Elements of

Modern Topology

Consulting Editor
Professor R. L. Goodstein
University of Leicester

Other titles in the European Mathematics Series

Flett: *Mathematical Analysis*
Goodstein: *Complex Functions*
Hawgood: *Numerical Methods in Algol*
Isaacs: *Real Numbers*
Jones: *Generalised Functions*
Nef: *Linear Algebra*
Nicol: *Elementary Programming and Algol*
Petersen: *Regular Matrix Transformations*

Elements of
Modern Topology

Ronald Brown
University of Hull
England

McGRAW-HILL Book Company
New York · St. Louis · San Francisco · Toronto · London · Sydney

Published by
McGRAW-HILL Publishing Company Limited
MAIDENHEAD · BERKSHIRE · ENGLAND

94059

PRINTED AND BOUND IN GREAT BRITAIN

Preface

This book is intended as a textbook on point set and algebraic topology at the undergraduate and immediate post-graduate levels. In particular, I have tried to make the point set topology commence in an elementary manner suitable for the student beginning to study the subject.

The choice of topics given here is perhaps unusual, but has the aim of presenting the subject with a geometric flavour and with a coherent outlook. The first consideration has led to the omission of a number of topics important more from the point of view of analysis, such as uniform spaces, convergence, and various alternative kinds of compactness. The second consideration, together with restriction of space, has led to the omission of homology theory, of the theory of manifolds, and of any complete account of simplicial complexes. It has also led to the omission of topics which are important but not exactly germane to this book; for example, I have not included accounts of paracompactness or the theory of continua, nor proofs of the Tychonoff theorem or the Tietze extension theorem.

I felt, on the other hand, that the general direction of this book should be towards homotopy theory, since this subject links naturally with the point set topology and also occupies a central role in modern developments. However, in homotopy theory there are a number of ideas and constructions which are important in many applications (particularly the ideas of adjunction space, cell complex, join, and homotopy extension property) but for which no elementary account has appeared. One of my aims, then, has been to cover this sort of topic, and so to supplement existing accounts, rather than provide a new reference book to contain everything a young geometer ought to know. At the same time, I have tried to show that point set topology has its main value as a language for doing 'continuous geometry'; I believe it is important that the subject be presented to the student in this way, rather than as a self-contained system of axioms, definitions, and theorems which are to be studied for their own sake, or for their interest as generalizations of facts about Euclidean space. For this reason, I have kept the point set topology to the minimum needed for later purposes, and at the same time emphasized general processes of construction which lead to interesting topological spaces, and for which other structures, such as metrics or uniformities, are largely irrelevant.

The language of point set topology is geometric. On the other hand, the notions of category and functor are algebraic: they play in modern mathematics the unifying role which has earlier been given to the notion of a group. Particular kinds of categories are the groupoids, of which in turn groups are special cases. The algebra of groupoids has been developed recently by P. J. Higgins. In 1965, I discovered the utility of this algebra for computing the fundamental group, and this has led me to include in the last four chapters of this book an account of the elements of the algebra of categories and groupoids. The treatment given in these chapters is quite novel, in that this algebra is used in an essential way and not just as a convenient, but not entirely necessary, general language. The writing of these chapters gave me great pleasure as I found the way in which the various topological notions of sum, adjunction space, homotopy, covering space, are modelled by the corresponding notions for groupoids. The most important feature here is probably the way in which the computations of the fundamental group derive from a general property of the fundamental groupoid (that it sends 'nice' pushouts to pushouts). This contrasts with the usual rather *ad hoc* computation via simplicial complexes.

The development of these last four chapters has meant that I have had to cut down on some other topics. For example, I have had to relegate the accounts of k-spaces, infinite cell complexes, and function spaces to the Exercises. A further reason for playing down these topics was that the recent development of the theory of quasi-topological spaces is likely to lead to a radical change in their exposition (cf. Spanier [2] and a forthcoming book by Dyer and Eilenberg). I have also relegated to the Exercises the discussion of group-like structures and exact sequences, and have omitted an account of the major properties of the Whitehead product. The reasons behind this were lack of space, the difficulty of presenting computations, and, finally, the feeling that these topics needed reconsideration in the light of the preceding work on groupoids.

The material in this book would more than cover a two-term undergraduate course in point set and algebraic topology. Such a course could include, for the point set topology, all of chapters 1 to 3 and some material from chapters 4 and 5. This could be followed by a course on the fundamental groupoid comprising chapter 6 and parts of chapters 8 or 9; naturally, more could be covered if this were linked with a course on categories and functors.

The last four chapters, with supporting parts of chapters 4 and 5, would probably be suitable for a two-term M.Sc. course. Alternatively, the purely algebraic parts of the last four chapters would cover the elements (but with not enough algebraic applications) of a one-term course on groupoids.

The Appendix contains an account of functions, cardinality, and some 'universal constructions' which are used from chapter 4 onwards. On pages *xii* to *xvi* is a Glossary of those terms from set theory which are used in the book but not defined in the main text.

This book is not self-contained—in particular, it will be assumed at certain points that the reader is familiar with the usual accounts of vector spaces and of groups.

On the other hand, a more 'topological' account of elementary analysis is by no means universal. For this reason, and in order to motivate the axioms for a topological space, I have started the book with an account of the elementary topological notions on the real line **R**.

In the earlier chapters, I have included among the exercises a fair number of straightforward verifications or simple tests of the reader's understanding. Apart from a few simple definitions and verifications, no results from the exercises are used in the text; this means that occasionally a result from an exercise is actually proved later in the text. However, there are some cross-references among the exercises themselves. Exercises from later chapters are usually more difficult than those from earlier ones—exercises of technical difficulty are marked with an asterisk.

Cross-references are always given in square brackets; thus [6.5.12] refers to 5.12 of §5 of chapter 6. The end of a Proof is denoted by the symbol □.

I would like to thank Dr W. F. Newns for reading through much of the manuscript and for many suggestions which resulted in improved exposition and in the removal of errors. I would also like to thank Professor G. Horrocks and Dr R. M. F. Moss for their comments on some of the chapters. I am indebted to Dr P. J. Higgins for conversations and letters; for sending me notes of a course on groupoids he gave at King's College, London in 1965; and for writing his paper, Higgins [1].

I would also like to thank Dr W. Gilbert, P. R. Heath, D. Pitt, T. Poston, Dr E. Rees, and E. E. Thompson for help in proofreading.

RONALD BROWN

Contents

Glossary of symbols

Standard spaces

$\mathbf{C}$	complex numbers	14
$\mathbf{I}$	unit interval	10
$\mathbf{H}$	quaternions	130
$\mathbf{L}$	$\{0\} \cup \{n^{-1} : n = 1, 2, \ldots\}$	57
$\mathbf{N}$	natural numbers (including 0)	336
$\mathbf{Q}$	rational numbers	336
$\mathbf{R}$	real numbers	336
$\mathbf{Z}$	integers	335
$\mathbf{K}$	one of **C**, **R**, **H**	42
$\dot{\mathbf{I}}$	$\{0, 1\}$	148
$\mathbf{K}^n$	standard n-dimensional space over **K**	43
$\mathbf{B}^n$	unit Euclidean ball	45
$\mathbf{E}^n$	unit Euclidean disc	45
$\mathbf{S}^{n-1}$	unit Euclidean sphere	45
$\mathbf{J}^n$	n-cube	45
T^2	torus	90
$M_n(\mathbf{K})$	$n \times n$ matrices	143
Δ^n	standard n-simplex	158
$\dot{\Delta}^n$	simplicial boundary of Δ^n	148
$P^n(\mathbf{K})$	projective space	136
$\mathrm{O}(n)$	orthogonal group	143
$\mathrm{SO}(n)$	special orthogonal group	143
$\mathrm{U}(n)$	unitary group	143
$\mathrm{SU}(n)$	special unitary group	143
$\mathrm{Sp}(n)$	symplectic group	143

Subsets

$[x, y]$, $]\leftarrow, x]$, $]x, y[$, $X^{\geqslant x}$	intervals in ordered set	335

$X^{\neq x}$	$X \setminus \{x\}$	
$X \setminus A$	complement of A in X	334
$[x, y]$	segment in normed space	74
$B(a, r)$	ball of radius r	44
$E(a, r)$	closed ball of radius r	45
$S(a, r)$	sphere of radius r	45
$B(V)$	unit ball	45
$E(V)$	closed unit ball	45
$S(V)$	unit sphere	45
V_*	$V^{\neq 0}$	136

Topological operations on subsets

Int A	interior	18
$\bar{A}$	closure	22
Ext A	exterior	23
Bd A	boundary	23
Fr A	frontier	23
$\hat{A}$	derived set	23
$\mathrm{Cl}_A\, C$	relative closure	29
$\mathrm{Int}_A\, C$	relative interior	30

Operations on spaces and sets

$X_1 \times X_2$	Cartesian product of sets (spaces)	333 (25)
$X_1 \sqcup X_2$	sum of sets (spaces)	329 (60)
$\prod_{\alpha \in A} X_\alpha$	product of a family of sets (spaces)	328 (154)
$\bigsqcup_{\alpha \in A} X_\alpha$	sum of a family of sets (spaces)	329 (94)
$\mathscr{P}(X)$	set of subsets of X	320
$\Delta(X)$	diagonal of X	55
SX	suspension of X	106
ΣX	reduced suspension of X	109, 185
CX	cone of X	99
ΓX	reduced cone of X	109
$X * Y$	join	157
$X \circledast Y$	smashed product	165
$X \vee Y$	wedge	165
$X_1\, {}_{i_1}\!\sqcup_{i_2} X_2$	pushout of i_1, i_2	111, 202
$B_f \sqcup X$	adjunction space	109
X/A	X with A shrunk to a point	99
$M(f)$	mapping cylinder	119
$C(f)$	mapping cone	119
$M(f) \cup X$	mapping cylinder union X	246

$X \cup e^n$	X union a cell	127
X^n	n-fold Cartesian product	
K^n	n-skeleton of a cell complex	121, 124
$C(x)$	component of the point x	66
$\mathscr{B}_r$	cover by balls of radius r	85
X^Y	function space	156
$X \times_W Y$	weak product	103
X_Σ	X with a weak topology	97
$\tilde{X}$	covering of X	290
$G(V)$	isometry group	143
$P(V)$	projective space	136

Functions

$f : A \to B$	function from A	313
$f : A \rightarrowtail B$	function out of A	152
$f \mid A', B'$	restriction of f	315
$f \mid A'$	restriction of f	315
$f[A]$	image of A by f	315
$f^{-1}[A]$	inverse image of A by f	315
$\operatorname{Im} f$	image of f	315
$x \rightsquigarrow fx$	function	313
1	identity function	314
$f \times g$	product function	317
(f, g)	function into product	32, 328
$p_\alpha : X_1 \times X_2 \to X_\alpha$	projection	27, 324
$i_\alpha : X_\alpha \to X_1 \sqcup X_2$	injection	328
$\Delta : X \to X \times X$	diagonal map	32
$\bar{i} : B \to B\,{}_f{\sqcup}\, X$		109
$\bar{f} : X \to B\,{}_f{\sqcup}\, X$		109
$p : V_* \to P(V)$	fundamental map	136
$h : S(V) \to P(V)$	Hopf map	138
$\xi_\alpha : X_1 * X_2 \to \mathbf{I}$	coordinate function of join	158
$\eta_\alpha : X_1 * X_2 \rightarrowtail X_\alpha$	coordinate function of join	158

Paths, homotopies

$\lvert a \rvert$	length of a	72
a_φ	final point of a	72
a_ι	initial point of a	72
$-a$	reverse of a	72
$b + a$	sum of b and a	73
PX	category of paths	171
πX	fundamental groupoid	180

πXA	fundamental groupoid on $A \cap X$	183
$\pi(X, x)$	fundamental group	183
$\mathscr{C}^{\mathrm{op}}$	opposite category	173
$\pi_n(Y, \cdot)$	nth homotopy group	220
F_t	homotopy	175, 192
$a \sim b$	homotopy of paths	175
$a \sim b$	equivalence of paths	178
$f \simeq g$	homotopy of maps	192
$G + F$	sum of homotopies	175, 218
$f \simeq g$ rel A	homotopy rel A	217
$P(X^Y \text{ rel } A)$	relative homotopy category	218
$\pi(X^Y \text{ rel } A)$	track groupoid	218
πX^Y	track groupoid	219
$\pi X^Y_{\cdot}$	pointed track groupoid	219
$X/\!/u$	classes of extensions	219
$X/\!/\gamma$	operation by γ	225
$f_* : \pi Y^A \to \pi Z^A$	induced morphism	227
HEP	homotopy extension property	223
WNDR	weak neighbourhood deformation retract	234
$\mathbf{f} \simeq \mathbf{g}$	homotopy of M-ads	239
$[X, Y]$	set of homotopy classes	230
$[X, Y]_{\cdot}$	pointed homotopy classes	220

Categories, groupoids

$\mathscr{S}$et	sets	171
$\mathscr{T}$op	topological spaces	171
$\mathscr{G}$	groups	171
$\mathscr{G}$d	groupoids	187
$\mathscr{T}$op.	pointed spaces	188
$\mathscr{H}\mathscr{T}$op	homotopy category	199
$\mathscr{C} \times \mathscr{D}$	product of categories	189
I	tree groupoid on two objects	185
$I^{\rightarrow}$	a subcategory of I	199
$\mathscr{F}$un $(\mathscr{C}, \mathscr{D})$	functor category	199
$G\{x\}$	object group at x	183
Cx	component of x	183
$\Gamma : \mathscr{C} \to \mathscr{D}$	functor	186
a_{*}	conjugacy function	184
$\mathscr{C}^x, \mathscr{C}_x$	functors to $\mathscr{S}$et	191
$\widehat{\mathscr{C}}$	functor to $\mathscr{S}$et	200
$U_\sigma(G)$	universal groupoid	266
σ^*	universal morphism	266

$U(G)$	universal group	268
FA	free group on A	268
$G * H$	free product of G and H	270
$\operatorname{Im} f$	Image of f	274
$\operatorname{Ker} f$	Kernel of f	278
G/N	quotient groupoid	278
$\underline{N}(R)$	groupoid of consequences	280
$\overline{\Gamma}$	$\Gamma \cup \Gamma^{-1}$	274
$\operatorname{St}_G x$	star of x in G	295
$\widetilde{G}$	covering groupoid	295
$\operatorname{Tr}(G, H)$	translation groupoid	305
χ_f	characteristic groupoid of f	306

Miscellaneous

$\operatorname{cls} x$	equivalence class of x	334
$x \in\in E$	$x \in \operatorname{cls} x \in E$	334
$\operatorname{Re}(q)$	Real part of a quaternion	130
$\operatorname{Ve}(q)$	Vector part of a quaternion	130
$\bar{q}$	conjugate of q	132
$\lvert x \rvert$	Euclidean norm of x	43
$\lVert x \rVert$	norm of x	43
$d(\mathbf{K})$	dimension of $\mathbf{K}$ over $\mathbf{R}$	134
$(x \mid y)$	inner product	142
$U^{\perp}$	orthogonal space of U	142
$U^{\dagger}$	subset of $P(V)$	137
$\operatorname{dist}(x, A)$	distance of x from A	53
※ X	cardinality of X	318

1. Some topology on the real line

In this chapter, we introduce the system of neighbourhoods of points on the real line **R**; this presents **R** with structure additional to its usual structures of addition, multiplication, and order. This additional structure makes **R** a *topological space*, and the study of this and other topological spaces is the subject matter of this book.

There are several reasons for giving this example of a topological space before a definition of such an object. First of all, we hope in this way to familiarize the reader with some of the techniques and terminology which will recur constantly. Secondly, the real line is central to mathematics, in the way that atoms are central to physics, and cells to biology. So time devoted to this example is well spent.

1.1 Neighbourhoods in R

Definition Let N be a subset of $\mathbf{R}$, and let $a \in \mathbf{R}$. Then N is a *neighbourhood of a* if there is a real number $\delta > 0$ such that the open interval about a of radius δ is contained in N; that is, if there is a $\delta > 0$ such that

$$]a - \delta, a + \delta[\subset N.$$

EXAMPLES 1. **R** itself is a neighbourhood of any a in **R**.
2. The closed interval $[0, 1]$ is a neighbourhood of $\frac{1}{2}$, but not of 0, nor of 1.
3. The set $\{0\}$ consisting of 0 alone is not a neighbourhood of 0.
4. $[0, 1[\cup]1, 2]$ is not a neighbourhood of 1.
5. $[0, 1] \cup [2, 3] \cup]3\frac{6}{7}, 8[$ is a neighbourhood of 2π.
6. The set **Q** of rationals is not a neighbourhood of any $a \in \mathbf{R}$: for any interval $]a - \delta, a + \delta[$ $(\delta > 0)$ must contain irrational as well as rational points, and so $]a - \delta, a + \delta[$ cannot be contained in **Q**.
7. This is a mildly pathological example. For each integer $n > 0$, let

$$X_n = \{x \in \mathbf{R} : (2n + 1)^{-1} < |x| < (2n)^{-1}\}.$$

Let X be the union of $\{0\}$ and the sets X_n for all $n > 0$ [cf. Fig. 1.1]. Then

X is not a neighbourhood of 0, since any interval about 0 contains points not in X (in addition, of course, to points of X).

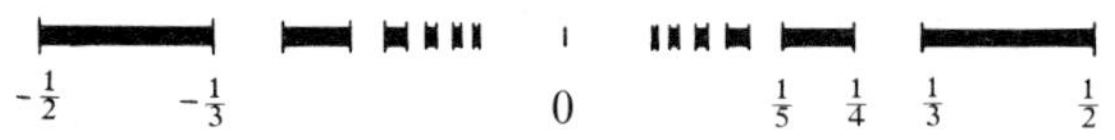

Fig. 1.1

The notion of neighbourhood is useful in formulating the definition of something being true 'near' a given point. In fact, let P be a property which applies to real numbers, and may or may not hold for any given real number. Let a be in $\mathbf{R}$. We say *P holds near a*, or *is valid near a*, if P holds for all points in some neighbourhood of a.

For example, let $f : \mathbf{R} \to \mathbf{R}$ be the function $x \rightsquigarrow x^2 + x^3$. Then f is positive near the point 1 (since $f(x) > 0$ for all $x > 0$); but it is not true that f is positive near 0, or near -1.

There is still no notion of absolute nearness, that is, of a point x being 'near a'. This is to be expected; the only definition of x being near a that makes sense is that x is near a if x is in some neighbourhood of a. But any x in $\mathbf{R}$ is then near a.

At this stage we could still dispense with arbitrary neighbourhoods and work entirely with intervals. However, the elegance and flexibility of the general notion will appear as we proceed.

We now derive some simple properties of neighbourhoods.

1.1.1 *Let $a \in \mathbf{R}$, and let M, N be neighbourhoods of a. Then $M \cap N$ is a neighbourhood of a.*

Proof Since M, N are neighbourhoods of a, there are real numbers δ, $\delta' > 0$ such that

$$]a - \delta, a + \delta[\subset M, \qquad]a - \delta', a + \delta'[\subset N.$$

Let $\delta'' = \min(\delta, \delta')$. Then

$$]a - \delta'', a + \delta''[\subset M \cap N$$

and so $M \cap N$ is a neighbourhood of a. □

1.1.2 *Let $a \in M \subset N \subset \mathbf{R}$. If M is a neighbourhood of a, then so is N.*

Proof If M is a neighbourhood of a, then there is a $\delta > 0$ such that $]a - \delta, a + \delta[\subset M$. Hence $]a - \delta, a + \delta[\subset N$, and the result follows. □

1.1.3 *An open interval is a neighbourhood of any of its points.*

Proof Suppose first that I is an open interval of the form $]a, b[$ where $a, b \in \mathbf{R}$. Let $x \in I$, and let $\delta = \min(x - a, b - x)$. Then δ is positive and $]x - \delta, x + \delta[\subset I$, whence I is a neighbourhood of x.

The proofs for the other kinds of open intervals are also simple. □

Clearly 1.1.3 is false if the word 'open' is removed.

A point a in $\mathbf{R}$ determines the sets N which are neighbourhoods of a. Also a set A determines the set of points of which A is a neighbourhood; this set is called the *interior of* A, and is written Int A; thus $x \in \operatorname{Int} A$ if and only if A is a neighbourhood of x. Since x belongs to any of its neighbourhoods, Int A is a subset of A.

EXAMPLES 8. If I is an open interval, then $\operatorname{Int} I = I$.

9. If A is finite, then Int A is empty.
10. If $A = [a, b]$, then $\operatorname{Int} A =]a, b[$.
11. If $A = \mathbf{Q}$ the set of rationals, then $\operatorname{Int} A = \varnothing$.
12. If $A = \mathbf{R} \setminus \mathbf{Q}$, then $\operatorname{Int} A = \varnothing$.

1.1.4 *If* $A \subset B$, *then* $\operatorname{Int} A \subset \operatorname{Int} B$.

Proof If A is a neighbourhood of x, then so also, by 1.1.2, is B. □

1.1.5 *If N is a neighbourhood of a, then so also is* Int N.

Proof Let N be a neighbourhood of a, and let $\delta > 0$ be such that $]a - \delta, a + \delta[\subset N$. Then $]a - \delta, a + \delta[\subset \operatorname{Int} N$, by 1.1.4 and since the interior of an open interval is the same interval. So Int N is a neighbourhood of a. □

EXERCISES

1. Let $a \in \mathbf{R}$ and let N be a subset of $\mathbf{R}$. Prove that the following conditions are equivalent.
(*a*) N is a neighbourhood of a.
(*b*) There is a $\delta > 0$ such that $[a - \delta, a + \delta] \subset N$.
(*c*) There is an integer $n > 0$ such that $[a - n^{-1}, a + n^{-1}] \subset N$.
2. Prove that if A is a countable set of real numbers, then $\operatorname{Int} A = \varnothing$.
3. Prove that $\operatorname{Int}(A \cap B) = \operatorname{Int} A \cap \operatorname{Int} B$.
4. Does $\operatorname{Int}(A \cup B) = \operatorname{Int} A \cup \operatorname{Int} B$?
5. Let (A_n) be a sequence of subsets of $\mathbf{R}$. Does

$$\operatorname{Int}(\cap_n A_n) = \cap_n \operatorname{Int} A_n \ ?$$

6. Let C be a neighbourhood of $c \in \mathbf{R}$, and let $a + b = c$. Prove that there are neighbourhoods A of a, B of b such that $x \in A$ and $y \in B$ implies $x + y \in C$.
7. Write down and prove a similar result to that of Exercise 6, but with $c = ab$.
8. Let C be a neighbourhood of c, where $c \neq 0$. Prove that there is a neighbourhood C' of c^{-1} such that if x is in C', then x^{-1} is in C.

9. Prove that Int (Int A) = Int A.
10. Let $A_1 = [-1, 1[\setminus \{0\}$, $A_2 = \{2\}$, $A_3 = \mathbf{Q} \cap [3, 4]$, and let $A = A_1 \cup A_2 \cup A_3$. Show that exactly fourteen distinct subsets of $\mathbf{R}$ may be constructed from A by means of the operations Int and complementation with respect to $\mathbf{R}$.

1.2 Continuity

In this section, we define continuity of real functions (that is, functions whose domain and range are subsets of $\mathbf{R}$). This definition is entirely in terms of neighbourhoods. In this section, we shall usually take the range of a real function to be $\mathbf{R}$ itself, but the definitions and results are the same as for the general case, when the range is any subset of $\mathbf{R}$.

Let $f : A \to \mathbf{R}$ be a function, where A is a subset of $\mathbf{R}$. Let $a \in A$.

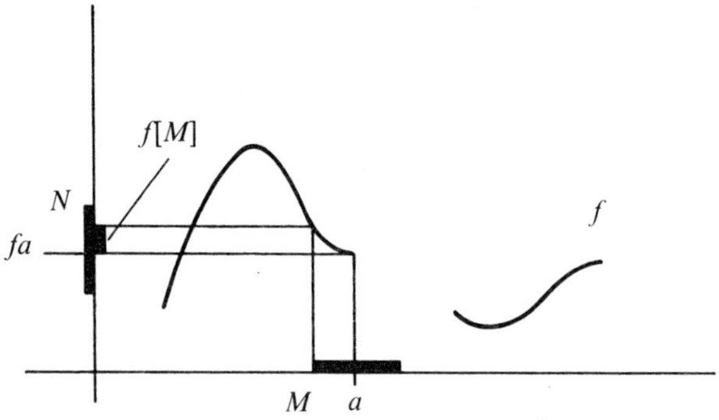

Fig. 1.2

Definition C The function f is *continuous at a* if for each neighbourhood N of $f(a)$ there is a neighbourhood M of a such that $f[M] \subset N$. Further, f is *continuous* if f is continuous at each a in A.

In this definition, $f[M]$ is the image† of the set M by f. So the definition can be restated:

Definition C′ The function f is *continuous at a* if, for each neighbourhood N of $f(a)$, there is a neighbourhood M of a such for all $x \in \mathbf{R}$

$$x \in M \cap A \Rightarrow f(x) \in N.$$

The statement $f[M] \subset N$ is equivalent to $M \cap A \subset f^{-1}[N]$. Suppose now that A is a neighbourhood of a. Then $M \cap A$ is a neighbourhood of a, so that $f[M] \subset N$ implies that $f^{-1}[N]$ is a neighbourhood of a. On the other hand, we always have

$$ff^{-1}[N] \subset N.$$

† See A.1.5 of the Appendix.

So if $f^{-1}[N]$ is a neighbourhood of a, then $f^{-1}[N]$ is itself a neighbourhood M of a such that $f[M] \subset N$. This shows that if A is a neighbourhood of a, we can restate Definition C as:

Definition C″ The function f is *continuous at a* if, for every neighbourhood N of $f(a)$, $f^{-1}[N]$ is a neighbourhood of a.

This last definition has only one quantifier, whereas Definition C has two, and Definition C′ has *three*. Thus Definition C″ is the easiest to understand, but we emphasize that it only applies to the case when A, the domain of f, is a neighbourhood of a.

Another advantage of Definition C″ is that it is easy to negate. We suppose A is a neighbourhood of a: then f is not continuous at a if, for some neighbourhood N of a, $f^{-1}[N]$ is not a neighbourhood of a. This is illustrated in the examples which follow.

EXAMPLES 1. Let $l \in \mathbf{R}$ and let $f: \mathbf{R} \to \mathbf{R}$ be the constant function $x \rightsquigarrow l$. Let $a \in \mathbf{R}$. The domain of f is $\mathbf{R}$, which is a neighbourhood of a. If N is a neighbourhood of l, then $f^{-1}[N] = \mathbf{R}$, which is a neighbourhood of a. Therefore, f is continuous at a, and since a is arbitrary, f is a continuous function.

2. Let $f: \mathbf{R} \to \mathbf{R}$ be the identity function $x \rightsquigarrow x$. Let $a \in \mathbf{R}$. The domain of f is a neighbourhood of a, and $f(a) = a$. If N' is a neighbourhood of $f(a)$, then $f^{-1}[N] = N$, so that $f^{-1}[N]$ is a neighbourhood of a. Thus f is a continuous function.

3. Consider the function

$$f: \mathbf{R} \to \mathbf{R}$$

$$x \rightsquigarrow \begin{cases} 0, & x < 1 \\ 2, & x \geqslant 1. \end{cases}$$

Here again, $\mathbf{R}$ is a neighbourhood of 1. In this case, f is not continuous at 1. For let $N = [1, \to[$. Then N is a neighbourhood of $f(1) = 2$, but $f^{-1}[N] = [1, \to[$ which is not a neighbourhood of 1.

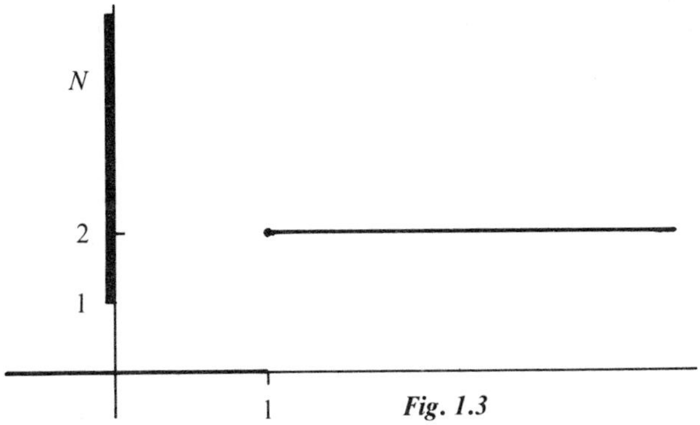

Fig. 1.3

4. Consider the function

$$f : [1, \rightarrow[\;\rightarrow \mathbf{R}$$
$$x \rightsquigarrow 2.$$

Here the domain of f is $[1, \rightarrow[$, which is not a neighbourhood of 1; so to prove continuity at 1, we must use Definition C. Let N be a neighbourhood of 2. Then

$$f[\mathbf{R}] = \{2\}$$

which is a subset of N. Since $\mathbf{R}$ is a neighbourhood of 1, we have proved continuity at 1. This example should be compared carefully with Example 3.

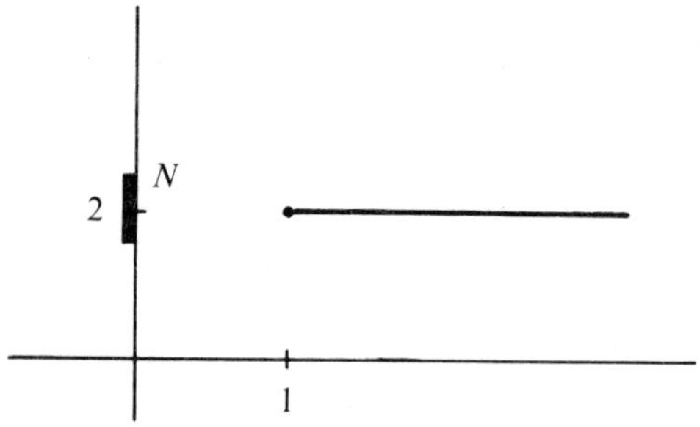

Fig. 1.4

5. Let $f : \mathbf{R} \rightarrow \mathbf{R}$ be defined as follows. If α is a rational number and is, in its lowest terms, p/q where p, q are integers such that $q > 0$, then $f(\alpha) = 1/q$ (in particular, $f(0) = 1$). If β is irrational, then $f(\beta) = 0$. We prove that f is not continuous at any rational number, but is continuous at any irrational number.

Let $\alpha = p/q$ be such that $f(\alpha) = 1/q$. Let $N =]0, \rightarrow[$. Then N is a neighbourhood of $f(\alpha)$, but $f^{-1}[N]$ contains no irrational numbers and so is not a neighbourhood of α. Therefore f is not continuous at α.

In order to prove that f is continuous at an irrational number β, we use the fact that rational numbers of high denominators are needed to approximate closely an irrational number.

Let N be a neighbourhood of $0 = f(\beta)$. We prove that $M = f^{-1}[N]$ is a neighbourhood of β. Certainly, M contains all irrational numbers, since these are all sent to 0 by f, and we show that all rational numbers close enough to β are also contained in M.

Let n be a positive integer such that

$$[-n^{-1}, n^{-1}] \subset N;$$

such an integer exists since N is a neighbourhood of 0 [cf. Exercise 1 of 9.1]. Let m be an integer such that $\beta \in [m, m+1]$. There are only a finite number of rational numbers p/q which are in $[m, m+1]$ and are such that $0 < q \leqslant n$. One of these, say r, will be closest to β. Let

$$|\beta - r| = \delta, \qquad I =]\beta - \delta, \beta + \delta[.$$

Since β is irrational, δ is positive. Also, all rational numbers p/q in I satisfy $q > n$. Hence

$$f[I] \subset]-n^{-1}, n^{-1}[, \quad \text{whence } I \subset M.$$

Therefore M is a neighbourhood of β, and f is continuous at β.

Examples 1, 2, and 5 show that some functions are proved continuous by working directly with the definition. More usually, though, we construct continuous functions by taking a basic stock of continuous functions and giving rules for making more complicated continuous functions from those of the basic stock.

We take from analysis the definitions and continuity of the functions sin and log. These, with the identity function and the constant function, form our basic stock. Let $f : A \to \mathbf{R}$, $g : B \to \mathbf{R}$ be functions; then $f + g$, $f.g$, f/g are respectively the functions $x \rightsquigarrow f(x) + g(x)$, $x \rightsquigarrow f(x).g(x)$, $x \rightsquigarrow f(x)/g(x)$ with respective domains $A \cap B$, $A \cap B$, $(A \cap B) \setminus \{x \in B : g(x) = 0\}$. Let A, B be subsets of $\mathbf{R}$, and let $a \in A \cap B$.

1.2.1 (*Sum, product, and quotient rule.*) *If f, g are continuous at a, so also are $f + g, f.g$; if f/g is defined at a, then it is continuous at a.*

We omit the proof, which is given, for example, in Flett [1] [cf. also Exercise 1 following, and sections 2.5, 2.8].

1.2.2 (*Restriction rule*) *Let $f : A \to \mathbf{R}$ be a real function continuous at $a \in A$. Let B be a subset of A containing a. Then $f \mid B$ is continuous at a.*

Proof Let $g = f|B$. Let N be a neighbourhood of $g(a) = f(a)$. Since f is continuous at a, there is a neighbourhood M of a such that $f[M] \subset N$. Then

$$g[M] = g[M \cap B] = f[M \cap B] \subset f[M] \subset N.$$

So $g[M] \subset N$, and g is continuous at a. □

The restriction rule can be used to deduce Example 4 from Example 1.

Let $f : A \to \mathbf{R}$, $g : B \to \mathbf{R}$ be real functions, and let $h : A \cap f^{-1}[B] \to R$ be the composite function $x \rightsquigarrow g(f(x))$.

1.2.3 (*Composite rule*) *If $a \in A \cap f^{-1}[B]$, f is continuous at a and g is continuous at $f(a)$, then h is continuous at a.*

Proof Let N be a neighbourhood of $h(a)$. Since g is continuous at $f(a)$, there is a neighbourhood M of $f(a)$ such that $g[M] \subset N$. Since f is continuous at a, there is a neighbourhood L of a such that $f[L] \subset M$. It is easy to verify that

$$h[L] = g[f[L]],$$

whence $h[L] \subset N$. Therefore h is continuous at a. □

Let $f : A \to \mathbf{R}$ be a real function which is injective, so that f has an inverse $f^{-1} : f[A] \to A$.

1.2.4 (*Inverse rule*) *If A is an interval and f is continuous, then f^{-1} is continuous.*

For a proof, see Flett [1; 3.10.5]. This result is also a consequence of general theorems on connectivity and compactness [cf. chapter 3]. The assumption that A is an interval is essential [cf. Exercise 4].

Examples 6. By repeated application of 1.2.1 to the identity function and constant functions, we can prove in turn the continuity of $x \rightsquigarrow x^n$ ($n \geqslant 0$), polynomial functions, and rational functions.

7. The continuity of cos, tan, sec, cosec follows easily from 1.2.1 and 1.2.3. For example, cos is the composite of sin and $x \rightsquigarrow \pi/2 - x$, and tan is the quotient sin/cos.

By 1.2.2, $\sin \mid [-\pi/2, \pi/2]$ is continuous; but this function is injective, so its inverse $\sin^{-1}$ is continuous by 1.2.5. Similarly, we derive the continuity of $\cos^{-1}$, $\tan^{-1}$. The function log is injective and continuous; therefore, its inverse exp is continuous. The function $\sqrt[n]{}$ is continuous, since it is the inverse of $x \rightsquigarrow x^n$ if n is odd, and of $x \rightsquigarrow x^n$ ($x \geqslant 0$) if n is even. So we can prove continuity of functions such as

$$x \rightsquigarrow (\sin x)^{1/n} + \log x + x^{17}.$$

EXERCISES

1. Prove 1.2.1 by use of Exercises 6, 7 of 1.1.

2. Prove the following 'Sandwich Rule' (also called the Squeeze Rule). Let $\lambda, \mu : A \to \mathbf{R}$ be two functions continuous at $a \in A$ and such that $\lambda(a) = \mu(a)$. Let $f : A \to \mathbf{R}$ be a function such that for some neighbourhood M of a

$$x \in M \cap A \Rightarrow \lambda(x) \leqslant f(x) \leqslant \mu(x).$$

Then f is continuous at a.

Use this rule to prove continuity of the function

$$x \rightsquigarrow \begin{cases} x \sin x^{-1} & x \neq 0 \\ 0 & x = 0. \end{cases}$$

3. Let $f : A \to \mathbf{R}$ be a function, let $a \in A$, and let N be a neighbourhood of a. Prove that if $f \mid N \cap A$ is continuous at a, then so also is f.
4. Let f be the function

$$x \rightsquigarrow \begin{cases} x & 0 \leqslant x \leqslant 1 \\ x - 1 & 2 < x \leqslant 3. \end{cases}$$

Prove that f is continuous and injective, but that f^{-1} is not continuous at 1.
5. Let $f : [a, b] \to [c, d]$ be a monotonic bijection. Prove that f is continuous.
6. Let $f : \mathbf{R} \to \mathbf{R}$. Prove that f is continuous if and only if for every subset A of $\mathbf{R}$

$$f^{-1}[\text{Int } A] \subset \text{Int } f^{-1}[A].$$

7. Let $f : A \to \mathbf{R}$ be a real function and let $a \in A$. Prove the equivalence of the following statements:
(*a*) f is continuous at a.
(*b*) for all $\varepsilon > 0$, there is a $\delta > 0$ such that

$$f]a - \delta, a + \delta[\subset]a - \varepsilon, a + \varepsilon[.$$

(*c*) For all positive integers m there is a positive integer n such that

$$f]a - n^{-1}, a + n^{-1}[\subset]a - m^{-1}, a + m^{-1}[.$$

8. Prove the following 'Glueing Rule'. Let $A = A_1 \cup A_2 \subset \mathbf{R}$, $a \in A_1 \cap A_2$. Let $f : A \to \mathbf{R}$ be a function such that $f \mid A_1, f \mid A_2$ are continuous at a. Then f is continuous at a. Prove also that f is continuous if $f \mid A_1, f \mid A_2$ are continuous and $A_1 \setminus A_2 \subset \text{Int } A_1$, $A_2 \setminus A_1 \subset \text{Int } A_2$.
9. Prove that the set of all continuous functions $[0, 1] \to \mathbf{R}$ is uncountable.

1.3 Open sets, closed sets, closure

In this section, we introduce some more topological concepts on the real line $\mathbf{R}$.

First we consider the open sets. A subset U of $\mathbf{R}$ is *open* if U is a neighbourhood of each of its points. Now for any set U, Int U is the set of points of which U is a neighbourhood. So U is open if and only if $U = \text{Int } U$.

Examples of open sets are the empty set (which has no points and so is a neighbourhood of each of them) and, by 1.1.3, any open interval. Other examples may be constructed by means of the following results.

1.3.1 *The union of any family of open sets is open.*

Proof Let $(U_i)_{i \in J}$ be a family of subsets of $\mathbf{R}$ such that each U_i is open, and let $U = \bigcup_{i \in J} U_i$. If U is empty, it is open, if not, let $u \in U$; we prove that U is a neighbourhood of u.

First, $u \in U_i$ for some i. Since U_i is open, it is a neighbourhood of u. Therefore U, which contains U_i, is also a neighbourhood of u. □

1.3.2 *A subset of* **R** *is open if and only if it is the union of a countable set of disjoint open intervals.*

Proof The union of any family of open intervals is open by 1.1.3 and 1.3.1.

To prove the converse, let U be an open set. If U is empty the result is true since U is the union of the empty family of intervals.

Suppose U is not empty. Two points x, y of U are called equivalent, written $x \sim y$, if the closed interval with end points x, y is contained in U. It is easily verified that $\sim$ is an equivalence relation on U. By the definition of $\sim$, the equivalence classes are intervals of $\mathbf{R}$ (that is, if x, y belong to an equivalence class E, and $x < y$, then any point z such that $x < z \leqslant y$ also belongs to E). They are also disjoint and cover U.

Now the open intervals of $\mathbf{R}$ are exactly those intervals of $\mathbf{R}$ which do not contain any of their end points—this is a non-trivial fact about $\mathbf{R}$ being a consequence of the completeness of the order relation (see the Glossary under *bounded* and *interval*). So to prove the theorem let E be one of the above intervals and let a be an end point of E. If $a \in U$ then, for some $\delta > 0$, $]a - \delta, a + \delta[$ is contained in U and hence also in E, and this is absurd. Therefore $a \notin U$ and so $a \notin E$. Thus E is an open interval.

Let φ be the function which sends each element of $\mathbf{Q} \cap U$ to its equivalence class. Then φ is a surjection to the set of equivalence classes since each non-empty open interval of $\mathbf{R}$ contains a rational number. Since $\mathbf{Q} \cap U$ is countable it follows that the number of equivalence classes is countable. □

The simple criterion 1.3.2 allows some pathological and complicated examples. For example, in Fig. 1.1 the union of all the sets X_n, that is, the set $X \setminus \{0\}$, is an open set.

EXAMPLE *The Cantor Set.* This is a subset K of $\mathbf{I} = [0, 1]$ such that $\mathbf{I} \setminus K$ is open.

The *middle-third* of a closed interval $[a, b]$ is the open interval,

$$]a + (b - a)/3, b - (b - a)/3[.$$

If (I_α) is a family of disjoint closed intervals, and $U = \bigcup I_\alpha$, then the *middle-third* of U is the union of the middle-thirds of each I_α.

Now let $\mathbf{I} = [0, 1]$; we define sets X_n, I_n by induction. First, X_1 is the middle-third of $\mathbf{I}$ and

$$I_1 = \mathbf{I} \setminus X_1.$$

Suppose X_n, I_n have been defined, and I_n is a finite union of closed intervals. Then X_{n+1} is defined to be the union of X_n and the middle-third of I_n, and we set $I_{n+1} = \mathbf{I} \setminus X_{n+1}$. The construction is illustrated in Fig. 1.5.

It is easy to prove by induction that I_n is the union of 2^n closed intervals each of length 3^{-n}, that X_n is a union of disjoint open intervals, and that

$$\mathbf{I} \supset I_1 \supset I_2 \supset \cdots; \quad X_1 \subset X_2 \subset X_3 \subset \cdots.$$

Let $X = \bigcup_{n \geqslant 1} X_n$ so that X is open. The *Cantor set* is

$$K = \mathbf{I} \setminus X = \bigcap_{n \geqslant 1} I_n.$$

There is a convenient representation of the points of K by ternary decimals. We recall [cf. for example, Flett [1], pp. 240–1] that each point of **I** can be represented as

$$.a_1a_2a_3\ldots = \sum_{n=1}^{\infty} a_n 3^{-n}, \quad a_n = 0, 1, \text{ or } 2.$$

The points of X_1 have $a_1 = 1$, and those of I_1 have $a_1 = 0$ or 2. The points of X_2 have $a_1 = 1$ or $a_2 = 1$, while for I_2 neither a_1 nor a_2 can be 1. In fact, it is not hard to prove by induction that the points of I_n are exactly those real numbers whose representation as ternary decimals $.a_1a_2\ldots$ have no 1's in the first n places. So the points of K are represented by ternary decimals $.a_1a_2\ldots$ with $a_n \neq 1$ for all $n \geqslant 1$; and the points of X are represented by ternary decimals $.a_1a_2\ldots$ in which at least one of the a_n is 1.

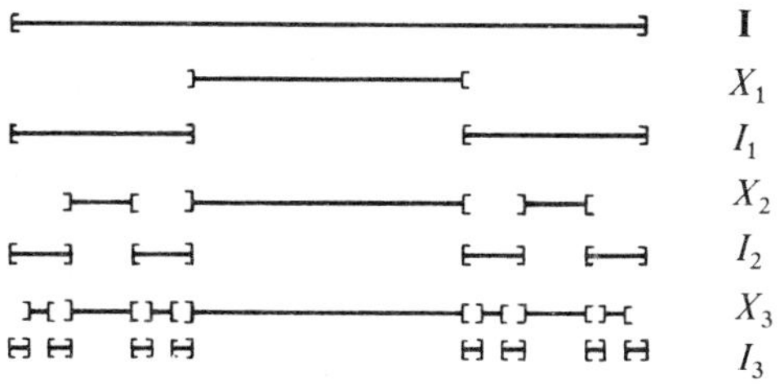

Fig. 1.5

The open sets in **R** generalize the open intervals. The corresponding generalizations of the closed intervals are the closed sets: a subset C of **R** is *closed* if $\mathbf{R} \setminus C$ is open. Thus a closed interval $[a, b]$ is closed since its complement is $]\leftarrow, a[\cup]b, \leftarrow[$ which is open. Corresponding to 1.3.1, we have,

1.3.3 *The intersection of any family of closed sets is closed.*

Proof Let (C_α) be a family of closed sets, and let $C = \bigcap C_\alpha$. We must prove that $\mathbf{R} \setminus C$ is open.

By the De Morgan laws,

$$\mathbf{R} \setminus C = \bigcup (\mathbf{R} \setminus C_\alpha).$$

But C_α is closed, so that $\mathbf{R} \setminus C_\alpha$ is open, and $\bigcup (\mathbf{R} \setminus C_\alpha)$ is open by 1.3.1. □

A corollary of 1.3.3 is that the Cantor set K is closed: for $K = \bigcap_{n \geqslant 1} I_n$, and $\mathbf{R} \setminus I_n$ is open, since it is the union of open intervals.

There are subsets of **R** which are neither open nor closed, for example, the half-open interval $[0, 1[$. A natural question is: which subsets of **R** are both open and closed?

1.3.4 *The only subsets of* **R** *which are both open and closed are* $\varnothing$ *and* **R**.

Proof Let U be a non-empty, open, proper subset of **R**.

By 1.3.2, U is the union of a family of disjoint open intervals. Since U is neither $\varnothing$ nor **R**, one of these intervals has an end point, say a. Now $a \in \mathbf{R} \setminus U$, but $\mathbf{R} \setminus U$ cannot be a neighbourhood of a since a is the end point of an open interval contained in U. Therefore $\mathbf{R} \setminus U$ is not open, and so U is not closed. □

The open sets U and **R** are characterized by the property: $U = \text{Int } U$. There is another operation on subsets of **R** which characterizes the closed sets.

Let A be a subset of **R**. We divide the points of **R** into three sets. First, we have the interior points of A; these form a set Int A which has already been discussed. Second, we have the *exterior* points of A, which form a set Ext A; these are the points which are interior to $\mathbf{R} \setminus A$, so that $x \in \text{Ext } A$ if and only if $\mathbf{R} \setminus A$ is a neighbourhood of x. Finally, the points which remain form the *frontier* Fr A.

For example, let $A = [0, 1[$. The interior of A is $]0, 1[$, the exterior of A is $]\leftarrow, 0[\cup]1, \rightarrow[$, and the frontier of A consists of the points 0 and 1.

The *closure* $\bar{A}$ of a set A is obtained by adding to Int A the points of the frontier of A; that is,

$$\bar{A} = \text{Int } A \cup \text{Fr } A.$$

In fact $A \subset \bar{A}$, so that $\bar{A} = A \cup \text{Fr } A$. We prove $A \subset \bar{A}$ as follows: Let $x \in A$. If $x \in \text{Int } A$, then certainly $x \in \bar{A}$. Suppose $x \in A \setminus \text{Int } A$: then A is not a neighbourhood of x, and neither is $\mathbf{R} \setminus A$; hence $x \in \text{Fr } A$, and so $x \in \bar{A}$.

We set out the definitions of these operators in terms of neighbourhoods:

(*a*) $x \in \text{Int } A \Leftrightarrow A$ is a neighbourhood of $x \Leftrightarrow$ some neighbourhood of x does not meet $\mathbf{R} \setminus A$.

(*b*) $x \in \operatorname{Ext} A \Leftrightarrow \mathbf{R} \setminus A$ is a neighbourhood of $x \Leftrightarrow$ some neighbourhood of x does not meet A.
(*c*) $x \in \operatorname{Fr} A \Leftrightarrow$ every neighbourhood of x meets both A and $\mathbf{R} \setminus A$.
(*d*) $x \in \bar{A} \Leftrightarrow$ every neighbourhood of x meets A.

It is necessary to explain why (*d*) defines $\bar{A}$. Suppose first that $x \in \bar{A}$, and that N is a neighbourhood of x. If $x \in A$, then $x \in N \cap A$ and so N meets A; if $x \in \operatorname{Fr} A$, then N meets A by (*c*). So if $x \in \bar{A}$ then every neighbourhood of x meets A.

Conversely, suppose every neighbourhood of x meets A. If $x \in A$, then $x \in \bar{A}$; and if $x \in \mathbf{R} \setminus A$, then every neighbourhood of x meets both A and $\mathbf{R} \setminus A$, so that $x \in \operatorname{Fr} A$. In either case, $x \in \bar{A}$.

The closure operation is probably the most important of these topological operators—when we define closure for subsets of a general topological space, we shall take (*d*) above as the definition.

We conclude this section by proving:

1.3.5 *A subset A of* $\mathbf{R}$ *is closed if and only if* $A = \bar{A}$.

Proof Suppose A is closed. Then $\mathbf{R} \setminus A$ is open, so that $\operatorname{Ext} A = \mathbf{R} \setminus A$. Hence

$$\bar{A} = \operatorname{Int} A \cup \operatorname{Fr} A = \mathbf{R} \setminus \operatorname{Ext} A = A.$$

Conversely, suppose $A = \bar{A}$. Then

$$\mathbf{R} \setminus A = \operatorname{Ext} A = \operatorname{Int} (\mathbf{R} \setminus A).$$

Therefore, $\mathbf{R} \setminus A$ is open, and A is closed. □

EXERCISES

1. Prove that the Cantor set K is uncountable.
2. Prove that the function

$$f : K \to \mathbf{I}$$
$$\sum_1^\infty a_n 3^{-n} \rightsquigarrow \sum_1^\infty a_n 2^{-n}$$

is continuous, increasing, and surjective.
3. Let X_n be as in the construction of the Cantor set K. Prove that $X_n \setminus X_{n-1}$ is the union of 2^{n-1} open intervals $X_{n,p}$ each of length 3^{-n}. Prove that the function f of Exercise 2 takes the same value at the end points of each $X_{n,p}$. Deduce that f extends to a continuous function $g : \mathbf{I} \to \mathbf{I}$ which is constant on each $X_{n,p}$. Sketch the graph of g.
4. Determine the exterior, frontier, and closure of the following subsets of $\mathbf{R}$: (i) $\mathbf{Q}$, (ii) $\mathbf{R} \setminus \mathbf{Q}$, (iii) $\{0\}$, (iv) $\{n^{-1} : n \text{ a positive integer}\}$, (v) $\mathbf{Z}$.

5. Let A be a subset of $\mathbf{R}$. Prove that

$$\mathbf{R} \setminus \bar{A} = \operatorname{Int}(\mathbf{R} \setminus A).$$

6. Let A be a subset of $\mathbf{R}$. Prove that if C is a closed set containing A, then C contains $\bar{A}$.
7. Let A be an open subset of $\mathbf{R}$ and $f: A \to \mathbf{R}$ a function. Prove that f is continuous if and only if $f^{-1}[U]$ is open for each open set U of $\mathbf{R}$.
8. Prove that any closed set of $\mathbf{R}$ is the intersection of countably many open sets. [Use 1.3.2].

1.4 Some generalizations

The idea of neighbourhood on which much of the previous sections was based, is applicable to other situations and this will of course be discussed in detail in later chapters. Here we wish to prepare the reader for the full scale axiomatics of the next chapter by showing some other examples of neighbourhoods.

There is an easy generalization to the Euclidean plane $\mathbf{R}^2 = \mathbf{R} \times \mathbf{R}$. Let us identify $\mathbf{R}^2$ with $\mathbf{C}$, the set of complex numbers. Then, we can define in $\mathbf{C}$ a neighbourhood of a complex number a to be any subset N of $\mathbf{C}$ which contains a set

$$B(a, \delta) = \{z \in \mathbf{C} : |z - a| < \delta\}$$

for some $\delta > 0$. Here the 'open ball' $B(a, \delta)$ replaces what in $\mathbf{R}$ was the open interval $]a - \delta, a + \delta[$. As will be clear later, most of the previous discussions and definitions go through without change. In particular, we can define continuity for functions with domain and range subsets of either $\mathbf{R}$ or $\mathbf{C}$, since all that is needed for the definition of continuity is the notion of neighbourhood.

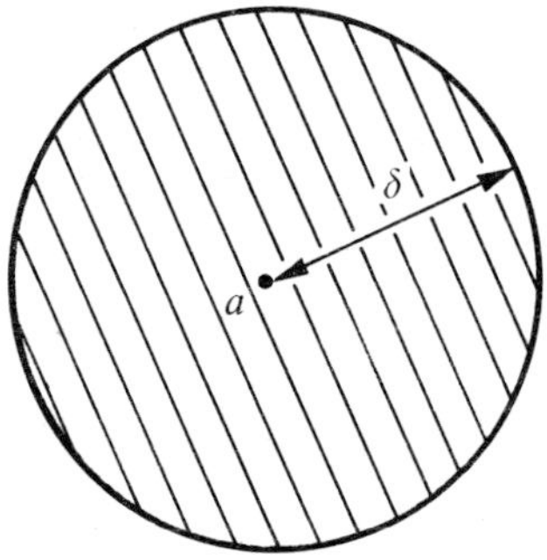

Fig. 1.6

Similarly, there is the notion of an open ball $B(a, \delta)$ in $\mathbf{R}^3$, namely, the set of points z whose Euclidean distance from a is less than δ. Given the notion of open ball, we can again define a neighbourhood of a in $\mathbf{R}^3$ to be any subset of $\mathbf{R}^3$ containing an open ball $B(a, \delta)$ for some $\delta > 0$. These definitions of neighbourhood find their proper place in the definition of neighbourhoods in metric spaces and in normed vector spaces.

These ideas also lead to definitions of neighbourhoods for subsets A of $\mathbf{R}^3$; viz., if $a \in A$, then a *neighbourhood in A* of a is a set $N \cap A$ where N is a neighbourhood of a in the above sense of neighbourhoods for points of $\mathbf{R}^3$. For example, in Fig. 1.7 which pictures the *2-sphere* (a) and *Möbius band* (b), a neighbourhood in A of a is any subset of A containing a 'disc' about a such as that shown.

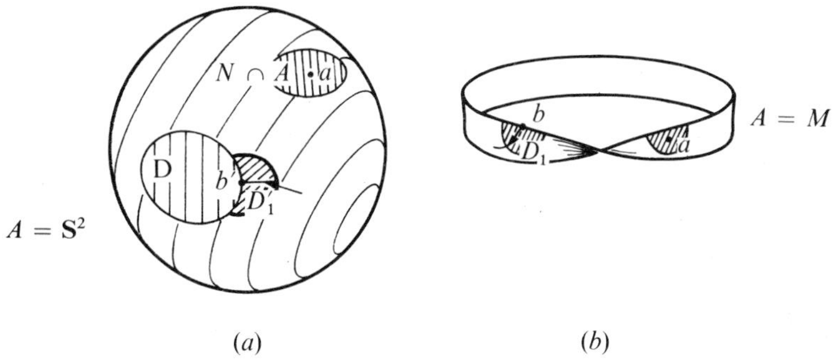

Fig. 1.7

These examples are simple. But we can also give examples of neighbourhoods where it is not easy to visualize the whole set.

Consider the Möbius band M. This has only one edge, and this edge E can be considered as a (somewhat twisted) circle. If we cut a disc D out of the 2-sphere $\mathbf{S}^2$, then the edge of this disc is again a circle, E' say. Let us suppose that these are models (in cloth perhaps) in $\mathbf{R}^3$ and that we have arranged the models so that E and E' have the same length. It would then seem reasonable to produce a new model by stitching E to E' and so joining the two models. Unfortunately, as experiment will show, the whole thing gets hopelessly tangled. The point is, that this sort of model making is impossible in $\mathbf{R}^3$—an extra dimension is needed. (The *proof* of this assertion is very difficult and we will not give it.)

The question is: can we anyway say what we mean by this stitching process without having to produce the result as a subset of $\mathbf{R}^3$? One of the properties of the model we should like is that if in Fig. 1.7 the point b' of $\mathbf{S}^2$ is identified with b in M, then the curve shown should be continuous.

This can be arranged by defining neighbourhoods suitably. First, the model K should be as a set the union of M and $\mathbf{S}^2 \setminus D$ (we suppose $\mathbf{S}^2$ and M are disjoint). If $a \in K \setminus E$, then $a \in \mathbf{S}^2 \setminus D$, or $a \in M \setminus E$; about a we can find discs contained in $\mathbf{S}^2 \setminus D$ or $M \setminus E$ as the case may be, and a neighbourhood of a in K shall be any subset of K containing some such disc. For a point b on E the situation is different. Suppose b is identified with b' on E'. About b and b' consider 'half-discs' D_1 and D_1' (Fig. 1.8). Then any

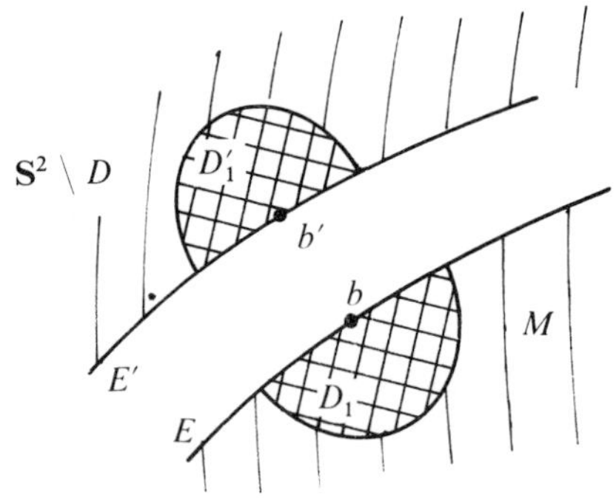

Fig. 1.8

subset of K containing $D_1 \cup (D_1' \setminus E')$ is to be a neighbourhood in K of b, and all neighbourhoods of b in K are to be obtained by this construction for some half-discs D_1, D_1'. It is easy to see that this definition gives the required continuity for curves through b and b'.

We must now leave the particular examples and go forward to the general theory. Examples of the type of the last one will be discussed again in chapter 4.

2. Topological spaces

In 2.1 of this chapter, we give the axioms for a topological space: the topological spaces are the objects of study of the rest of this book.

A topological space consists of a set X and a 'topology' on X. The importance for science of the notion of a topology is that it gives a precise but general sense to the intuitive ideas of nearness and continuity. As we saw in chapter 1 the basis of both these ideas is that of neighbourhood, and so it is this idea of neighbourhood which we axiomatize at first. It will appear quickly that there are equivalent ways of defining a topology, for example by means of open sets, or of closed sets. These definitions do not have the same intuitive appeal as the neighbourhood definition, but they are logically simpler and in some cases give the best method of defining a topology.

The type of structure of the real line **R** with its neighbourhoods of points is that for each r in **R** we have a set of subsets of **R** called neighbourhoods of r. More precisely, we have a function $r \rightsquigarrow \mathcal{N}(r)$ where $\mathcal{N}(r)$ is the set of neighbourhoods of r.

The way to generalize this structure is apparent. We consider a set X and a function $\mathcal{N} : x \rightsquigarrow \mathcal{N}(x)$ assigning to each x in X a set $\mathcal{N}(x)$ of subsets of X called neighbourhoods of x. The function $\mathcal{N}$ will be a topology on X if it satisfies suitable axioms.

The question of what axioms are suitable can not be decided on *a priori* grounds. The decision rests on the character of (*a*) the theory derived from the axioms, and (*b*) the examples of objects which satisfy the axioms. We shall see in later chapters how the theory presented here gives a language which is essential for certain aspects (in fact the topological aspects!) of geometry.

2.1 Axioms for neighbourhoods

Let X be a set and $\mathcal{N}$ a function assigning to each x in X a non-empty set $\mathcal{N}(x)$ of subsets of X. The elements of $\mathcal{N}(x)$ will be called *neighbourhoods of x with respect to $\mathcal{N}$* (or, simply, *neighbourhoods of* x). The function $\mathcal{N}$

is called a *neighbourhood topology* if Axioms N1–N4 below are satisfied; and then X with $\mathcal{N}$ is called a *topological space.*

The following axioms must hold for each x in X.

N1 *If N is a neighbourhood of x, then $x \in N$.*

N2 *If N is a subset of X containing a neighbourhood of x, then N is a neighbourhood of x.*

N3 *The intersection of two neighbourhoods of x is again a neighbourhood of x.*

N4 *Any neighbourhood N of x contains a neighbourhood M of x such that N is a neighbourhood of each point of M.*

We allow the *empty* topological space in which $X = \varnothing$ and $\mathcal{N}$ is the empty function from $\varnothing$ to $\{\varnothing\}$, the set of subsets of $\varnothing$.

EXAMPLE The real line **R** with its neighbourhoods of points is a topological space. The first three axioms have been verified in 1.1, and we now verify N4. Let N be a neighbourhood of $x \in R$; there is a $\delta > 0$ such that $]x - \delta, x + \delta[\subset N$. Let $M =]x - \delta, x + \delta[$. Then M is a neighbourhood of each of its points, and so N is a neighbourhood of each point of M.

This neighbourhood topology on **R** is called the *usual topology* on **R**.

The first three axioms for neighbourhoods have a clear meaning. The fourth axiom has a very important use in the structure of the theory, that of linking together the neighbourhoods of different points of X.

Intuitively, Axiom N4 can be expressed as follows: *a neighbourhood of x is also a neighbourhood of all points sufficiently close to x.* Another way of expressing the axiom is that *each point is inside any of its neighbourhoods.* To explain what is meant by this, we introduce the notion of the interior of a subset of X.

Any point x in X determines its neighbourhoods, which are subsets of X. On the other hand, given a subset A of X we can find those points of which A is a neighbourhood—the set of all such points is called the *interior* of A and is written Int A.

2.1.1 Int $A \subset A$.

2.1.2 *If $A \subset B \subset X$, then* Int $A \subset$ Int B.

2.1.3 Int A *is a neighbourhood of any of its points.*

2.1.4 Int (Int A) = Int A.

Proof If Int $A = \varnothing$, then 2.1.1–2.1.3 (and so 2.1.4) are trivially satisfied. Suppose then $x \in$ Int A, so that A is a neighbourhood of x.

By Axiom N1, $x \in A$. Hence Int $A \subset A$. Further, if $A \subset B$ then B is also a neighbourhood of x (Axiom N2), so that $x \in$ Int B. Hence Int $A \subset$ Int B.

Let M be a neighbourhood of x such that $M \subset A$ and A is a neighbourhood of each point of M (such an M exists by Axiom N4). Then $M \subset \text{Int } A$, and so Int A is a neighbourhood of x, by Axiom N2. This proves 2.1.3.

By 2.1.3, $\text{Int } A \subset \text{Int}(\text{Int } A)$. By 2.1.1 and 2.1.2 $\text{Int}(\text{Int } A) \subset \text{Int } A$. □

The sentence 'each point is inside any of its neighbourhoods' means that if N is a neighbourhood of x, then x belongs to Int N. Also, Int N is again a neighbourhood of x. Thus Axioms N1–N4 imply:

N4′ *If N is a neighbourhood of x, then so also is* Int N.

This axiom implies Axiom N4, since, by definition N is a neighbourhood of each point of Int N.

EXERCISES

1. Let X be a set with a neighbourhood topology. Prove that if A, B are subsets of X then,

$$\text{Int } A \cap \text{Int } B = \text{Int}(A \cap B)$$
$$\text{Int } A \cup \text{Int } B \subset \text{Int}(A \cup B).$$

Prove also that if $(X_\lambda)_{\lambda \in L}$ is a family of subsets of X then,

$$\text{Int} \bigcap_{\lambda \in L} X_\lambda \subset \bigcap_{\lambda \in L} \text{Int } X_\lambda .$$

2. Let $\leqslant$ be an order relation on the set X. Let $x \in X$ and $N \subset X$. We say that N is a neighbourhood of x if there is an open interval I of X such that

$$x \in I \subset N.$$

Prove that these neighbourhoods of points of X form a neighbourhood topology on X. This topology is called the *order topology* on X. What is the order topology on **R**?

3. Prove that the following are neighbourhood topologies on a set X.
(*a*) The *discrete topology*: N is a neighbourhood of x if and only if $x \in N \subset X$.
(*b*) The *indiscrete topology*: N is a neighbourhood of $x \in X$ if and only if $N = X$.
For what X do these topologies coincide? Let $x \in X$. What is Int $\{x\}$ if X has the discrete topology?, the indiscrete topology?

4. Prove that the order topology on **Z** is the discrete topology, but that the order topology on **Q** is not discrete.

5. Let $\leqslant$ be a partial order on the set X. Discuss the possibility, or impossibility, of using $\leqslant$ to define a neighbourhood topology on X.

6. Let X be an uncountable set. Prove that the following define distinct neighbourhood topologies on X.

(*a*) N is a neighbourhood of $x \in X$ if $x \in N \subset X$ and $X \setminus N$ is finite.
(*b*) N is a neighbourhood of $x \in X$ if $x \in N \subset X$ and $X \setminus N$ is countable.

Can either of these topologies be the discrete topology?

7. Let $X = \mathbf{Z}$ and let p be a fixed integer. A set $N \subset \mathbf{Z}$ is a *p-adic neighbourhood* of $n \in \mathbf{Z}$ if N contains the integers $n + mp^r$ for some r and all $m = 0, \pm 1, \pm 2, \ldots$

(so that in a given neighbourhood r is fixed but m varies). Prove that the p-adic neighbourhoods form a neighbourhood topology on $\mathbf{Z}$, the *p-adic topology*. Is this topology the same as the order topology?, the discrete topology?, the indiscrete topology?

The reader familiar with ring theory should develop two generalizations of the p-adic topology on $\mathbf{Z}$. First, replace the ring $\mathbf{Z}$ by an arbitrary ring R, so that now $p \in R$. Second, replace the element p by any ideal P of R. What is the P-adic topology in R if (*a*) $P = R$, (*b*) $P = \{0\}$?

8. Prove that Axioms N1–N4 are independent.

2.2 Open sets

Let $\mathcal{N}$ be a neighbourhood topology on the set X. A subset U of X is *open* (with respect to $\mathcal{N}$) if U is a neighbourhood of each of its points. Thus U is open if and only if $U = \text{Int } U$.

2.2.1 *Let $x \in X$ and $N \subset X$; N is a neighbourhood of x if and only if there is an open set U such that*

$$x \in U \subset N.$$

Proof If N is a neighbourhood of x, then Int N is an open set such that $x \in \text{Int } N \subset N$. Conversely, if U is an open set such that $x \in U \subset N$, then U is a neighbourhood of x and hence so also is N. □

The most important properties of open sets are given by:

2.2.2 *The open sets of X satisfy*

O1 *X and $\varnothing$ are open sets.*

O2 *If U, V are open sets, then $U \cap V$ is open.*

O3 *If $(U_\lambda)_{\lambda \in L}$ is any family of open sets, then $\bigcup_{\lambda \in L} U_\lambda$ is open.*

Proof The relation Int $\varnothing \subset \varnothing$ implies that Int $\varnothing = \varnothing$; thus $\varnothing$ is open. If $x \in X$, then x has at least one neighbourhood N; but $N \subset X$ and so X is a neighbourhood of x. Thus X is open.

If $U \cap V$ is empty, then it is open. If it is not empty, let $x \in U \cap V$. Then U and V are both neighbourhoods of x, and hence $U \cap V$ is a neighbourhood of x. Thus $U \cap V$ is open.

Let $U = \bigcup_{\lambda \in L} U_\lambda$. If U is empty, it is open. If not, let $x \in U$. Then $x \in U_\lambda$ for some $\lambda \in L$, and U_λ, being open, is a neighbourhood of x. But $U_\lambda \subset U$. So U also is a neighbourhood of x. □

We now show that the innocent seeming properties O1, O2, O3 suffice to axiomatize topological spaces in terms of open sets.

Let $\mathcal{U}$ be a set of subsets of X, called open sets, satisfying O1, O2, O3.

For each $x \in X$ a set $\mathcal{M}(x)$ of *$\mathcal{U}$-neighbourhoods* of x is defined by:

$$N \in \mathcal{M}(x) \Leftrightarrow N \subset X \text{ and there is a } U \in \mathcal{U} \text{ such that } x \in U \subset N.$$

(Compare 2.2.1.) The function $x \rightsquigarrow \mathcal{M}(x)$ is said to be *associated with* $\mathcal{U}$.

$\mathcal{M}$ *is* a neighbourhood topology on X. The Axioms N1, N2 are immediately verified while Axiom N3 follows from O2. Also, if $x \in U \in \mathcal{U}$, then $U \in \mathcal{M}(x)$, and this implies Axiom N4.

We now prove:

2.2.3 *$\mathcal{U}$ is the set of open sets of $\mathcal{M}$.*

Proof Certainly, each $U \in \mathcal{U}$ is open with respect to $\mathcal{M}$. Suppose, conversely, that $U \subset X$ and U is open with respect to $\mathcal{M}$. If $U = \varnothing$, then $U \in \mathcal{U}$. If $U \neq \varnothing$, then for each $x \in U$ there is a set U_x in $\mathcal{U}$ such that $x \in U_x \subset U$. Let U' be the union of these U_x for all x in U. Then $U' \subset U$ since each $U_x \subset U$; and $U \subset U'$ since each x in U belongs to U_x. So $U = U'$. Hence, $U = \bigcup_{x \in U} U_x$ belongs to $\mathcal{U}$ by O3. □

Suppose $\mathcal{N}$ is a neighbourhood topology on X, and $\mathcal{U}$ is the set of open sets of $\mathcal{N}$. It is immediate from 2.2.1 that $\mathcal{N}$ is the neighbourhood topology associated with $\mathcal{U}$. Since 2.2.1 is a consequence of Axiom N4, we have shown another use for this axiom—it ties together the neighbourhoods and the open sets.

A set of subsets of X called open sets and satisfying O1, O2, O3 is called an *open set topology* on X. We have proved that the structures of open set topology and neighbourhood topology determine one another: so topology may be developed using either as a starting point.

As we shall see later, topological spaces may be axiomatized in terms of other structures, for example, closed sets, closure, interior, or the relation $A \subset \operatorname{Int} B$. We shall use the word *topology* to denote the set of these equivalent structures, and shall be more specific when necessary. A topological space will be a set with a topology; thus a topological space carries all these structures, and may be defined by any one of them.

A topological space is really a pair $(X, \mathcal{T})$ where $\mathcal{T}$ is a topology on X. It is often convenient to use the symbol X to denote also this pair. Such a notation causes confusion only when we are considering two topologies on the same set X. In such case, we shall write $X_{\mathcal{T}}$ for the topological space consisting of the set X and the topology $\mathcal{T}$. We call X the *underlying set* of $X_{\mathcal{T}}$.

Closed sets, closure

Let X be a topological space. A subset C of X is *closed* if $X \setminus C$ is open.

2.2.4 *The closed sets of X satisfy*

C1 *X and $\varnothing$ are closed sets.*

C2 *If C, D are closed sets, then $C \cup D$ is closed.*

C3 *If $(C_\lambda)_{\lambda \in L}$ is a family of closed sets of X, then $\bigcap_{\lambda \in L} C_\lambda$ is a closed set.*

This is immediate from O1, O2, O3 and the De Morgan laws.

If $\mathscr{C}$ is a set of subsets of a set X, called closed sets, which satisfy C1, C2, C3, then the set $\mathscr{U}$ of complements with respect to X of the elements of $\mathscr{C}$ is a set of open sets satisfying O1, O2, O3; thus $\mathscr{C}$ determines a topology on X, and $\mathscr{C}$ is exactly the set of closed sets of this topology. Thus topological spaces may be axiomatized in terms of closed sets.

Let X be a topological space and let $A \subset X$. The *closure* of A is the set $\bar{A}$ of points x in X such that every neighbourhood of x meets A.

2.2.5 $X \setminus \bar{A} = \text{Int}\,(X \setminus A)$.

Proof Each of the following statements is obviously equivalent to its successor.

(*a*) $x \in X \setminus \bar{A}$.
(*b*) There is a neighbourhood N of x not meeting A.
(*c*) There is a neighbourhood N of x such that $N \subset X \setminus A$.
(*d*) $X \setminus A$ is a neighbourhood of x.
(*e*) $x \in \text{Int}\,(X \setminus A)$. □

2.2.6 $A \subset \bar{A}$.
2.2.7 *If $A \subset B$, then $\bar{A} \subset \bar{B}$.*
2.2.8 *$\bar{A}$ is a closed set.*
2.2.9 *If A is a closed set, then $A = \bar{A}$.*
2.2.10 $\bar{\bar{A}} = \bar{A}$.

Proofs Let $x \in A$. Then any neighbourhood N of x meets A (since $x \in N$). So $x \in \bar{A}$, and 2.2.6 is proved.

2.2.7 is obvious. For 2.2.8, $X \setminus \bar{A} = \text{Int}\,(X \setminus A)$, and $\text{Int}\,(X \setminus A)$ is open by 2.1.4. Hence $\bar{A}$ is closed.

Suppose A is a closed set and $x \notin A$. Then $X \setminus A$ is a neighbourhood of x not meeting A. So $x \notin \bar{A}$. Thus $\bar{A} \subset A$ and so $A = \bar{A}$.

Finally, 2.2.10 follows from 2.2.8 and 2.2.9. □

2.2.11 *If $A, B \subset X$, then $\overline{A \cup B} = \bar{A} \cup \bar{B}$.*

Proof This can be deduced from 2.2.5 and Exercise i of 2.1. Alternatively, we argue as follows.

$\bar{A} \cup \bar{B}$ is closed and contains $A \cup B$. Hence $\overline{A \cup B} \subset \bar{A} \cup \bar{B}$. On the other hand, $A \subset A \cup B$ implies $\bar{A} \subset \overline{A \cup B}$; similarly, $\bar{B} \subset \overline{A \cup B}$, whence $\bar{A} \cup \bar{B} \subset \overline{A \cup B}$. □

EXERCISES

1. What are the open sets of X when X is discrete, that is, has the discrete topology?, is indiscrete, that is, has the indiscrete topology? What is the closure of $\{x\}$, $x \in X$, in these cases?
2. Let X be a topological space and let $A \subset X$. Prove that Int A is the union of all open sets U such that $U \subset A$, and $\bar{A}$ is the intersection of all closed sets C such that $A \subset C$.
3. Let X be a topological space, and let $A \subset X$. A point x in X is called a *limit point* of A if each neighbourhood of x contains points of A other than x. The set of limit points of A is written $\hat{A}$. Prove that $\bar{A} = A \cup \hat{A}$, and that A is closed $\Leftrightarrow$ $\hat{A} \subset A$. Give examples of non-empty subsets A of $\mathbf{R}$ such that (i) $\hat{A} = \varnothing$, (ii) $\hat{A} \neq \varnothing$ and $\hat{A} \subset A$, (iii) A is a proper subset of $\hat{A}$, (iv) $\hat{A} \neq \varnothing$ but $A \cap \hat{A} = \varnothing$.
4. Let X be a topological space and let $A \subset B \subset X$. We say A is *dense in* B if $B \subset \bar{A}$, and A is *dense* if $\bar{A} = X$. Prove that, if A is dense in X and U is open then

$$U \subset \overline{A \cap U}.$$

5. Let $\mathbf{I} = [0, 1]$. Define an order relation $\leqslant$ on $\mathbf{I}^2 = \mathbf{I} \times \mathbf{I}$ by

$$(x, y) \leqslant (x', y') \Leftrightarrow y < y' \text{ or } (y = y' \text{ and } x \leqslant x').$$

The *television topology* on $\mathbf{I}^2$ is the order topology with respect to $\leqslant$ (the name is due to E. C. Zeemann). Let A be the set of points $(2^{-1}, 1 - n^{-1})$ for positive integral n. Prove that in the television topology on $\mathbf{I}^2$

$$\bar{A} = A \cup \{(0, 1)\}.$$

6. A topological space is *separable* if it contains a countable, dense subset. Which of the following topological spaces are separable? (i) $\mathbf{Q}$ with the order topology, (ii) $\mathbf{R}$ with the usual topology, (iii) $\mathbf{I}^2$ with the television topology, (iv) an uncountable set with the indiscrete topology, (v) the spaces defined in Exercise 6 of 2.1.

7. Prove that if A is the closure of an open set, then $A = \overline{\text{Int } A}$. Prove that at most fourteen distinct sets can be constructed from A by the operations of closure and complementation.
8. For any subset A of a topological space X, define Ext A (the *exterior* of A), Bd A (the *boundary* of A) and Fr A (the *frontier* of A) as follows:

$$\begin{aligned} \text{Ext } A &= \text{Int } (X \setminus A) \\ \text{Bd } A &= A \setminus \text{Int } A \\ \text{Fr } A &= \text{Bd } A \cup \text{Bd } (X \setminus A). \end{aligned}$$

Prove that the following relations hold:

(i) $\bar{A} = \text{Int } A \cup \text{Fr } A = A \cup \text{Fr } A = A \cup \text{Bd } (X \setminus A)$.
(ii) $\text{Int } (\text{Bd } A) = \varnothing$.
(iii) $\text{Bd } (\text{Int } A) = \varnothing$.
(iv) $\text{Bd } (\text{Bd } A) = \text{Bd } A$.

(v) $\text{Fr } A = \bar{A} \cap \overline{(X \setminus A)}$.
(vi) Fr A is closed. If A is closed then Bd A = Fr A.
(vii) $\text{Fr}(\text{Fr}(\text{Fr } A)) = \text{Fr}(\text{Fr } A) \subset \text{Fr } A$.
(viii) $\text{Fr } A = \varnothing \Leftrightarrow A$ is both open and closed.
(ix) $\text{Bd } A = A \cap \overline{(X \setminus A)}$.
(x) Bd A is closed $\Leftrightarrow$ A is the union of a closed and an open set.
(xi) $\text{Ext}(\text{Ext } A) = \text{Int } \bar{A}$.
(xii) Ext Ext Ext Ext A = Ext Ext A.

9. A topological space H is defined as follows. The underlying set of H is **R**, and for each $x \in H$ and $N \subset H$, N is a neighbourhood of $x \Leftrightarrow$ there are real numbers x', x'' such that

$$x \in [x', x''[\subset N.$$

Prove that H is a topological space and that (i) each interval $[a, b[$ is both open and closed, (ii) H is separable, (iii) if $A \subset H$, then $A \setminus \hat{A}$ is countable. (This topology on **R** is the *half-open topology*.)

10. Let X be a non-empty set and $i : \mathscr{P}(X) \to \mathscr{P}(X)$ a function such that for all $A, B \in \mathscr{P}(X)$

(i) $i(A) \subset A$
(ii) $i(i(A)) = i(A)$
(iii) $i(X) = X$
(iv) $i(A \cap B) = i(A) \cap i(B)$.

For each x in X, define A to be a neighbourhood of x if $x \in i(A)$. Prove that these neighbourhoods form a topology $\mathscr{N}$ on X. Which of the axioms (i)–(iv) are essential in the proof?

*11. With the notation of Exercise 10, prove that i is the interior operator for the topology $\mathscr{N}$.

*12. Let X be a non-empty set and $\lhd$ a relation on subsets of X such that

(i) $\varnothing \lhd \varnothing$, $X \lhd X$
(ii) $A \subset A'$, $A' \lhd B'$ and $B' \subset B$ imply $A \lhd B$
(iii) $A \lhd B$ implies $A \subset B$
(iv) $A \lhd B$ and $A' \lhd B'$ imply $A \cap A' \lhd B \cap B'$
(v) $A_i \lhd B_i$ for all $i \in I$ implies $\cup_{i \in I} A_i \lhd \cup_{i \in I} B_i$.

For each $x \in X$, $A \subset X$ define A to be a neighbourhood of x if $\{x\} \lhd A$. Prove that these neighbourhoods define a topology on X for which $A \lhd B \Leftrightarrow A \subset \text{Int } B$.

*13. Show how to axiomatize topologies using the closure operator. [Use 2.2.5 and Exercise 10.]

2.3 Product spaces

Let X, Y be topological spaces. We consider the problem of defining a reasonable topology on the set $X \times Y$. For example, if $X = Y = \mathbf{R}$, this is the problem of finding the notion of 'nearness' in the Euclidean plane $\mathbf{R}^2 = \mathbf{R} \times \mathbf{R}$.

We consider the abstract situation. Let $x \in X$, $y \in Y$; we wish to define neighbourhoods of (x, y) in $X \times Y$ in such a way that Axioms N1–N4 are satisfied.

An obvious first attempt is to say that the neighbourhoods of (x, y) shall be the sets $M \times N$ for M, N neighbourhoods of x, y respectively. This would correspond to the intuitive idea '(x, y) is near to (x', y') if x is near to x' and y is near to y''. However, with this definition Axiom N2 is not satisfied: the set P in the following figure is not of the form $M \times N$ for any sets $M \subset X$, $N \subset Y$.

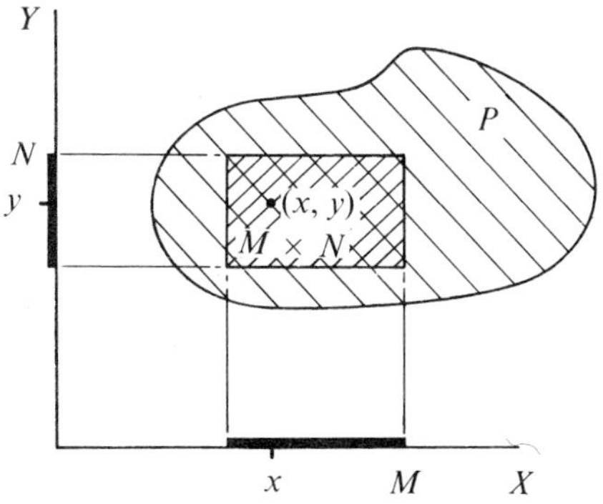

Fig. 2.1

We therefore make a virtue of necessity. The sets $M \times N$ as above we call the *basic neighbourhoods* of (x, y), and we define a *neighbourhood of* (x, y) *in* $X \times Y$ to be any subset of $X \times Y$ containing a basic neighbourhood of (x, y). We show that with these neighbourhoods $X \times Y$ is a topological space.

Let P be a neighbourhood of (x, y), and $M \times N$ a basic neighbourhood of (x, y) contained in P. Clearly, $(x, y) \in P$, and if $P \subset Q$ then Q is a neighbourhood of (x, y). This verifies Axioms N1, N2.

Let $M^0 = \text{Int } M$, $N^0 = \text{Int } N$. Then $M^0 \times N^0$ is a basic neighbourhood of (x, y) and of any $(x', y') \in M^0 \times N^0$. Hence, P is a neighbourhood of any $(x', y') \in M^0 \times N^0$. This verifies Axiom N4.

Finally, let P' be a neighbourhood of (x, y) containing the basic neighbourhood $M' \times N'$ of (x, y). Then $(M \cap M') \times (N \cap N') = (M \times N) \cap (M' \times N') \subset P \cap P'$, and so $P \cap P'$ is a neighbourhood of (x, y). This verifies Axiom N3 and completes the proof that $X \times Y$ is a topological space.

The product $X_1 \times \cdots \times X_n$ of n topological spaces $X_1, \ldots, X_n$ is defined inductively by

$$X_1 \times \cdots \times X_n = (X_1 \times \cdots \times X_{n-1}) \times X_n.$$

It is easily shown that a set $P \subset X_1 \times \cdots \times X_n$ is a neighbourhood of

$(x_1, \ldots, x_n)$ if and only if there are neighbourhoods M_i of x_i, $i = 1, \ldots, n$, such that $M_1 \times \cdots \times M_n \subset P$. In particular, the product topology of $\mathbf{R}^n = \mathbf{R} \times \cdots \times \mathbf{R}$ is called the *usual topology* on $\mathbf{R}^n$.

Let X, Y be topological spaces.

2.3.1 *If U, V are open in X, Y respectively, then $U \times V$ is open in $X \times Y$.*

2.3.2 *If C, D are closed in X, Y respectively, then $C \times D$ is closed in $X \times Y$.*

Proofs U is a neighbourhood of each x in U, V is a neighbourhood of each y in V. So $U \times V$ is a neighbourhood of each (x, y) in $U \times V$.

That $C \times D$ is closed is immediate from the formula

$$(X \times Y) \setminus (C \times D) = (X \times (Y \setminus D)) \cup ((X \setminus C) \times Y). \quad \square$$

2.3.3 *A set U is open in $X \times Y \Leftrightarrow$ there are sets U_λ, $V_\lambda (\lambda \in L)$ open in X, Y respectively such that $U = \bigcup_{\lambda \in L} U_\lambda \times V_\lambda$.*

Proof The implication $\Leftarrow$ is clear from 2.3.1 and property O3 of open sets.

In order to prove the implication $\Rightarrow$, let U be open in $X \times Y$. For each $\lambda \in U$ there is a basic neighbourhood $M \times N$ of λ such that $M \times N \subset U$. Let $U_\lambda = \operatorname{Int} M$, $V_\lambda = \operatorname{Int} N$. Then U_λ, V_λ are open and $U = \bigcup_{\lambda \in U} U_\lambda \times V_\lambda$. $\square$

EXAMPLE Let $\alpha = (a, b) \in \mathbf{R}^2$, and let $r > 0$. The *open ball about α of radius r* is the set

$$B(\alpha, r) = \{(x, y) \in \mathbf{R}^2 : (x - a)^2 + (y - b)^2 < r^2\}.$$

This open ball is an open set: For, let $\alpha' = (a', b') \in B(\alpha, r)$ and let $s = \sqrt{\{(a' - a)^2 + (b' - b)^2\}}$. Then $s < r$. Let $0 < \delta < (r - s)/\sqrt{2}$, $M =]a' - \delta, a' + \delta[$, $N =]b' - \delta, b' + \delta[$. Then $M \times N \subset B(\alpha, r)$ and so $B(\alpha, r)$ is a neighbourhood of α'.

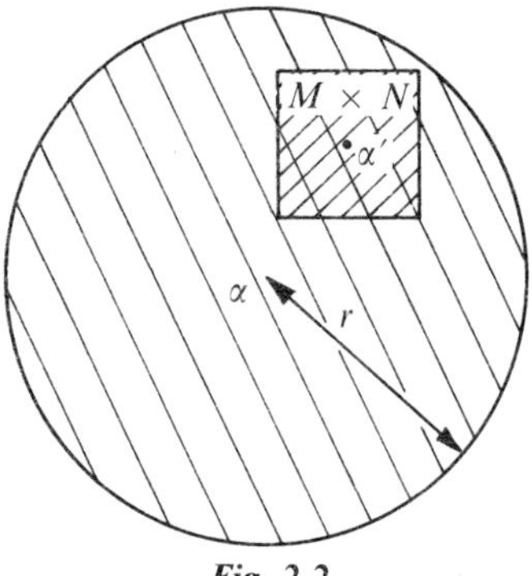

Fig. 2.2

2.3.4 *Let $A \subset X$, $B \subset Y$. Then*

$$\operatorname{Int}(A \times B) = \operatorname{Int} A \times \operatorname{Int} B$$

$$\overline{A \times B} = \bar{A} \times \bar{B}.$$

Proof The relation $\operatorname{Int} A \times \operatorname{Int} B \subset \operatorname{Int}(A \times B)$ follows from the fact that $\operatorname{Int} A \times \operatorname{Int} B$ is an open set contained in $A \times B$. On the other hand, suppose $(x, y) \in \operatorname{Int}(A \times B)$. Then $A \times B$ is a neighbourhood of (x, y) and so contains a basic neighbourhood $M \times N$ of (x, y). Thus, $M \subset A$, $N \subset B$ and so A, B are neighbourhoods of x, y respectively. Hence, $(x, y) \in \operatorname{Int} A \times \operatorname{Int} B$.

The proof of the second relation is left as an exercise to the reader. □

We recall that the projections $p_1 : X \times Y \to X$, $p_2 : X \times Y \to Y$ are the functions $(x, y) \rightsquigarrow x$, $(x, y) \rightsquigarrow y$.

2.3.5 *If U is open in $X \times Y$, then $p_1[U]$ is open in X, $p_2[U]$ is open in Y.*

Proof We prove only that $p_1[U]$ is open in X. This is clear if $p_1[U]$ (and so U) is empty.

Suppose $x \in p_1[U]$. Then there is a y in Y such that $(x, y) \in U$. Since U is open, it contains a basic neighbourhood $M \times N$ of (x, y). So $M = p_1[M \times N] \subset p_1[U]$. Hence, $p_1[U]$ is a neighbourhood of x. □

It is not true that the projection of a closed set of $X \times Y$ is closed [Exercise 3].

EXERCISES

1. State which of the following sets of points (x, y) of $\mathbf{R}^2$ are (*a*) open, (*b*) closed, (*c*) neither open nor closed. (i) $|x| < 1$ and $|y| < 1$, (ii) $|x| + |y| \leqslant 1$, (iii) $x^2 + y^2 > 1$, (iv) $|xy| < 1$, (v) $\sin \pi x = 0$, (vi) one of x, y is rational, (vii) $y = \sin(1/x)$, $x \neq 0$, (viii) there is an integer $n > 0$ such that $xy = 1/n$, (ix) there are integers p, q such that $q > 0$, p/q is in its lowest terms, and $x = p/q$, $y = 1/q$.
2. Find the closure and interior of each of the sets of Exercise 1.
3. Prove that the set $\{(x, y) \in \mathbf{R}^2 : xy = 1\}$ is closed in $\mathbf{R}^2$ but that its projections in $\mathbf{R}$ are not closed.
4. Prove the second relation in 2.3.4.
5. Let $A \subset X \times Y$, $x \in X$, and let $A_x = \{y \in Y : (x, y) \in A\}$. Prove that A_x is open (closed) in Y if A is open (closed) in $X \times Y$. Give examples of subsets A of $\mathbf{R}^2$ such that (i) A is not open but A_x is open for each x in $\mathbf{R}$, (ii) A is not closed but A_x is closed for each x in $\mathbf{R}$.
6. Let $A \subset \mathbf{R}^3$ be the set of points

$$\{((2 + \cos \alpha t) \cos t, (2 + \cos \alpha t) \sin t, \sin \alpha t) : t \in \mathbf{R}\}.$$

Prove that A is closed if and only if α is a rational multiple of π, and that if A is not closed then $\bar{A}$ is the *anchor ring*

$$\{(x, y, z) \in \mathbf{R}^3 : ((x^2 + y^2)^{\frac{1}{2}} - 2)^2 + z^2 = 1\}.$$

7. Prove directly that the open sets given by 2.3.3 satisfy the Axioms O1, O2, O3 for an open set topology on $X \times Y$.
8. Prove that if X, Y are separable spaces, then so also is $X \times Y$.
9. Prove that $X \times Y$ is discrete (indiscrete) if and only if both X and Y are discrete (indiscrete).

2.4 Relative topologies and subspaces

Let X be a topological space and A a subset of X. We consider the problem of defining a topology on A so that A becomes a topological space. The theory here works out slightly simpler if the topology is defined in terms of open sets rather than of neighbourhoods.

The *induced*, or *relative*, *topology* on A (with respect to X) is that in which the open sets are the sets $U \cap A$ where U is an open set of X. These sets $U \cap A$ are called *open in* A. (Thus *open in* X means the same as open.)

We must verify that this does define a topology.

(a) Axiom O1 is trivially verified, since $A \cap \varnothing = \varnothing$, $A \cap X = A$.
(b) Let V, V' be open in A. Then $V = U \cap A$, $V' = U' \cap A$ where U, U' are open in X. Hence, $V \cap V' = U \cap U' \cap A$ is open in A.
(c) Let $(V_\lambda)_{\lambda \in L}$ be a family of sets open in A, and let V be the union of this family. For each $\lambda \in L$, there is an open set U_λ such that $V_\lambda = U_\lambda \cap A$. So

$$V = \bigcup_{\lambda \in L} (U_\lambda \cap A) = \Big(\bigcup_{\lambda \in L} U_\lambda\Big) \cap A$$

is open in A.

EXAMPLE Let $X = \mathbf{R}$, $A = [0, 1[$. Then A itself is open in A, and $[0, \tfrac{1}{2}[$ is open in A; neither of these sets are open in X. On the other hand $]\tfrac{1}{2}, 1[$ is open in A and open in X.

Let X be a topological space. A topological space A which is a subset of X and whose topology is the relative topology as a subset of X is called a *subspace* of X. A subset of X is usually assumed to have the relative topology (if it has a topology at all) and so to be a subspace of X. The relative topologies on $\mathbf{N}$, $\mathbf{Z}$, $\mathbf{Q}$ as subsets of $\mathbf{R}$ are called the *usual topologies* on these sets. In the case of $\mathbf{N}$, $\mathbf{Z}$, the usual topologies are the discrete topologies: for, if $n \in \mathbf{Z}$, then

$$\{n\} = \mathbf{Z} \cap \,]n - \tfrac{1}{2}, n + \tfrac{1}{2}[;$$

hence $\{n\}$ is open in $\mathbf{Z}$ and so any subset of $\mathbf{Z}$ is open in $\mathbf{Z}$. A similar argument applies to $\mathbf{N}$.

For the rest of this section, we suppose that A is a subspace of the topological space X.

Let $a \in A$. The neighbourhoods of a for the topology of A are called *neighbourhoods in* A of a.

2.4.1 *Let $a \in A$. A set $N \subset A$ is a neighbourhood in A of $a \Leftrightarrow$ there exists M, a neighbourhood in X of a, such that $N = M \cap A$.*

Proof $\Leftarrow$ Let $N = M \cap A$, where N is a neighbourhood in X of a. Then Int N is open in X and so (Int N) $\cap A$ is open in A. Hence M is a neighbourhood in A of a since $a \in (\text{Int } N) \cap A \subset M$.

$\Rightarrow$ Let N be a neighbourhood in A of a. Then there is a set V open in A such that $a \in V \subset N$. Also, $V = U \cap A$ where U is open in X. Then, $M = U \cup N$ is a neighbourhood in X of a such that $M \cap A = N$. □

For example, if $X = \mathbf{R}$, $A = [0, 1]$, then $[0, \frac{1}{2}[$ is a neighbourhood in A of 0; and if $A = \{0\}$ then A itself is a neighbourhood in A of 0.

Again, let $X = \mathbf{R}$, $A = \{(x, y) \in \mathbf{R}^2 : x^2 + y^2 = 1\}$. The thickened part of A in Fig. 2.3 is a neighbourhood in A of P, but is not a neighbourhood in A of Q.

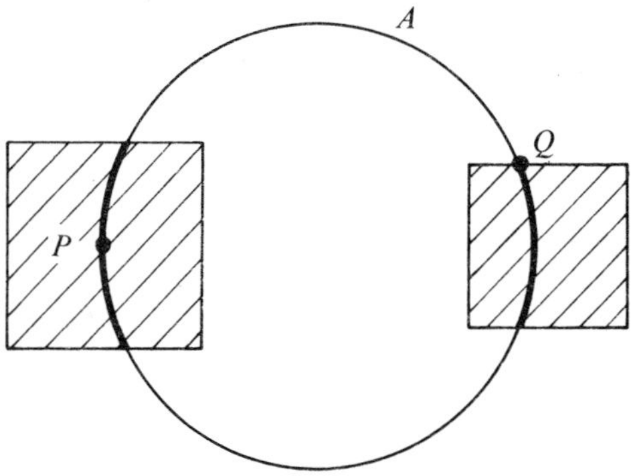

Fig. 2.3

A subset C of A is *closed in A* if C is closed in the topology of A, that is, if $A \setminus C$ is open in A. Also, if $C \subset A$, then we denote by

$$\text{Cl}_A\, C$$

the closure of C with respect to the relative topology of A. This operation is very simply related to the closure $\bar{C}$ of C in X.

2.4.2 *If $C \subset A$, then*

$$\text{Cl}_A\, C = \bar{C} \cap A.$$

Proof Suppose M is a subset of X. Since C is contained in A, we have M meets C if and only if $M \cap A$ meets C. It follows from this and 2.4.1 that if $x \in A$, then all neighbourhoods in A of x meet C if and only if all neighbourhoods in X of x meet C. □

2.4.3 *If $C \subset A$, then C is closed in A if and only if there is a set D closed in X such that $C = D \cap A$.*

Proof If C is closed in A, then $C = \mathrm{Cl}_A\, C$ and so $C = \bar{C} \cap A$. The result follows (with $D = \bar{C}$).

Conversely, if $C = D \cap A$ where D is closed in X, then $C \subset D$ and so $\bar{C} \subset D$. It follows easily that $C = \bar{C} \cap A$ and so $C = \mathrm{Cl}_A\, C$. This shows that C is closed in A. □

2.4.4 *If A is closed in X, then any set closed in A is also closed in X. The same holds with the word closed replaced by open.*

Proof The first part follows from 2.4.2 and the fact that the intersection of two closed sets is closed. The second part is similar. □

EXERCISES

1. Prove that the relation 'X is a subspace of Y' is a partial order relation for topological spaces.
2. Prove that the set $\{x \in \mathbf{Q}: -\sqrt{2} \leqslant x \leqslant \sqrt{2}\}$ is both open and closed in $\mathbf{Q}$.
3. Let A be the subspace of $\mathbf{R}$ of points $1/n$ for $n \in \mathbf{Z} \setminus \{0\}$. Prove that A is discrete, but that the subspace $A \cup \{0\}$ of $\mathbf{R}$ is not discrete.
4. Prove that a subspace of a discrete space is discrete, and a subspace of an indiscrete space is indiscrete.
5. Let A be the subspace $[0, 2] \setminus \{1\}$ of $\mathbf{R}$. Prove that $[0, 1[$ is both open and closed in A.
6. Let $x \in X$ and let A be a neighbourhood (in X) of x. Prove that the neighbourhoods in A of x are exactly the neighbourhoods in X of x which are contained in A.
7. Let $\leqslant$ be an order relation on the set X. If $A \subset X$ then the restriction of $\leqslant$ is an order relation on A. Show that it is not necessarily true that if A, X have the order topologies, then A is a subspace of X. What is the order topology on $\mathbf{Q}$?
8. Let A, B be subspaces of X, Y respectively. Prove that $A \times B$ is a subspace of $X \times Y$.
9. Let A be a subspace of X, and let Int, Int_A denote respectively the interior operators for X, A. Prove that if $B \subset X$, then

$$(\mathrm{Int}\, B) \cap A \subset \mathrm{Int}_A\,(B \cap A).$$

Give an example for which $(\mathrm{Int}\, B) \cap A \neq \mathrm{Int}_A\,(B \cap A)$.

10. Let A be a subspace of X, and let Cl, Cl_A denote respectively the closure operators for X, A. Prove that if $B \subset X$, then

$$\mathrm{Cl}_A\,(B \cap A) \subset (\mathrm{Cl}\, B) \cap A.$$

Give an example for which $\mathrm{Cl}_A\,(B \cap A) \neq (\mathrm{Cl}\, B) \cap A$.

*11. A set $A \subset X$ is *locally closed* if, for each a in A, there exists N, a neighbourhood in X of a, such that $N \cap A$ is closed in N. Prove that A is locally closed $\Leftrightarrow$ A is the intersection of a closed set and an open set of X.

12. Write an account of relative topologies inverting the order of the present section; that is, define neighbourhoods in A using 2.4.1, prove that these neighbourhoods form a topology on A, and show that the sets open in A are as defined here.

*13. Let H be the real line with the half-open topology [Exercise 9 of 2.2]. Prove that the subspace $\{(x, x) \in H \times H : x \in H\}$ of $H \times H$ is not separable.

2.5 Continuity

One of the main reasons for studying the concept of a topological space is that it provides the most natural context for dealing with continuity. In this section, we define continuity for functions $X \to Y$, where X, Y are topological spaces, and we give also a number of important rules for constructing continuous functions.

Let $X_{\mathscr{S}}$, $Y_{\mathscr{T}}$ be topological spaces with underlying sets X, Y and topologies $\mathscr{S}$, $\mathscr{T}$ respectively. By a *function* $X_{\mathscr{S}} \to Y_{\mathscr{T}}$ is meant the triple $(f, \mathscr{S}, \mathscr{T})$ consisting of a function $f : X \to Y$ and the two topologies $\mathscr{S}$, $\mathscr{T}$. The purpose of this notation is that two functions $(f, \mathscr{S}, \mathscr{T})$, $(f', \mathscr{S}', \mathscr{T}')$ are equal if and only if $f = f'$, $\mathscr{S} = \mathscr{S}'$, $\mathscr{T} = \mathscr{T}'$. However, we make an abuse of language and denote such a function $(f, \mathscr{S}, \mathscr{T})$ also by f.

Let X, Y be topological spaces, and $f : X \to Y$ a function. We say f is *continuous* if, for all x in X, N is a neighbourhood of $f(x)$ implies $f^{-1}[N]$ is a neighbourhood of x. This condition is obviously equivalent to: for all x in X, if N is a neighbourhood of $f(x)$, then there is a neighbourhood M of x such that $f[M] \subset N$.

A *map* $X \to Y$ is simply a continuous function $X \to Y$.

2.5.1 *Let X, Y be topological spaces. Any constant function $X \to Y$ is continuous.*

Proof Let $f : X \to Y$ be a constant function and let $x \in X$. If N is a neighbourhood of $f(x)$, then $f^{-1}[N] = X$ which is a neighbourhood of x. □

2.5.2 *Let X be a topological space. The identity $1_X : X \to X$ is continuous.*

This follows easily from the rule $1_X^{-1}[N] = N$ for $N \subset X$.

2.5.3 *Let $f : X \to Y$ be a map and let $A \subset X$, $B \subset Y$ be such that $f[A] \subset B$. Then $f \mid A, B$, the restriction† of f, is a map.*

Proof Let $g = f \mid A, B$, let $a \in A$ and let M be a neighbourhood in B of

† See A.1.4 of the Appendix.

$f(a)$. Then there exists N, a neighbourhood in Y of $f(a)$, such that $M = N \cap B$. Since f is a map, $f^{-1}[N]$ is a neighbourhood in X of a. Hence $g^{-1}[M] = f^{-1}[N] \cap A$ is a neighbourhood in A of a. □

A corollary of 2.5.3 is that, if $A \subset X$, then the inclusion function $i : A \to X$ is a map: for $i = 1_X|A$.

2.5.4 *If $f : X \to Y$, $g : Y \to Z$ are maps, so also is $gf : X \to Z$.*

Proof Let $x \in X$ and let N be a neighbourhood of $gf(x)$. Then

$$(gf)^{-1}[N] = f^{-1}g^{-1}[N]$$

and so $(gf)^{-1}[N]$ is a neighbourhood of x. □

2.5.5 *If X, Y are topological spaces, then the projections $p_1 : X \times Y \to X$, $p_2 : X \times Y \to Y$ are maps.*

Proof Let N be a neighbourhood of $x = p_1(x, y)$. Then $p_1^{-1}[N] = N \times Y$ is a neighbourhood of (x, y). □

Now, functions $f : Z \to X$, $g : Z \to Y$ determine uniquely a function $(f, g) : Z \to X \times Y$ whose components are f, g, that is, (f, g) is $z \rightsquigarrow (fz, gz)$.

2.5.6 *Let $f : Z \to X$, $g : Z \to Y$ be maps. Then $(f, g) : Z \to X \times Y$ is a map.*

Proof Let $h = (f, g)$, so that h sends $z \rightsquigarrow (f(z), g(z))$. Let P be a neighbourhood of $h(z)$, and let $M \times N$ be a basic neighbourhood of $h(z)$ contained in P. Then $h^{-1}[P]$ contains the set

$$\begin{aligned} h^{-1}[M \times N] &= \{z \in Z : f(z) \in M, g(z) \in N\} \\ &= f^{-1}[M] \cap g^{-1}[N]. \end{aligned}$$

It follows that $h^{-1}[P]$ is a neighbourhood of z. □

This result can also be expressed: *a function $h : Z \to X \times Y$ is continuous $\Leftrightarrow$ p_1h, p_2h are continuous.* The implication $\Rightarrow$ follows from 2.5.4 and 2.5.5, while the converse implication follows from 2.5.6 since p_1h, p_2h are the components f, g of h.

There are a number of useful corollaries of 2.5.6.

2.5.7 *The diagonal map $\Delta : X \to X \times X$ is continuous.*

2.5.8 *The twisting function $T : X \times Y \to Y \times X$ which sends $(x, y) \rightsquigarrow (y, x)$ is continuous.*

2.5.9 *If $f : X \to X'$, $g : Y \to Y'$ are continuous, then so is $f \times g : X \times Y \to X' \times Y'$.*

Proofs The diagonal map Δ is simply $(1_X, 1_X)$, and so is continuous. The twisting map T is (p_2, p_1), where p_1, p_2 are the projections of $X \times Y$. Finally, $f \times g = (fp_1, gp_2)$, and so $f \times g$ is continuous. □

EXAMPLES 1. We use some of these results to prove a sum and product rule for maps $X \to \mathbf{R}$. Let $f, g : X \to \mathbf{R}$ be maps. Then $f + g, f.g$ are the functions

$$X \xrightarrow{(f,g)} \mathbf{R} \times \mathbf{R} \xrightarrow{+} \mathbf{R}, X \xrightarrow{(f,g)} \mathbf{R} \times \mathbf{R} \xrightarrow{\cdot} \mathbf{R}$$

respectively where $+, .$ denote the addition and multiplication functions $(x, y) \rightsquigarrow x + y$, $(x, y) \rightsquigarrow xy$. We shall prove later that these latter functions are continuous—the continuity of $f + g, f.g$ follows.

2. The following type of example will occur frequently in later chapters. Suppose $F : X \times X \to X$ is a map, and consider the function

$$G : X \times X \to X$$
$$(x, y) \rightsquigarrow F(y, F(x, y)).$$

Then G is a map since it is the composite of the maps

$$X \times X \xrightarrow{1 \times \Delta} X \times X \times X \xrightarrow{T \times 1} X \times X \times X \xrightarrow{1 \times F} X \times X \xrightarrow{F} X$$
$$(x, y) \rightsquigarrow (x, y, y) \rightsquigarrow (y, x, y) \rightsquigarrow (y, F(x, y)) \rightsquigarrow G(x, y).$$

3. Let $\mathbf{S}^1 = \{(x, y) \in \mathbf{R}^2 : x^2 + y^2 = 1\}$, and consider the function

$$f : [0, 1[\to \mathbf{S}^1$$
$$t \rightsquigarrow (\cos 2\pi t, \sin 2\pi t).$$

Then f is a bijection, and is continuous since its components are continuous. However f^{-1} is not continuous since $M = [0, \frac{1}{2}]$ is a neighbourhood in $[0, 1[$ of 0, but $f[M]$ is not a neighbourhood in $\mathbf{S}^1$ of $f(0) = (1, 0)$. This confirms the intuitive idea that breaking a loop of string is a non-continuous process.

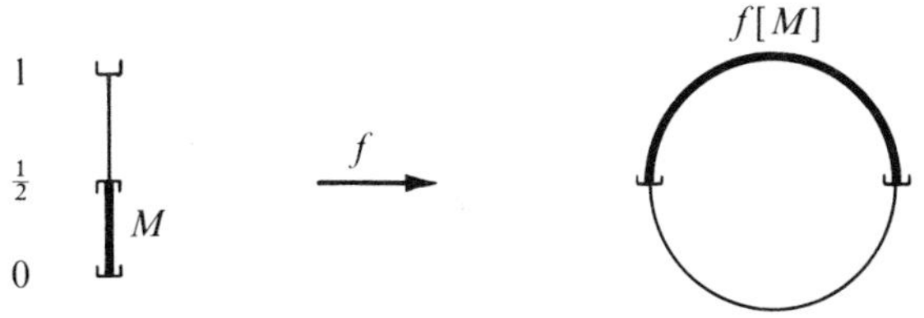

Fig. 2.4

Continuity of a function is a 'local' property in the following sense.

2.5.10 *Let X, Y be topological spaces and $f : X \to Y$ a function such that each $x \in X$ has a neighbourhood N such that $f \mid N$ is continuous. Then f is continuous.*

Proof Let $x \in X$ and let M be a neighbourhood of $f(x)$. Let N be a neighbourhood of x such that $f \mid N$ is continuous. Then

$$(f \mid N)^{-1}[M] = f^{-1}[M] \cap N \subset f^{-1}[M].$$

Since $f \mid N$ is continuous, $(f \mid N)^{-1}[M]$ is a neighbourhood in N of x, and so is also a neighbourhood in X of x (since N is a neighbourhood of x). Therefore, $f^{-1}[M]$ is a neighbourhood in X of x. □

Finally, we prove two versions of the 'glueing rule'.

2.5.11 *Let X, Y be topological spaces and $f : X \to Y$ a function. Let $X = A \cup B$ where $A \setminus B \subset \text{Int } A$, $B \setminus A \subset \text{Int } B$. If $f \mid A, f \mid B$ are continuous, then f is continuous.*

Proof Let $x \in X$ and let P be a neighbourhood of $f(x)$.

Suppose first $x \in A \cap B$. By continuity of $f \mid A, f \mid B$ and 2.4.1 there are neighbourhoods M, N of x in X such that

$$f^{-1}[P] \cap A = (f \mid A)^{-1}[P] = M \cap A$$

$$f^{-1}[P] \cap B = (f \mid B)^{-1}[P] = N \cap B.$$

So $M \cap N \subset (M \cap A) \cup (N \cap B) = f^{-1}[P]$, and hence $f^{-1}[P]$ is a neighbourhood of x.

Suppose next $x \in A \setminus B$. Then A is a neighbourhood of x and hence so also is $M \cap A$ where M is constructed as above. *A fortiori*, $f^{-1}[P]$ is a neighbourhood of x. A similar argument applies if $x \in B \setminus A$. □

2.5.12 *Let X, Y be topological spaces and $f : X \to Y$ a function. Let A, B be closed subsets of X such that $X = A \cup B$. If $f \mid A, f \mid B$ are continuous, so also is f.*

Proof Since A, B are closed and have union X, the set $A \setminus B = X \setminus B$ is open, and so $A \setminus B \subset \text{Int } A$. Similarly, $B \setminus A \subset \text{Int } B$. So the result follows from 2.5.11. □

EXERCISES

1. Prove continuity of the following functions $\mathbf{R}^3 \to \mathbf{R}$.

(i) $(x, y, z) \rightsquigarrow P(x, y, z)$ where P is a polynomial.
(ii) $(x, y, z) \rightsquigarrow \sin xz + \cos (x + y + z)$.

2. Let $F, G : X \times X \to X$ be maps. Prove continuity of the functions

(i) $X \times X \times X \to X, (x, y, z) \rightsquigarrow F(G(y, z), x)$.
(ii) $X \times X \times X \to X \times X, (x, y, z) \rightsquigarrow (F(x, z), G(y, y))$.

3. Let $X = A \cup B$, Y be topological spaces and let $f : X \to Y$ be a function

such that $f \mid A, f \mid B$ are continuous. Prove that f is continuous if

$$\overline{(A \setminus B)} \cap (B \setminus A) = \varnothing, \qquad (A \setminus B) \cap \overline{(B \setminus A)} = \varnothing.$$

4. Let X be a topological space and let $f, g : X \to \mathbf{R}$ be maps. Prove that the following functions $X \to \mathbf{R}$ are maps.
(i) $x \rightsquigarrow |f(x)|$
(ii) $x \rightsquigarrow f(x) / g(x)$ (if $g(x)$ is never 0)
(iii) $x \rightsquigarrow \max\{f(x), g(x)\}$, $x \rightsquigarrow \min\{f(x), g(x)\}$.

5. Prove that X has the discrete topology $\Leftrightarrow$ for all spaces Y any function $X \to Y$ is continuous. Find a similar characterization of the indiscrete topology.
6. Let $x_0 \in X$, $y_0 \in Y$ and let $X \vee Y$ be the subspace $X \times \{y_0\} \cup \{x_0\} \times Y$ of $X \times Y$. Let $i_1 : X \to X \vee Y$, $i_2 : Y \to X \vee Y$ be the functions $x \rightsquigarrow (x, y_0)$, $y \rightsquigarrow (x_0, y)$ respectively. Prove that a function $f : X \vee Y \to Z$ is continuous $\Leftrightarrow f\, i_1$, $f\, i_2$ are continuous.

2.6 Other conditions for continuity

The set of neighbourhoods at a point x in a topological space X contains 'large' neighbourhoods, for example X itself. However, for many purposes, such as deciding continuity, it is only necessary to look at 'small' neighbourhoods of x, or at neighbourhoods of a particular type. For example, in $X \times Y$ it is the basic neighbourhoods $M \times N$ which are important. The precise way of expressing these notions is in terms of a *base* for the neighbourhoods of x.

A *base for the neighbourhoods* at $x \in X$ is a set $\mathscr{B}(x)$ of neighbourhoods of x such that if N is a neighbourhood of x then N contains some B of $\mathscr{B}(x)$.

EXAMPLES 1. The set of all neighbourhoods of x is a base for the neighbourhoods at x. So also is the set of open neighbourhoods of x.
2. The intervals $]-1/n, 1/n[$ for positive integral n form a base for the neighbourhoods of 0 in $\mathbf{R}$. So also do the closed intervals $[-1/n, 1/n]$.
3. The basic neighbourhoods $M \times N$ of (x, y) in $X \times Y$ form a base for the neighbourhoods of (x, y).
4. Let M be a neighbourhood of x. The neighbourhoods N of x such that $N \subset M$ form a base for the neighbourhoods of x.

If we have such a base $\mathscr{B}(x)$ for each x in X, then the function $\mathscr{B} : x \rightsquigarrow \mathscr{B}(x)$ is called a *base for the neighbourhoods of* X. Our main result on bases is the following.

Let $\mathscr{B}$, $\mathscr{B}'$ be bases for the neighbourhoods of X, X' respectively. Let $f : X \to X'$ be a function.

2.6.1 *f is continuous $\Leftrightarrow$ for each x in X and $N \in \mathscr{B}'(f(x))$, there is an $M \in \mathscr{B}(x)$ such that $f[M] \subset N$.*

Proof The proof is simple.

$\Rightarrow$ Let $x \in X$, $N \in \mathscr{B}'(f(x))$. Then $f^{-1}[N]$ is a neighbourhood of x and so there is an $M \in \mathscr{B}(x)$ such that $M \subset f^{-1}[N]$. This implies $f[M] \subset N$.

$\Leftarrow$ Let $x \in X$ and let P be a neighbourhood of $f(x)$. Then there exists $N \in \mathscr{B}'(f(x))$ such that $N \subset P$. By assumption there is an $M \in \mathscr{B}(x)$ such that $M \subset f^{-1}[N]$. So $f^{-1}[P]$, which contains $f^{-1}[N]$, is a neighbourhood of x. □

The argument here is similar to that of 2.5.6.

The continuity of a function can also be described in terms of open sets, closed sets, or closure. This fact, which is of vital importance later, is contained in the following omnibus theorem (in which *(a)*–*(d)* are the important conditions).

Let X, Y be topological spaces and $f : X \to Y$ a function.

2.6.2 *The following conditions are equivalent.*

(a) f is continuous.

(b) If U is open in Y, then $f^{-1}[U]$ is open in X.

(c) If C is closed in Y, then $f^{-1}[C]$ is closed in X.

(d) If A is a subset of X, then

$$f[\bar{A}] \subset \overline{f[A]}.$$

(e) If B is a subset of Y, then

$$\overline{f^{-1}[B]} \subset f^{-1}[\bar{B}].$$

(f) If D is a subset of Y, then

$$f^{-1}[\operatorname{Int} D] \subset \operatorname{Int} f^{-1}[D].$$

Proof *(a)* $\Rightarrow$ *(b)* If $f^{-1}[U]$ is empty, then it is open. Otherwise, let $x \in f^{-1}[U]$. Then $f(x) \in U$ and so U is a neighbourhood of $f(x)$. Hence $f^{-1}[U]$ is a neighbourhood of x.

(b) $\Rightarrow$ *(a)* This follows easily from the fact that if N is a neighbourhood of $f(x)$, then Int N is open.

(b) $\Leftrightarrow$ *(c)* This is a simple consequence of

$$f^{-1}[Y \setminus C] = X \setminus f^{-1}[C].$$

(a) $\Rightarrow$ *(d)* Let $y \in f[\bar{A}]$ so that $y = f(x)$ where $x \in \bar{A}$. Let N be a neighbourhood of y. Then $f^{-1}[N]$ meets A since $f^{-1}[N]$ is a neighbourhood of x.

Hence N meets $f[A]$, and so $y \in \overline{f[A]}$.

(d) $\Rightarrow$ *(e)* Let $A = f^{-1}[B]$ so that $f[A] \subset B$. Then

$$f[\bar{A}] \subset \overline{f[A]} \subset \bar{B}$$

whence $\overline{f^{-1}[B]} = \bar{A} \subset f^{-1}[\bar{B}]$.

$(e) \Leftrightarrow (f)$ This is an immediate consequence of the rules

$$f^{-1}[Y \setminus D] = X \setminus f^{-1}[D]$$
$$X \setminus \overline{f^{-1}[D]} = \operatorname{Int}(X \setminus f^{-1}[D]).$$

$(f) \Rightarrow (b)$ Let U be open in Y, so that $U = \operatorname{Int} U$. Then

$$f^{-1}[U] = f^{-1}[\operatorname{Int} U] \subset \operatorname{Int} f^{-1}[U].$$

So $f^{-1}[U] = \operatorname{Int} f^{-1}[U]$ and $f^{-1}[U]$ is open. □

It should be confessed that 2.6.2 is useful to solve rigorously some earlier exercises, particularly those of 2.3. For example, to prove that the set $A = \{(x, y) \in \mathbf{R}^2 : |xy| < 1\}$ is open in $\mathbf{R}^2$, we consider the function $f : \mathbf{R}^2 \to \mathbf{R}$ which sends $(x, y) \rightsquigarrow |xy|$. Then f is continuous and $A = f^{-1}]-1, 1[$. Since $]-1, 1[$ is open in $\mathbf{R}$, A is open in $\mathbf{R}^2$. The reader should work again through the Exercises in 2.3 to show how 2.6.2 can be used.

EXAMPLES 5. Let $f, g : X \to \mathbf{R}^n$ be maps. Then the set A of points on which f, g agree is closed in X. For let $h = f - g : X \to \mathbf{R}^n$; then h is continuous (as we shall prove later), $\{0\}$ is closed in $\mathbf{R}^n$ and so, by 2.6.2(c), $A = h^{-1}[\{0\}]$ is closed in X.

6. Let $f : X \to \mathbf{R}$ be a map and let $A \subset X \times \mathbf{R}$ be the graph of f. Then A is closed in $X \times \mathbf{R}$ since A is the set of points of $X \times \mathbf{R}$ on which the maps p_2, fp_1 agree.

For example, the set $\{(x, y) : y = x^2 + e^x + \sin x\}$ is closed in $\mathbf{R}^2$.

7. The function $x \rightsquigarrow x/|x|$ is a continuous function $\mathbf{R}^{\neq 0} \to \mathbf{R}$. The graph of this function is closed in $\mathbf{R}^{\neq 0} \times \mathbf{R}$, but not in $\mathbf{R}^2$.

8. Let X be the graph of the function $x \rightsquigarrow \sin \pi/x$ $(x \neq 0)$. Then X is closed in $Z = \mathbf{R}^{\neq 0} \times \mathbf{R}$, but not in $\mathbf{R}^2$. In fact, let $J = \{(0, y) \in \mathbf{R}^2 : -1 \leqslant y \leqslant 1\}$; we prove that $\overline{X}$, the closure of X in $\mathbf{R}^2$, is $X \cup J$.

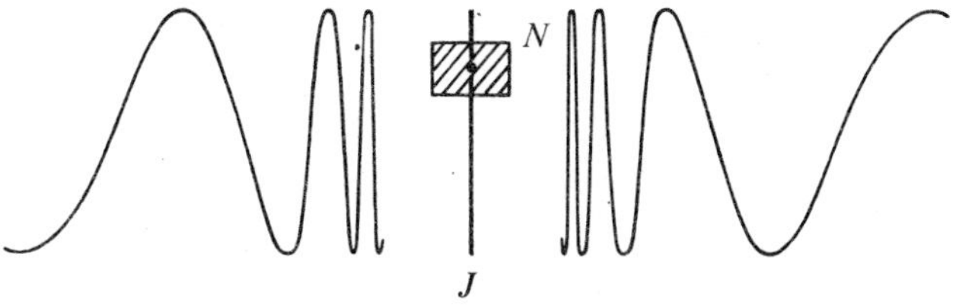

Fig. 2.5

First, $\sin \pi/x$ assumes all values in $[-1, 1]$ for x in any interval $[1/(n + 1), 1/n]$ (n a positive integer); therefore, if $z \in J$, then any neighbourhood N of z meets X. That is, $X \cup J \subset \overline{X}$.

Second, $|\sin \pi/x| \leqslant 1$ $(x \neq 0)$; therefore, if $(x, y) \in \bar{X}$, then $|y| \leqslant 1$.

Third, X is closed in Z and hence $X = Z \cap \bar{X}$ [2.4.2]. So the only points of $\bar{X}$ which are not in X are the points of J.

EXERCISES

1. Let A be a subspace of X and let $\mathscr{B}$ be a base for the neighbourhoods of X. Construct from $\mathscr{B}$ a base for the neighbourhoods of A.
2. Let $\mathscr{B}(x)$, $\mathscr{B}'(x')$ be bases for the neighbourhoods of $x \in X$, $x' \in X'$ respectively. Prove that the sets $M \times N$ for $M \in \mathscr{B}(x)$, $N \in \mathscr{B}'(x')$ form a base for the neighbourhoods of $(x, x') \in X \times X'$, and that the sets $M \times M$ for $M \in \mathscr{B}(x)$ form a base for the neighbourhoods of $(x, x) \in X \times X$.
3. A topological space X is said to satisfy the *first axiom of countability* if there is a base $\mathscr{B}$ for the neighbourhoods of X such that $\mathscr{B}(x)$ is countable for each x in X. Prove that the following satisfy the first axiom of countability: $\mathbf{R}$, $\mathbf{Q}$, a discrete space, a space with a countable number of open sets.
4. Prove that subspaces and (finite) products of spaces satisfying the first axiom of countability also satisfy the first axiom of countability.
5. A topological space X has a countable base for the neighbourhoods at x. Prove that there is a base for the neighbourhoods of x of sets B_n, $n \in \mathbf{N}$, such that $B_n \supset B_{n+1}$, $n \in \mathbf{N}$.
6. Deduce Exercises 9 and 10 of 2.4 from 2.6.2(*e*), (*f*).
7. Prove that the continuity of $f : X \to Y$ is not equivalent to the condition: if $A \subset X$, then $\operatorname{Int} f[A] \subset f[\operatorname{Int} A]$.
8. Give an example of X, Y, a map $f : X \to Y$ and subsets A of X, and B, D of Y such that $f[\bar{A}] \neq \overline{f[A]}$, $\overline{f^{-1}[B]} \neq f^{-1}[\bar{B}]$, $f^{-1}[\operatorname{Int} D] \neq \operatorname{Int} f^{-1}[D]$. [Let X have the discrete topology.]
9. Let $f, g : X \to \mathbf{R}$ be maps. Prove that the sets

$$\{x \in X : f(x) \geqslant g(x)\}, \qquad \{x \in X : f(x) \leqslant g(x)\}$$

are closed in X.
10. Deduce the version 2.5.12 of the glueing rule from 2.6.2(*c*). Generalize this rule to the case when X is the union of n closed sets $C_1, \ldots, C_n$.

Is this rule true if the sets C_i need not be closed? if their number is not finite?
11. Let $f : X \to \mathbf{R}$ be a map and let $A = \{(x, 1/f(x)) \in X \times \mathbf{R} : f(x) \neq 0\}$. Prove that A is closed in $X \times \mathbf{R}$.
12. A map $f : X \to Y$ is *closed* if f maps closed sets of X to closed sets of Y, and is *open* if f maps open sets of X to open sets of Y. Give examples of X, Y, f for which (i) f is neither open nor closed, (ii) f is open but not closed, (iii) f is closed but not open, (iv) f is both open and closed.

2.7 Comparison of topologies, homeomorphism

There is a partial order on the set of topologies on a set X defined as follows. Let $\mathscr{S}$, $\mathscr{T}$ be topologies in X. We say $\mathscr{S}$ is *finer* than $\mathscr{T}$ (and $\mathscr{T}$ is *coarser* than $\mathscr{S}$) if the identity function $i : X_{\mathscr{S}} \to X_{\mathscr{T}}$ is continuous. In such case we write $\mathscr{S} \geqslant \mathscr{T}$ (or $\mathscr{T} \leqslant \mathscr{S}$). Clearly, this relation is transitive and reflexive.

Let $\mathscr{S}, \mathscr{T}$ be topologies on X. If $N \subset X$, then $i^{-1}[N] = N$. So we have: *$\mathscr{S}$ is finer than $\mathscr{T}$ if and only if for each x in X each $\mathscr{T}$-neighbourhood of x is also an $\mathscr{S}$-neighbourhood of x.* It follows easily that $\leqslant$ is also an anti-symmetric relation, which is to say that if $\mathscr{S} \geqslant \mathscr{T}$ and $\mathscr{T} \geqslant \mathscr{S}$ then $\mathscr{S} = \mathscr{T}$, and the neighbourhoods of any x in X are the same for $\mathscr{S}$ as for $\mathscr{T}$.

By 2.6.2 this relation can also be described in terms of open, or of closed, sets: $\mathscr{S} \geqslant \mathscr{T} \Leftrightarrow$ each set open for $\mathscr{T}$ is also open for $\mathscr{S} \Leftrightarrow$ each set closed for $\mathscr{T}$ is also closed for $\mathscr{S}$.

The finest topology on X is the discrete topology and the coarsest topology is the indiscrete topology.

Two topologies $\mathscr{S}$, $\mathscr{T}$ on X may be *incomparable*, in which case $\mathscr{S}$ is neither finer nor coarser than $\mathscr{T}$. For example, let $X = \{x, y\}$, let $\mathscr{S}$ be the topology whose open sets are $\varnothing, \{x\}, X$ and let $\mathscr{T}$ be the topology whose open sets are $\varnothing$, $\{y\}$, X. Clearly $\mathscr{S}$ and $\mathscr{T}$ are incomparable.

Let X, Y be topological spaces. A function $f : X \to Y$ is a *homeomorphism* if (i) f is a bijection, and (ii) for all x in X, M is a neighbourhood of $x \Leftrightarrow f[M]$ is a neighbourhood of $f(x)$. Clearly (i) and (ii) are equivalent to (i) and (ii′): for all x in X, N is a neighbourhood of $f(x) \Leftrightarrow f^{-1}[N]$ is a neighbourhood of x.

This definition, though possibly the most intuitive, is not the most elegant or useful. Better conditions for homeomorphism are given in 2.7.1: let X, Y be topological spaces and $f : X \to Y$ a function.

2.7.1 *The following conditions are equivalent.*
(*a*) *f is a homeomorphism.*
(*b*) *f is continuous, a bijection, and $f^{-1} : Y \to X$ is continuous.*
(*c*) *f is continuous and there is a continuous function $g : Y \to X$ such that $gf = 1_X, fg = 1_Y$.*

Proof (*b*) is obviously equivalent to (*a*). Given (*b*) then $g = f^{-1}$ satisfies $gf = 1_X, fg = 1_Y$. Conversely, if g exists as in (*c*), then f must be a bijection and g must be f^{-1}. So (*c*) implies (*b*). □

This shows that we may replace (ii) in the definition of homeomorphism by (ii″): *U is open in $X \Leftrightarrow f[U]$ is open in Y*, or similar conditions involving f^{-1}, or closed sets.

It is important to note that a continuous bijection need not be a homeomorphism. For example, let $\mathscr{S}, \mathscr{T}$ be topologies on a set X. The identity $i : X_{\mathscr{S}} \to X_{\mathscr{T}}$ is a homeomorphism $\Leftrightarrow \mathscr{S} = \mathscr{T}$; and i is continuous $\Leftrightarrow \mathscr{S} \geqslant \mathscr{T}$. The relation $\mathscr{S} \geqslant \mathscr{T}$ does not imply $\mathscr{S} = \mathscr{T}$.

If $f : X \to Y$ is a homeomorphism we write $f : X \approx Y$, and if a homeomorphism $f : X \approx Y$ exists we say that X is homeomorphic to Y or X is of the same homeomorphism type as Y and write $X \approx Y$. We leave the reader to check that the relation $X \approx Y$ is an equivalence relation.

2.7.2 *Any two open intervals of* **R** *are homeomorphic.*

Proof First let $a, b \in \mathbf{R}$ $(a < b)$ and consider the function

$$f :]0, 1[\to]a, b[$$
$$t \rightsquigarrow a(1 - t) + bt.$$

Then f is continuous, a bijection, and with continuous inverse $s \rightsquigarrow (s - a)/(b - a)$. Thus, f is a homeomorphism. It follows that all bounded open intervals in **R** are homeomorphic.

We now show that any interval in **R** is homeomorphic to a bounded interval. Consider the function

$$g : \mathbf{R} \to]-1, +1[$$
$$r \rightsquigarrow r/(1 + |r|).$$

Then g is continuous, a bijection and with inverse $s \rightsquigarrow s/(1 - |s|)$, which is continuous. Let I be any interval of **R**. Then $g[I]$ is a bounded interval and $g \mid I, g[I]$ is a homeomorphism. □

This example illustrates that to prove two spaces are homeomorphic, one simply constructs a homeomorphism from one to the other. It is usually more difficult to prove that two given spaces are not homeomorphic.

Topology is often characterized as the study of those properties of spaces which are not changed under homeomorphism. For this reason, homeomorphic spaces are also called *topologically equivalent*.

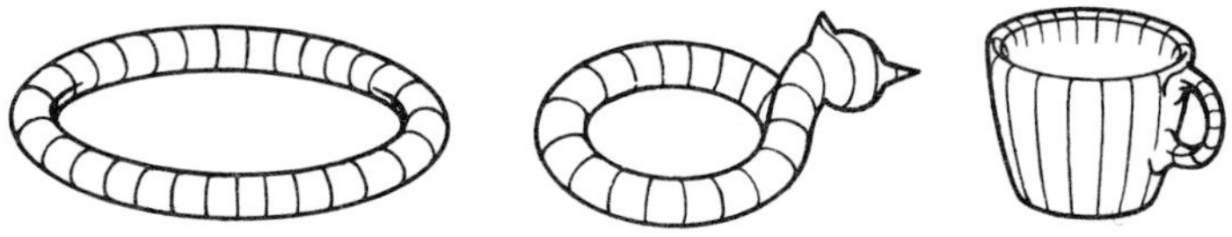

Fig. 2.6

Topology has also been called 'rubber sheet geometry', because if a surface X is constructed from sheets of rubber, then elastic deformations such as pulling and squashing, do not change the homeomorphism type of X. As an example, Fig. 2.6 illustrates surfaces in $\mathbf{R}^3$ which are all topologically equivalent.

One final definition will be needed later. Let X, Y be topological spaces. A function $f : X \to Y$ is a *homeomorphism into* if its restriction $f \mid X, f[X]$

is a homeomorphism. For example, any inclusion mapping of a subspace into a total space is a homeomorphism and, in general, a homeomorphism into is the composite of a homeomorphism and an inclusion map.

EXAMPLE Let X, Y be topological spaces, and let $y \in Y$. The function $f : X \to X \times Y$, $x \rightsquigarrow (x, y)$, is a homeomorphism into. First, f is continuous, since its components are the identity map and the constant map $x \rightsquigarrow y$. Second, the inverse of $f \mid X, f[X]$ is simply the restriction of the projection $p_1 : X \times Y \to Y$, and so this inverse is continuous.

The following glueing rule is often useful for constructing homeomorphisms (cf. section 4 of Chapter 4).

2.7.3 *Let $X = X_1 \cup X_2$, $Y = Y_1 \cup Y_2$ be topological spaces such that X_1, X_2 are closed in X and Y_1, Y_2 are closed in Y. Let $f_1 : X_1 \to Y_1$, $f_2 : X_2 \to Y_2$ be homeomorphisms which restrict to the same homeomorphism $f_0 : X_1 \cap X_2 \to Y_1 \cap Y_2$. Then the function*

$$f : X \to Y$$
$$x \rightsquigarrow \begin{cases} f_1 x, & x \in X_1 \\ f_2 x, & x \in X_2 \end{cases}$$

is well-defined and a homeomorphism.

Proof The function f is well-defined since f_1, f_2 agree on $X_1 \cap X_2$. The continuity of f follows from 2.5.12. A similar argument shows that f^{-1} is defined and continuous. □

EXERCISES

1. Construct homeomorphisms between the subspaces A, B of $\mathbf{R}^2$ in each of the following cases.

(i) $A = \mathbf{R}^2$, $B = \{(x, y) : y > 0\}$,
(ii) $A = \{(x, y) : x^2 + y^2 \leqslant 1\}$, $B = \{(x, y) : |x| \leqslant 1, |y| \leqslant 1\}$.
(iii) $A = \{(x, y) : y \geqslant 0\}$, $B = \{(x, y) : y \geqslant x^2\}$,
(iv) $A = \mathbf{R}^2$, $B = \{(x, y) : x^2 + y^2 < 1\}$,
(v) $A = \{(x, y) : y \geqslant 0 \text{ and } |x| \leqslant 1\}$, $B = \{(x, y) : y \geqslant 0 \text{ and } |x| \leqslant (1 + |y|)^{-1}\}$,
(vi) $A = \{(x, y) : y \neq 0\}$, $B = \{(x, y) : y < 0 \text{ or } y > 1\}$.

2. Let $\mathbf{S}^1 = \{z \in \mathbf{C} : |z| = 1\}$, let $\alpha \in \mathbf{R}$ be irrational. Prove that the function $f : \mathbf{R} \to \mathbf{S}^1 \times \mathbf{S}^1$, $t \rightsquigarrow (e^{2\pi i\alpha t}, e^{2\pi i t})$, is not a homeomorphism into.

3. Let f: [0, 1] $\to$ [a, b] be an order preserving bijection. Prove that f is a homeomorphism.

[In Exercises 4–7 it should be assumed that if $f : [a, b] \to \mathbf{R}$ is a map, then Im f is a closed, bounded interval.]

4. Let $f : [0, 1] \to [0, 1]$ be a homeomorphism. Prove that $f(0)$ is 0 or 1, and that $f]0, 1[=]0, 1[$.

5. Prove that $[0, 2]$ is not homeomorphic to $[-1, 0[\cup]1, 2]$.
6. Prove that there is no continuous surjection $[0, 1] \to]0, 1[$; construct a continuous surjection $]0, 1[\to [0, 1]$.
7. Let $X \subset \mathbf{R}$ be the union of the open intervals $]3n, 3n + 1[$ and the points $3n + 2$ for $n = 0, 1, 2, \ldots$. Let $Y = (X \setminus \{2\}) \cup \{1\}$. Prove that there are continuous bijections $f : X \to Y$, $g : Y \to X$, but that X, Y are not homeomorphic.
8. Is the half-open topology on $\mathbf{R}$ finer, coarser, or incomparable to the usual topology on $\mathbf{R}$?
9. Let A, the *anchor ring*, be the set of points in $\mathbf{R}^3$

$$(\cos \theta(2 + \cos \varphi), \sin \theta(2 + \cos \varphi), \sin \varphi), \quad \theta, \varphi \in \mathbf{R}.$$

Construct a homeomorphism from A to the *torus* $\mathbf{S}^1 \times \mathbf{S}^1$.
10. Let $\mathbf{E}^2 = \{(x, y) \in R^2 : x^2 + y^2 \leqslant 1\}$. The space $\mathbf{S}^1 \times \mathbf{E}^2$ is called the *solid torus*. Prove that the *3-sphere*

$$\mathbf{S}^3 = \{(x_1, \ldots, x_4) \in \mathbf{R}^4 : x_1^2 + \cdots + x_4^2 = 1\}$$

is the union of two spaces each homeomorphic to a solid torus and with intersection homeomorphic to a torus. [Consider the subspaces of $\mathbf{S}^3$ given by $x_1^2 + x_2^2 \leqslant x_3^2 + x_4^2$ and by $x_1^2 + x_2^2 \geqslant x_3^2 + x_4^2$.]
11. Construct a homeomorphism $f : \mathbf{I}^2 \to \mathbf{I}^2$ (where $\mathbf{I}^2 = \mathbf{I} \times \mathbf{I}$) such that f maps $\mathbf{I} \times \{0, 1\} \cup \{0\} \times \mathbf{I}$ onto $\{0\} \times \mathbf{I}$.

2.8 Metric spaces and normed vector spaces

Let $\mathbf{K}$ denote either the real numbers, the complex numbers or the quaternions. The reader not familiar with quaternions may simply not consider them at this stage—in fact the only property of $\mathbf{K}$ we use is that $\mathbf{K}$ is a field (with non-commutative multiplication if $\mathbf{K}$ is the quaternions) and that for each element $\alpha \in \mathbf{K}$ there is defined an *absolute value*, or *modulus*, $|\alpha| \in \mathbf{R}$ with the properties:

(*a*) $|\alpha| > 0$ if $\alpha \neq 0$, and $|0| = 0$,
(*b*) $|\alpha\beta| = |\alpha|\,|\beta|$,
(*c*) $|\alpha + \beta| \leqslant |\alpha| + |\beta|$, for all $\alpha, \beta \in \mathbf{K}$.

We shall be considering vector spaces V over $\mathbf{K}$. Now, it is usual in elementary work to write αx for the multiple of the vector x in V by the scalar α in $\mathbf{K}$—this is expressed by saying that V is considered as a *left* vector space over $\mathbf{K}$. However, it turns out that in the case $\mathbf{K}$ is non-commutative it is more convenient to write $x\alpha$ instead of αx—that is, to consider V as a *right* vector space over $\mathbf{K}$, with scalar multiplication a function $V \times \mathbf{K} \to V$. Given such a right vector space, we can always define left scalar multiplication by $\alpha x = x\alpha$, $x \in V$, $\alpha \in \mathbf{K}$. However, if $\alpha, \beta \in \mathbf{K}$, $x \in V$, then

$$\beta(\alpha x) = (\alpha x)\beta = (x\alpha)\beta = x(\alpha\beta) = (\alpha\beta)x.$$

So we obtain the usual associativity rule $\beta(\alpha x) = (\beta\alpha)x$ if and only if **K** is commutative. Thus a vector space over **R** or **C** can, and will, be considered as both a left and a right vector space, while a vector space over the quaternions **H** will be considered only as a right vector space. To cover all cases, we frame our axioms for normed spaces in terms of right vector spaces.

Let V be any right vector space over **K**. A *norm* on V is a function

$$\| \ \| : V \to \mathbf{R}$$

such that for any x, y in V and α in **K**:

NVS1 $\|x\| > 0$ if $x \neq 0$,

NVS2 $\|x\alpha\| = |\alpha| \, \|x\|$,

NVS3 $\|x + y\| \leqslant \|x\| + \|y\|$.

Then V with such a norm is called a *normed vector space*. Intuitively, a norm gives a measure of the size of elements of V. From the formal viewpoint, the three axioms tie in the norm with the addition and scalar multiplication of the vector space structure on V.

The following examples show the importance of this concept.

EXAMPLES 1. The field **K** itself is a normed vector space over **K** with norm $\|x\| = |x|$, $x \in \mathbf{K}$.

2. More generally, the n-dimensional vector space $\mathbf{K}^n$ over **K** has many norms of which two are particularly important. (i) The *Euclidean norm* or *modulus* on $\mathbf{K}^n$ is written $|x|$ and is defined by

$$|(x_1, \ldots, x_n)| = \left\{ \sum_{i=1}^{n} |x_i|^2 \right\}^{1/2}, \quad x_i \in \mathbf{K}.$$

The verification of axioms NVS1 and 2 is trivial, while axiom NVS3 follows from the well-known Cauchy–Schwarz inequality [cf. section 5.4]. The resulting normed vector space is called *n-dimensional Euclidean space* if $\mathbf{K} = \mathbf{R}$, *n-dimensional unitary space* if $\mathbf{K} = \mathbf{C}$, and *n-dimensional symplectic space* if $\mathbf{K} = \mathbf{H}$, the field of quaternions. (ii) The *Cartesian norm* on $\mathbf{K}^n$ is defined by

$$\|(x_1, \ldots, x_n)\| = \max \{|x_1|, \ldots, |x_n|\}, \quad x_i \in \mathbf{K}.$$

The verification that this is a norm is very simple and is left to the reader.

As a further example, let p be a real number such that $p \geqslant 1$, and let

$$\|(x_1, \ldots, x_n)\|_p = \left\{ \sum_{i=1}^{n} |x_i|^p \right\}^{1/p}, \quad x_i \in \mathbf{K}.$$

This defines a norm on $\mathbf{K}^n$, axiom NVS3 being the generalized Minkowski inequality. [cf. Flett [1] p. 182].

3. Let $\mathscr{C}$ be the vector space over $\mathbf{R}$ of all maps $[0, 1] \to \mathbf{R}$. There are two norms on $\mathscr{C}$ which are important in analysis, the *sup norm* defined by

$$\|f\|_s = \sup \{|f(x)| : x \in [0, 1]\}$$

and the *integral norm* defined by

$$\|f\|_I = \int_0^1 |f(x)| \, dx.$$

The verification that the sup norm is a norm is entirely trivial—the continuity of any f in $\mathscr{C}$ is used only to show that $\|f\|$ is well defined. On the other hand, the continuity of any f in $\mathscr{C}$ is used essentially in showing that for the integral norm, $\|f\|_I > 0$ if $f \neq 0$.

As we shall see in the next section, a norm on a vector space V induces in a natural way a topology on V, and also on subsets of V. However, a subset of V need not be a vector space, and so not a normed vector space. For this reason, we widen our outlook and consider the more general concept of a metric space.

Let X be any set. A metric on X is a function $d : X \times X \to \mathbf{R}^{\geqslant 0}$ with the following properties.

M1 $d(x, y) = 0 \Leftrightarrow x = y$
M2 $d(x, y) = d(y, x)$
M3 $d(x, z) \leqslant d(x, y) + d(y, x)$.

A set X with a metric d is called a *metric space*, and is denoted by X_d or simply X.

EXAMPLES 4. Let V be a normed vector space and X a subset of V. Then the function

$$d(x, y) = \|x - y\|, \quad x, y \in X$$

is a metric on X. In particular, V itself will always be taken to have this metric.

5. Let X be a set, and define $d(x, y)$ to be 0 if $x = y$ and 1 otherwise. This is the *discrete* metric on X.

6. Let X_d be a metric space. Then we can define a new metric on X by

$$d'(x, y) = \min \{1, d(x, y)\}.$$

We can in a metric space define generalizations of open and of closed intervals in $\mathbf{R}$. Let X be a metric space with metric d. Let $a \in X$, $r \geqslant 0$. The *open ball about a of radius r* is the set

$$B(a, r) = \{x \in X : d(x, a) < r\}.$$

The *closed ball about a of radius r* is the set

$$E(a, r) = \{x \in X : d(x, a) \leqslant r\}$$

and the *sphere about a of radius r* is

$$S(a, r) = \{x \in X : d(x, a) = r\}.$$

The closed ball is sometimes called a *cell*, or *disc*.

In a normed vector space V, the sets $B(0, 1)$, $E(0, 1)$, $S(0, 1)$ are called the standard ball, cell, and sphere and are written $B(V)$, $E(V)$, $S(V)$. In particular, if $V = \mathbf{R}^n$ with the Euclidean norm, these are denoted by $\mathbf{B}^n$, $\mathbf{E}^n$, $\mathbf{S}^{n-1}$. The standard 1-cell $\mathbf{E}^1$ is of course the interval $[-1,1]$. If $V = \mathbf{R}^n$ with the Cartesian norm, then $E(V)$ is the n-fold product of $[-1, 1]$ with itself, and is written $\mathbf{J}^n$.

The following diagram illustrates $E(V)$ when $V = \mathbf{R}^2$ with respectively the Cartesian, Euclidean, $\| \ \|_1$, and $\| \ \|_3$ norms.

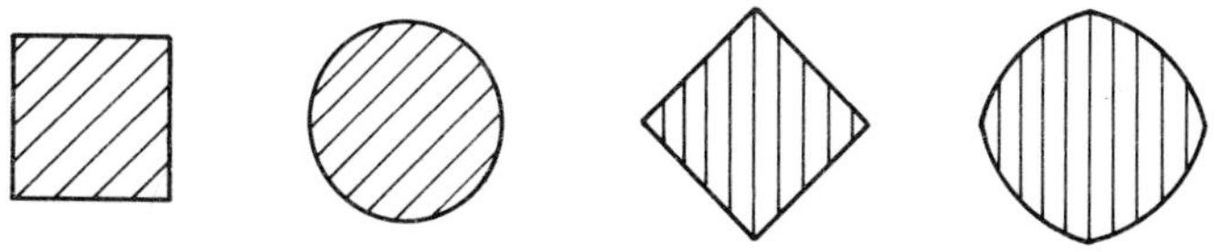

Fig. 2.7

Let X be a metric space and let $a \in X$, $r > 0$.

2.8.1 (*a*) *If $a' \in B(a, r)$ then there is a $\delta > 0$ such that*

$$B(a', \delta) \subset B(a, r).$$

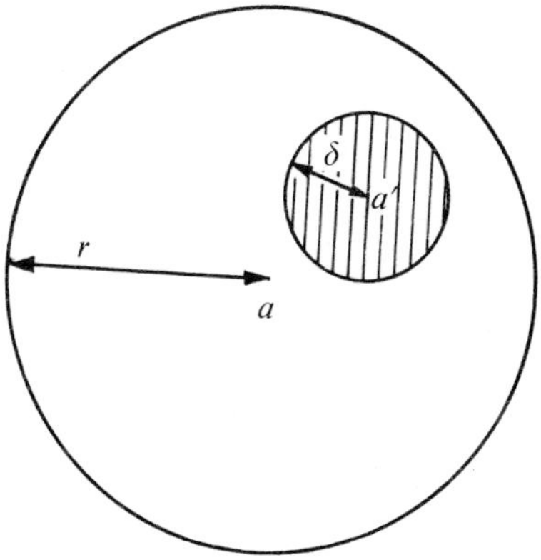

Fig. 2.8

(*b*) *If* $a' \in X \setminus E(a, r)$, *then there is a* $\delta > 0$ *such that*

$$B(a', \delta) \subset X \setminus E(a, r).$$

Proof Let $a' \in B(a, r)$. Then $\delta = r - d(a', a)$ is positive. Also if $x \in B(a', \delta)$ then $d(x, a) \leqslant d(x, a') + d(a', a) < \delta + r - \delta = r$. This proves (*a*). The proof of (*b*) is similar and is left to the reader. □

Metric topologies

We now show how a metric d on X defines a topology on X.

Definition A set $N \subset X$ is a *neighbourhood of* $a \in X$ if there is a real number $r > 0$ such that $B(a, r) \subset N$.

We verify axioms N1–N4. The relation $d(a, a) = 0$ implies $a \in B(a, r)$; this verifies Axiom N1. The verification of Axioms N2 and N3 is simple and is left to the reader. By 2.8.1 an open ball is a neighbourhood of each of its points; so the verification of axiom N4 is immediate.

Thus the neighbourhoods on X form a topology on X called the topology *induced by d* or the *metric topology*. Clearly, the open balls $B(a, r)$, all $r > 0$, form a base for the neighbourhoods of a; so also do the open balls $B(a, r)$ for r rational and positive, or for r of the form $1/n$, n a positive integer. This proves that X has a countable base for the neighbourhoods at each point. Notice also that $E(a, r/2) \subset B(a, r)$, so the closed balls of radius $r > 0$ also form a base for the neighbourhoods of r.

EXAMPLES 7. The *usual metric* on **R** is the metric $d(x, y) = |x - y|$. The topology induced by d is clearly the usual topology on **R**. Similarly, the fields **C**, **H** of complex numbers and quaternions have topologies induced by $d(x, y) = |x - y|$.

8. The Cartesian norm on $\mathbf{K}^n$ induces the product topology. This is proved as follows. If N_i is a neighbourhood in **K** of a_i, then $N_i \supset B(a_i, r_i)$ for some $r_i > 0$. Let $a = (a_1, \ldots, a_n)$, $\delta = \min \{r_1, \ldots, r_n\}$. Then

$$N_1 \times \cdots \times N_n \supset B(a, \delta).$$

This shows that each product neighbourhood of a is a metric neighbourhood of a. The converse is trivial since

$$B(a, r) = B(a_1, r) \times \cdots \times B(a_n, r).$$

9. Let X be a set and let d be the discrete metric on X. Then, if $a \in X$, the open balls about a are given by

$$B(a, r) = \begin{cases} X, & r > 1 \\ \{x\}, & 0 < r \leqslant 1. \end{cases}$$

Hence, any set containing a is a neighbourhood of a. That is, the discrete metric induces the discrete topology.

10. There is no metric on X inducing the indiscrete topology (unless $\divideontimes X \leqslant 1$). However, let us define a *pseudo-metric* to be a function $d : X \times X \to \mathbf{R}^{\geqslant 0}$ satisfying Axioms M2, M3 and the following weak form of Axiom M1.

M1′ $d(x, x) = 0.$

A pseudo-metric induces a topology on X in the same way as does a metric, since Axiom M1′ is enough to show that $a \in B(a, r)$. The indiscrete topology on X is induced by the *indiscrete pseudo-metric* given by $d(x, y) = 0$ for all $x, y \in X$.

It is clear from 2.8.1(a) that the open balls of X are open sets, and from 2.8.1(b) that the closed balls of X are closed sets. However, the closed ball is not necessarily the closure of the open ball, since in Example 9 above

$$B(a, 1) = \{a\},\ E(a, 1) = X.$$

Let X, Y be metric spaces and $f : X \to Y$ a function. We can use 2.6.1 to give another description of the continuity of f, namely, f is continuous $\Leftrightarrow$ for each $a \in X$ and each $\varepsilon > 0$, there is a $\delta > 0$ such that

$$fB(a, \delta) \subset B(fa, \varepsilon).$$

Further, we may replace here any open ball by a closed ball, and also restrict any or both of ε, δ to rational numbers, or to rational numbers $1/n$ for n a positive integer.

Let d, e be metrics on X. The metrics are *equivalent* if they induce the same topology on X or, what is the same thing, if the identity $1 : X_d \to X_e$ is a homeomorphism. Let subscripts (e.g., B_d, B_e) be used to distinguish the balls for the two metrics. Then *d, e are equivalent if and only if for each $a \in X$*

(a) *for each $\varepsilon > 0$ there is a $\delta > 0$ such that $B_d(a, \delta) \subset B_e(a, \varepsilon)$, and*
(b) *for each $\varepsilon > 0$ there is a $\delta > 0$ such that $B_e(a, \delta) \subset B_d(a, \varepsilon)$.*

EXAMPLE 11. Let $d : \mathbf{R} \times \mathbf{R} \to \mathbf{R}^{\geqslant 0}$ be defined by

$$d(x, y) = \frac{|x - y|}{1 + |x - y|}.$$

Then d is a metric on $\mathbf{R}$; the only verification which is non-trivial is of the triangle inequality M3, and this is proved as follows:

Let $x \neq z$. Then

$$\begin{aligned} d(x, z) &= (1 + |x - z|^{-1})^{-1} \\ &\leqslant (1 + (|x - y| + |y - z|)^{-1})^{-1} \\ &= |x - y|(1 + |x - y| + |y - z|)^{-1} \\ &\qquad + |y - z|(1 + |x - y| + |y - z|)^{-1} \\ &\leqslant d(x, y) + d(y, z). \end{aligned}$$

This metric d is equivalent to the usual metric. For it is easily seen that

$$|x - y| < \varepsilon \Leftrightarrow d(x, y) < \varepsilon(1 + \varepsilon)^{-1} \Rightarrow d(x, y) < \varepsilon.$$

Let $B(x, \varepsilon)$ be the open ball for the metric d. Then

$$]x - \varepsilon, x + \varepsilon[= B(x, \varepsilon(1 + \varepsilon)^{-1}) \subset B(x, \varepsilon)$$

and this implies the equivalence of the two metrics.

A metric d on a set X is *bounded* if there is a real number r such that $d(x, y) \leqslant r$ for all x, y in X. The last example shows that **R** admits an equivalent bounded metric, and a similar argument applies to any metric space X. It is also easy to check that the metric d' of Example 6 of 2.7 is a bounded metric equivalent to the given one.

There is a very useful continuity criterion for additive and bi-additive maps on normed vector spaces. Let U, V, W be normed vector spaces. A function $f : V \to W$ is *additive* if $f(x + y) = f(x) + f(y)$ for all $x, y \in V$; and $g : U \times V \to W$ is *bi-additive* if

$$\begin{aligned} g(x, y + y') &= g(x, y) + g(x, y'), \\ g(x + x', y) &= g(x, y) + g(x', y) \end{aligned}$$

for all $x, x' \in U$, $y, y' \in V$.

2.8.2(*a*) *An additive function $f : V \to W$ is continuous if there is a real number $r > 0$ such that*

$$\|f(x)\| \leqslant r\,\|x\| \quad \text{for all } x \in V.$$

(*b*) *A bi-additive function $g : U \times V \to W$ is continuous if there is a real number $r > 0$ such that*

$$\|g(x, y)\| \leqslant r\,\|x\|\,\|y\| \quad \text{for all } x \in U, y \in V.$$

Proof We prove (*b*) first. Let $\varepsilon > 0$, $(a, b) \in U \times V$. Let $\|x\|, \|y\| < \delta$ where $\delta < 1$. Then

$$\begin{aligned} \|g(a + x, b + y) - g(a, b)\| &\leqslant \|g(x, y)\| + \|g(a, y)\| + \|g(x, b)\| \\ &\leqslant r\{\|x\|\,\|y\| + \|a\|\,\|y\| + \|b\|\,\|x\|\} \\ &< r\delta\{1 + \|a\| + \|b\|\} \quad \text{since } \delta < 1 \\ &= k\delta \quad \text{say, where } k > 0. \end{aligned}$$

So $g[B(a, \delta) \times B(b, \delta)] \subset B(g(a, b), \varepsilon)$ if $\delta < \varepsilon/k$. This proves continuity of g.

We use (b) to prove (a). If $f : V \to W$ is additive and satisfies $\|f(x)\| \leqslant r\|x\|$, then $g : V \times \mathbf{K} \to W$ defined by $g(x, \lambda) = f(x)\lambda$ is bi-additive and satisfies $\|g(x, \lambda)\| \leqslant r\,|\lambda|\,\|x\|$. Since $f = gi$ where $i : V \to V \times \mathbf{K}$ is $x \leadsto (x, 1)$, the continuity of g implies that of f. □

It is easy to give a direct proof of (a), and we leave this to the reader.

Remark For f linear and g bilinear the converses of 2.8.2(a) and (b) are true. However we do not need this fact and so leave its proof as an exercise.

EXAMPLES 12. Let $V \times V$ be given the norm

$$\|(x, y)\| = \max\{\|x\|, \|y\|\}.$$

The axioms for a norm are easily verified. The addition map $V \times V \to V$ given by $(x, y) \leadsto x + y$ is additive and satisfies

$$\|x + y\| \leqslant \|x\| + \|y\| \leqslant 2\,\|(x, y)\|.$$

Therefore addition is continuous.

13. Let $f : \mathbf{R}^n \times \mathbf{R}^m \to V$ be bilinear, that is, f is additive and also

$$f(\lambda x, y) = f(x, \lambda y) = \lambda f(x, y) \quad \text{for } \lambda \in \mathbf{R}.$$

Let e_j, $j = 1, \ldots, n$ denote the standard basis elements of $\mathbf{R}^n$, and let $\mathbf{R}^n$, $\mathbf{R}^m$, $\mathbf{R}^l$ have the Euclidean or Cartesian norm. This implies that if $x = \sum \lambda_j e_j$, then $|\lambda_j| \leqslant \|x\|$. Hence, if $x = \sum_{j=1}^{n} \lambda_j e_j$, $y = \sum_{i=1}^{m} \mu_i e_i$, then

$$\begin{aligned} \|f(x, y)\| &\leqslant \sum_{i,j} |\lambda_j|\,|\mu_i|\,\|f(e_j, e_i)\| \\ &\leqslant \|x\|\,\|y\| \sum_{i,j} \|f(e_j, e_i)\|. \end{aligned}$$

It follows that f is continuous.

14. A similar, and simpler, argument to that of the last example shows that if $\mathbf{R}^n$ has the Euclidean or Cartesian norms, then any linear function $\mathbf{R}^n \to V$ is continuous. This implies that the Euclidean and Cartesian norms are equivalent (that is, define the same metric topology), since the identity $\mathbf{R}^n \to \mathbf{R}^n$ is continuous whichever of these norms we put on each $\mathbf{R}^n$. Actually it can be proved, as an application of compactness, that *on a finite dimensional normed vector space any two norms are equivalent.* [Exercise 11 of 3.3.]

15. The sup norm $\| \ \|_S$ and the integral norm $\| \ \|_I$ on the space $\mathscr{C}$ of continuous functions $[0, 1] \to \mathbf{R}$ are not equivalent. For let $0 < r \leqslant 1$ and

5

let $f_r : [0, 1] \to \mathbf{R}$ be the function whose graph is shown in Fig. 2.9. Then $\|f_r\|_S = 1$, but $\|f_r\|_I = r/2$. Hence $f_r \in E_I(0, r/2)$ but $f_r \notin E_S(0, \frac{1}{2})$. Thus $E_I(0, r/2) \not\subset E_S(0, \frac{1}{2})$ for any $0 < r \leqslant 1$. So the two norms are not equivalent.

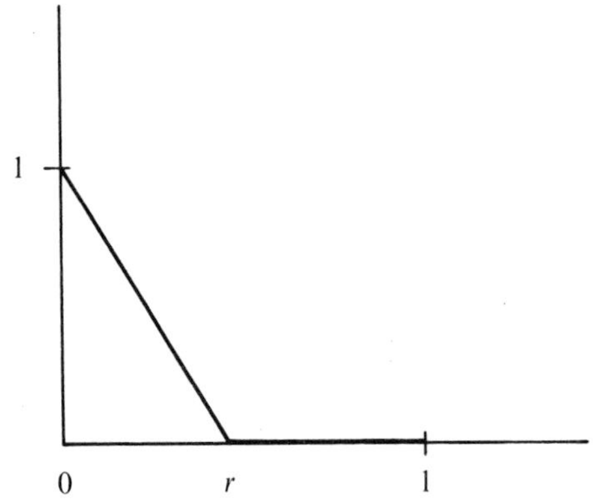

Fig. 2.9

16. Let V_1, V_2 be the same vector space V but with norms $\|\ \|_1$, $\|\ \|_2$ respectively. If we apply 2.8.2(*b*) to the two identity maps $V_1 \to V_2$, $V_2 \to V_1$ we see that these norms are equivalent if there are real numbers $r, s > 0$ such that

$$\|x\|_1 \leqslant r\,\|x\|_2, \qquad \|x\|_2 \leqslant s\,\|x\|_1.$$

(This sufficient condition is also necessary—cf. the Remark on p. 49). This gives another proof that in $\mathbf{R}^n$ the Euclidean norm $\|\ \|_2$ and the Cartesian norm $\|\ \|_\infty$ are equivalent, since for any x in $\mathbf{R}^n$

$$\|x\|_\infty \leqslant \|x\|_2 \leqslant \sqrt{n}\,\|x\|_\infty.$$

In this section, we have shown that every metric on a set X induces a topology on X. On the other hand, not every topology on X is induced by a metric. One example of this, the indiscrete topology, has been given already, and other examples will be given later. The characterization of metric topologies has been completely solved (for an account of this see Kelley [1]), but this kind of problem is outside the main line of direction of this book.

Products and subspaces

Let X, Y be metric spaces. On $X \times Y$ we can define a metric by

$$d((x, y), \ (x', y')) = \max\{d(x, x'), d(y, y')\}.$$

The verification of the axioms for a metric is simple, and is left to the reader. The topology on $X \times Y$ induced by this metric is simply the product topology, as is easily seen from the formula

$$B(a, r) \times B(b, r) = B((a, b), r) \subset B(a, s) \times B(b, t)$$

where $r = \min \{s, t\}$.

Let X be a metric space with metric d and let A be a subset of X. Let $d_A = d \mid A \times A$. It is easily verified that d_A is a metric on A. The open balls in X we write $B(a, r)$ and in A, $B_A(a, r)$. Clearly

$$B_A(a, r) = B(a, r) \cap A.$$

Figure 2.10 gives a picture of a subset A of $\mathbf{R}^2$ (with the Euclidean metric) and various open balls in A.

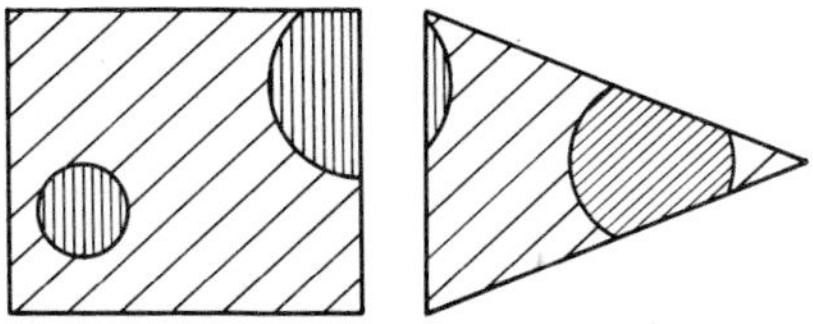

Fig. 2.10

2.8.3 *The metric topology on A is its relative topology as a subset of X.*

Proof Let $a \in A$. If N is a neighbourhood of a in X, then $N \supset B(a, r)$ for some $r > 0$ and so $N \cap A \supset B_A(a, r)$. Thus each subspace neighbourhood of a is also a metric neighbourhood.

Conversely, suppose M is a neighbourhood of a for the metric d_A. Then $M \supset B_A(a, r)$ for some $r > 0$. Let

$$N = B(a, r) \cup M.$$

Then N is a neighbourhood of a in X and $N \cap A = M$. Therefore, M is also a subspace neighbourhood of a. □

Because of this proposition, a subspace A of a metric space X will mean a subset with the metric d_A and the metric topology.

EXERCISES

1. Let X be a metric space. Prove that the topology of X is discrete if and only if for each x in X there is an $r > 0$ such that $B(x, r) = \{x\}$.
2. Let $f : \mathbf{R}^{\geqslant 0} \to \mathbf{R}^{\geqslant 0}$ be a function such that

(i) $f(x) = 0 \Leftrightarrow x = 0$,
(ii) $x \leqslant x' \Rightarrow f(x) \leqslant f(x')$,
(iii) $f(x + x') \leqslant f(x) + f(x')$.

Let d be a metric on X. Show that the composition $e = fd$ is a metric on X equivalent to d. Show also that the function $x \rightsquigarrow x/(1 + x)$ satisfies conditions (i), (ii), and (iii) above.

3. Let X, Y be metric spaces. Show that the following formulae define metrics on $X \times Y$ whose metric topology is the product topology.

(*a*) $d((x, y), (x', y')) = d(x, x') + d(y, y')$,

(*b*) $d((x, y), (x', y')) = [\{d(x, x')\}^2 + \{d(y, y')\}^2]^{1/2}$.

4. In Euclidean space two open balls meet if the distance between their centres is less than the sum of their radii. Is this true in an arbitrary metric space? in an arbitrary normed vector space?

5. Let $d : X \times X \to \mathbf{R}^{\geqslant 0}$ be a metric on X. Prove that d is continuous.

6. Let d be a metric on X. Prove that the metric topology is the coarsest topology $\mathcal{T}$ on X such that each function $d_x : X_{\mathcal{T}} \to \mathbf{R}^{\geqslant 0}$, $y \rightsquigarrow d(x, y)$, is continuous.

7. Let V be a normed vector space over $\mathbf{K}$, and let $a \in V$, $\alpha \in \mathbf{K}$ ($\alpha \neq 0$). Prove that the functions $V \to V$ given by $x \rightsquigarrow x + a$, $x \rightsquigarrow x\,\alpha$ are homeomorphisms. Prove also that any open ball in V is homeomorphic to V.

8. Let V be a normed vector space and let $a \in V$, $r > 0$. Prove that $B(a, r)$ is the interior of $E(a, r)$, and $E(a, r)$ is the closure of $B(a, r)$.

9. Let A, B be subsets of the normed vector space V. Prove that if one of A, B is open, then so also is

$$A + B = \{a + b : a \in A, b \in B\}.$$

10. Let $V_1, \ldots, V_n, V$ be normed vector spaces over $\mathbf{R}$. We say $u : V_1 \times \cdots \times V_n \to V$ is *multilinear* if, for any $\lambda, \mu \in \mathbf{R}$, and $x_i, x_i' \in V_i$ ($i = 1, \ldots, n$)

$$u(x_1, \ldots, \lambda x_i + \mu x_i', \ldots, x_n) = \lambda u(x_1, \ldots, x_i, \ldots, x_n) + \mu u(x_1, \ldots, x_i', \ldots, x_n).$$

Prove that such a multilinear map is continuous $\Leftrightarrow$ there is a real number $r > 0$ such that for all $x_i \in V_i$

$$\|u(x_1, \ldots, x_n)\| \leqslant r \|x_1\| \cdots \|x_n\|.$$

(This applies to normed vector spaces over $\mathbf{K}$, since such objects are also, by restriction of the field, normed vector spaces over $\mathbf{R}$.)

[Exercise 10 implies the necessity of the condition for equivalent norms given in Example 16, p. 50.]

11. Let V_1, V_2 be the same vector space V with distinct, but equivalent, norms $\| \ \|_1$, $\| \ \|_2$ respectively. Construct homeomorphisms

$$B(V_1) \to B(V_2),\ E(V_1) \to E(V_2),\ S(V_1) \to S(V_2).$$

12. Let $f : \mathbf{E}^n \to \mathbf{E}^n$ be a map such that $f \mid \mathbf{B}^n, \mathbf{B}^n$ is a homeomorphism. Prove that $f[\mathbf{S}^{n-1}] \subset \mathbf{S}^{n-1}$.

13. Brouwer has proved the following theorem known as the *Invariance of Domain* [cf. Nagata [1], Spanier [3]]. *Let A, B be subsets of $\mathbf{R}^n$ and $f : A \to B$ a homeomorphism. Then $f[\mathrm{Int}\, A] \subset \mathrm{Int}\, B$.*

Use the Invariance of Domain to prove (i) (*Invariance of Dimension*). If $f : \mathbf{R}^m \to \mathbf{R}^n$ is a homeomorphism, then $m = n$. (ii) If $f : \mathbf{E}^n \to \mathbf{E}^n$ is a homeomorphism, then $f \mid \mathbf{B}^n, \mathbf{B}^n$ and $f \mid \mathbf{S}^{n-1}, \mathbf{S}^{n-1}$ are defined and are homeomorphisms.

14. Let p be a prime number. For each $n \in \mathbf{N}$ define $v_p(n)$ to be the exponent of p in the decomposition of n into prime numbers. If $x = \pm\, m/n$ is any non-zero rational number, with $m, n \in \mathbf{N}$, define

$$v_p(x) = v_p(m) - v_p(n).$$

Finally, if x, y are rational numbers define

$$d(x, y) = \begin{cases} p^{-v_p(x-y)}, & x \neq y \\ 0, & x = y. \end{cases}$$

(i) Prove that d is a metric on $\mathbf{Q}$ and that d satisfies the following strong form of the triangle inequality

$$d(x, z) \leqslant \max \{d(x, y), d(y, z)\}.$$

The topology induced by d is called the *p-adic topology*.

(ii) Prove that the topology induced by d on $\mathbf{Z}$ is the p-adic topology of Exercise 7 of 2.1.

(iii) Justify the following statement: in the p-adic topology on $\mathbf{Q}$, small rational numbers are those which are multiples of large powers of p.

15. A subset A of $\mathbf{R}^m$ is *convex* if for any x, y in A the line segment joining x to y (that is, the set of points $(1 - t)x + ty, 0 \leqslant t \leqslant 1$) is contained in A. This exercise outlines a proof that any two open convex subsets A, B of $\mathbf{R}^m$ are homeomorphic. The steps are as follows:

(i) There is a homeomorphism $f : \mathbf{R}^m \to \mathbf{B}^m$ such that $f[A], f[B]$ are convex. So we may suppose A, B bounded. Let $a \in A$. Then, there is a real number $\delta > 0$ such that $B(a, \delta) \subset A$.

(ii) For each x in A, let

$$r(x) = \sup \{\lambda \in \mathbf{R} : \lambda(x - a) \in A\}.$$

Then $r(x)$ is well-defined and non-zero, and the function $x \rightsquigarrow r(x)$ is continuous.

(iii) The function $x \rightsquigarrow a + (\delta/r(x))\,(x - a)$ is a homeomorphism $A \to B(a, \delta)$.

(iv) A, B are homeomorphic.

2.9 Distance from a subset

Let X be a metric space with metric d, and let A be a (non-empty) subset of X. For each $x \in X$ we define the *distance of x from A* to be

$$\text{dist}\,(x, A) = \inf \{d(x, a) : a \in A\}.$$

2.9.1 *The function $x \rightsquigarrow$ dist (x, A) is a continuous function $X \to \mathbf{R}^{\geqslant 0}$.*

Proof We prove that for any $\varepsilon > 0$ there is a $\delta > 0$ such that

$$d(x, y) \leqslant \delta \Rightarrow |\text{dist}\,(x, A) - \text{dist}\,(y, A)| \leqslant \varepsilon. \qquad (*)$$

For any a in A

$$d(x, a) \leqslant d(x, y) + d(y, a),$$
$$d(y, a) \leqslant d(y, x) + d(x, a).$$

We apply $\inf_{a \in A}$ to each of these inequalities to obtain

$$\text{dist}\,(x, A) \leqslant d(x, y) + \text{dist}\,(y, A)$$
$$\text{dist}\,(y, A) \leqslant d(x, y) + \text{dist}\,(x, A)$$

whence

$$|\text{dist}\,(x, A) - \text{dist}\,(y, A)| \leqslant d(x, y).$$

This proves (*) with $\delta = \varepsilon$. □

2.9.2 $x \in \bar{A} \Leftrightarrow \text{dist}\,(x, A) = 0$.

Proof The inverse image of $\{0\}$ under $x \rightsquigarrow \text{dist}\,(x, A)$ is a closed set containing A, and so containing $\bar{A}$. Thus $x \in \bar{A} \Rightarrow d(x, A) = 0$.

On the other hand, if $x \notin \bar{A}$, then there is a closed ball $E(x, r)$, $r > 0$, not meeting A. Hence $\text{dist}\,(x, A) \geqslant r$. □

2.9.3 *Let A, B be disjoint closed sets in X. There are disjoint open sets U, V in X such that $A \subset U, B \subset V$.*

Proof Let $f : X \to \mathbf{R}$ be the function $x \rightsquigarrow \text{dist}\,(x, A) - \text{dist}\,(x, B)$, and let $U = f^{-1}[\mathbf{R}^{<0}]$, $V = f^{-1}[\mathbf{R}^{>0}]$. By 2.9.1, U and V are open, and they are clearly disjoint. If $x \in A$, then $\text{dist}\,(x, B) > 0$ since B is closed and A, B are disjoint. Therefore, $f(x) < 0$ and so $x \in U$. Thus $A \subset U$ and, similarly, $B \subset V$. □

In a topological space, we say a subset N is a *neighbourhood* of a subset A if there is an open set U such that $A \subset U \subset N$. We can express 2.9.3 succintly as: in a metric space, disjoint closed sets have disjoint neighbourhoods. A topological space with this property is called *normal*—examples of non-normal spaces are given in the Exercises.

EXERCISES

1. A topological space X is called T_1 if, for each x in X, the set $\{x\}$ is closed; and X is called *Hausdorff* if distinct points of X have disjoint neighbourhoods. Prove that a metric space is Hausdorff, and that a Hausdorff space is T_1.
2. A subset A of a topological space X is called a G_δ-set if A is the intersection of a countable number of open sets of X. Prove that a closed subset of a metric space is a G_δ-set.
3. Give examples of (i) a metric space X and a subset A of X which is not a G_δ-set, (ii) a topological space X and a closed subset A of X which is not a G_δ-set.

4. Let X be the unit interval $[0, 1]$ with the following topology. The neighbourhoods of t for $0 < t \leqslant 1$ are the usual ones. The neighbourhoods of 0 are the usual ones and also the sets $N \setminus A$ where N is a usual neighbourhood of 0 and A is a set $\{x_1, x_2, x_3, \ldots\}$ of points x_n such that $x_n \neq 0$ for any n, and $x_n \to 0$ as $n \to \infty$. Prove that this defines a topology on X, and that X is Hausdorff but not normal.

5. Prove that a topological space X is $T_1 \Leftrightarrow$ for each x, y in X there is a neighbourhood of x not containing y.

6. Let X be a metric space, A a subset of X, and $r > 0$. We define

$$B(A, r) = \{x \in X : \text{dist}\,(x, A) < r\}.$$

If $B(A, r) \subset N \subset X$ for some $r > 0$, then N is a neighbourhood of A. Prove that the converse of this implication is false.

7. Generalize the notion of a base for the neighbourhoods of a point to the notion of a base for the neighbourhoods of a set. Give an example of a subset A of **R** such that A does not have a countable base for its neighbourhoods.

2.10 Hausdorff spaces

We recall [Exercise 1 of 2.9] that a topological space X is Hausdorff if distinct points of X have disjoint neighbourhoods.† The following characterization of this property is more aesthetic, and often more useful.

2.10.1 *A topological space X is Hausdorff if and only if the diagonal*

$$\Delta(X) = \{(x, x) \in X \times X : x \in X\}$$

is closed in $X \times X$.

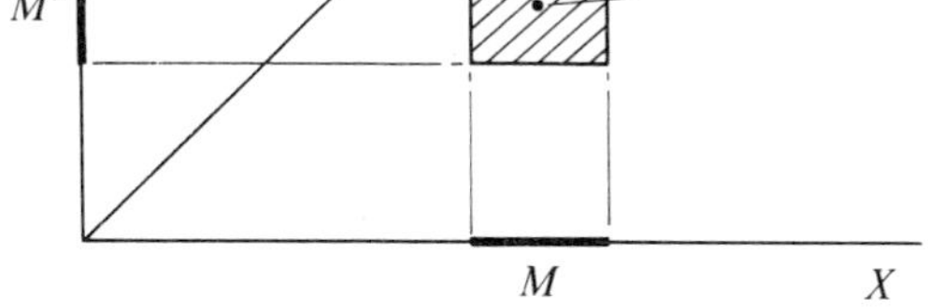

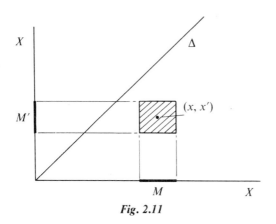

Fig. 2.11

† A standard joke is that X is Hausdorff if any two points can be housed off from each other.

Proof Let $\Delta = \Delta(X)$. The following statements are each equivalent to their successors (since $x \neq x' \Leftrightarrow (x, x') \notin \Delta$).

(*a*) X is Hausdorff.
(*b*) if $x \neq x'$, then there exist neighbourhoods M, M' of x, x' such that $M \cap M' = \emptyset$.
(*c*) if $x \neq x'$, then there exist neighbourhoods M, M' of x, x' such that $(M \times M') \cap \Delta = \emptyset$.
(*d*) if $x \neq x'$, then $(X \times X) \setminus \Delta$ is a neighbourhood of (x, x').
(*e*) Δ is closed in $X \times X$. □

2.10.2 *Let $f, g : Y \to X$ be maps of the topological space Y to the Hausdorff space X. Then the set A of points on which f, g agree is closed in Y.*

Proof $A = (f, g)^{-1}[\Delta(X)]$. □

2.10.3 *Subspaces, and products, of Hausdorff spaces are Hausdorff.*

Proof Let A be a subspace of the Hausdorff space X. Then

$$\Delta(A) = \Delta(X) \cap (A \times A).$$

Therefore, $\Delta(A)$ is closed in $A \times A$, and so A is Hausdorff.

Let X, Y be Hausdorff spaces and let

$$T : X \times X \times Y \times Y \to X \times Y \times X \times Y$$
$$(x, x', y, y') \rightsquigarrow (x, y, x', y').$$

Then T is a homeomorphism and

$$T[\Delta(X) \times \Delta(Y)] = \Delta(X \times Y).$$

Therefore, $\Delta(X \times Y)$ is closed in $X \times Y \times X \times Y$. □

2.10.4 *Let $f : Y \to X$ be a continuous injection and let X be Hausdorff. Then Y is Hausdorff.*

Proof Since f is an injection

$$\Delta(Y) = (f \times f)^{-1}[\Delta(X)].$$

Therefore, $\Delta(Y)$ is closed in $Y \times Y$. □

This result has two important special cases.

(*a*) A finer topology than a Hausdorff topology is also Hausdorff.
(*b*) If X and Y are homeomorphic and X is Hausdorff, then so also is Y.

In this book, the spaces of main interest will all be Hausdorff. But the reader should beware of thinking that non-Hausdorff spaces are of little importance. In the Exercises we sketch the construction of two important classes of non-Hausdorff spaces, firstly the Zariski topology and, secondly, the sheaf of germs of functions.

EXERCISES

1. Write down proofs of 2.10.2, 2.10.3, and 2.10.4 using directly the definition of Hausdorff spaces.
2. The *Zariski topology* on $\mathbf{R}^n$ (or on $\mathbf{C}^n$) is that in which C is closed if and only if there is a set of polynomials in n variables such that C is the set of points on which all these polynomials vanish. In the case $n = 1$, this is the topology in which C is closed $\Leftrightarrow C = \mathbf{R}$ or C is finite. The Zariski topology on $\mathbf{R}$ is not Hausdorff. (The proof that the Zariski topology on $\mathbf{R}^n$ is not Hausdorff requires knowledge of the ideal theoretic properties of polynomial rings—cf. Zariski–Samuel [1] Ch. VII §3.)
3. *The sheaf of germs of functions.* Let X, Y be topological spaces. For each $x \in X$ let $F(x)$ denote the set of continuous functions from some neighbourhood of x to Y. An equivalence relation $\sim$ is defined in $F(x)$ by $f \sim g \Leftrightarrow f, g$ agree on some neighbourhood of x. The set of equivalence classes is written $G(x)$, and $G = \cup_{x \in X} G(x)$.

An element of $G(x)$ is called a *function-germ* or *germ* at x, and G is the *sheaf of germs of continuous functions.* If $f \in F(x)$, the germ of f (that is the equivalence class of f) is written f^x. The value of f^x at x is well-defined by $f^x(x) = f(x)$. Let U be an open set containing x, and let $f: U \to Y$ be continuous. Then f defines a germ f^y for each $y \in U$ and the set $f^U = \{f^y : y \in U\}$ is defined to be a basic neighbourhood of f^x. The topology on G is that in which a neighbourhood of f^x is any set containing a basic neighbourhood.

Prove that G is in fact a topological space, and that G is non-Hausdorff even if $X = Y = \mathbf{R}$. (If $f, g: \mathbf{R} \to \mathbf{R}$ are such that $f(x) = g(x)$ for $x \leqslant a$, $f(x) \neq g(x)$ for $x > a$, then f^a, g^a are germs which do not have disjoint neighbourhoods.)

Prove that if $X = Y = \mathbf{R}$ the above construction can be varied by replacing the word continuous by (i) integrable, (ii) differentiable, (iii) of class C^∞, (i.e., with derivatives of all orders), (iv) polynomial, (v) analytic (i.e., expressible locally by power series). Prove that in the last two cases the corresponding sheaf of germs is a Hausdorff space.
4. Let X be a topological space. Let $(x_n)_{n>0}$ be a sequence of points of X. If $Y \subset X$, we say (x_n) is *eventually in* Y if there is a number n_0 such that $n \geqslant n_0$ implies $x_n \in Y$. If $x \in X$, we say $x_n \to x$ *as* $n \to \infty$, or (x_n) *has limit* x, (or, briefly, $(x_n) \to x$) if for any neighbourhood N of x, (x_n) is eventually in N. Let $\mathbf{L} = \{0\} \cup \{n^{-1} : n \in \mathbf{N}^{>0}\}$ have its relative topology as a subset of $\mathbf{R}$. Prove that $(x_n) \to x$ if and only if the function $g: \mathbf{L} \to X$ which sends $n^{-1} \rightsquigarrow x_n$, $0 \rightsquigarrow x$, is continuous. Prove also that if X is Hausdorff, then the conditions $(x_n) \to x$, $(x_n) \to y$ imply $x = y$.
5. Consider the following conditions on a space X: (*a*) X is *Fréchet*, that is, if $x \in X$ and $A \subset X$, then $x \in \bar{A}$ if and only if there is a sequence of points (x_n) of A such that $(x_n) \to x$. (*b*) X is *sequential*, that is a subset U of X is open if and only if every sequence converging to a point of U is eventually in U. (*c*) A subset U of X

is open if and only if $U \cap A$ is open in A for every countable subset A of X. Prove that (*a*) is satisfied if X satisfies the first axiom of countability [Exercise 3 of 2.6], and that (*a*) $\Rightarrow$ (*b*) $\Rightarrow$ (*c*). [Use Exercise 5 of 2.6 for the proof that first countable implies Fréchet.]

6. Let X be a sequential space, let Y be a space and $f: X \to Y$ a function. Prove that the following conditions are equivalent, (*a*) f is continuous, (*b*) $fg: \mathbf{L} \to Y$ is continuous for all continuous functions $g: \mathbf{L} \to X$, (*c*) for all x in X and sequences (x_n) in X, $(x_n) \to x$ implies $(fx_n) \to fx$.

7. Let X be the space of Exercise 4 of 2.9 but defined using $\mathbf{Q} \cap [0, 1]$ instead of $[0, 1]$. Prove that X is not sequential, but that X satisfies (*c*) of Exercise 5.

8. Let X be the set $[0, 1]$ retopologized as follows. The neighbourhoods of t in $]0, 1]$ are the usual neighbourhoods. The neighbourhoods of 0 are the usual neighbourhoods and also any set containing $\{0\} \cup U$ where U is a usual open neighbourhood of $\mathbf{L}^* = \mathbf{L} \setminus \{0\}$. Prove that X is sequential but not Fréchet. Prove also that $X \setminus \mathbf{L}^*$ is not sequential.

9. Give an example of a topological space X which is not indiscrete, and in which limits of sequences are not unique.

10. Let X be an uncountable set with the topology that $C \subset X$ is closed if C is countable or if $C = X$. Let $g: \mathbf{L} \to X$ be continuous. Prove that there is an integer n_0 such that $n \geqslant n_0 \Rightarrow g(n^{-1}) = g(0)$. Prove also that in X limits of sequences are unique. Let Y be the underlying set of X with the discrete topology. Let $f: X \to Y$ be the identity function. Prove that f is not continuous, but fg is continuous for all continuous functions $g: \mathbf{L} \to X$.

11. Prove that the space X of Exercise 10 is not a pseudometric space. [A pseudometric space satisfies the first axiom of countability.]

12. Let X be an uncountable set and let $x_0 \in X$. Let X have the topology in which a subset C of X is closed if C is countable or if $x_0 \in C$. Prove that this is a topology and that X with this topology is a Hausdorff, non-metric space.

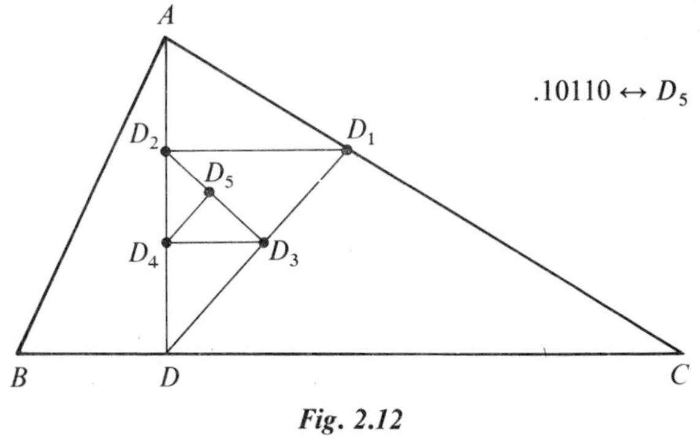

Fig. 2.12

13. Let $\Delta \subset \mathbf{R}^2$ be the set of points inside and on the right-angled triangle ABC, which we suppose has a right-angle at A and satisfies $AC > AB$. This exercise outlines the construction of a continuous surjection $f: [0, 1] \to \Delta$. Let D on

BC be such that AD is perpendicular to BC. Let $a = .a_1a_2a_3\ldots$ be a binary decimal, so that each a_n is 0 or 1. Then we construct a sequence (D_n) of points of Δ as follows: D_1 is the foot of the perpendicular from D onto the hypotenuse of the larger or smaller of the triangles ADB, ADC according as $a_1 = 1$ or 0 respectively. This construction is now repeated using D_1 in place of D and the appropriate triangle of ADB, ADC in place of ABC. For example, Fig. 2.12 illustrates the points D_1 to D_5 for the binary decimal .10110.... The reader should give a decent inductive definition of the sequence (D_n) and prove in turn: (i) the sequence (D_n) tends to a limit $D(a)$ in Δ; (ii) if $\lambda \in [0, 1]$ is represented by distinct binary decimals a, a' then $D(a) = D(a')$; hence, the point $D(\lambda)$ in Δ is uniquely defined; (iii) if $f : [0, 1] \to \Delta$ is the function $\lambda \rightsquigarrow D(\lambda)$ then f is surjective. (iv) f is continuous.

14. Using the previous exercise, prove the existence of continuous surjections from [0, 1] to the sets (i) $\mathbf{E}^2$, (ii) $\mathbf{I}^2$, (iii) $\mathbf{I}^n$.

15. Prove that there is a continuous surjection $\mathbf{R} \to \mathbf{R}^m$.

NOTES

A variety of axiomatizations for topological spaces are listed in the Exercises of Vaidyanathaswamy [1]. Topological spaces are not adequate to deal with the notion of uniform continuity—for this, there is needed the concept of uniform space which may be found for example in Kelley [1], Bourbaki [1]. A general account of the type of axiom system of which uniform spaces and topological spaces form a particular example is given in Cszazar [1]. The relation between convergence and topologies is best shown by means of the filters of Bourbaki [1]. Spaces whose topology can be defined by sequences are discussed, for example, by Franklin [1]. For an account of the general theory of sheaves see Godement [1]; applications of sheaves to algebraic topology are given in Swan [1] and to algebraic geometry in Hirzebruch [1] (but none of these books on sheaves is for the beginner in topology).

3. Connected spaces, compact spaces

3.1 The sum of topological spaces

Let X_1, X_2 be disjoint subspaces of a topological space X, and suppose $X = X_1 \cup X_2$. In general, it is not possible to recover the topology of X from the topologies of X_1, X_2. For example, the set $\{a, b\}$ has four distinct topologies, while the sets $\{a\}$, $\{b\}$ have each only one topology.

A case when the topology of X is determined by the topologies of X_1, X_2 is when U is open in X if and only if both $U \cap X_1$ is open in X_1 and $U \cap X_2$ is open in X_2. In this case, we say X is a *topological sum* of X_1, X_2 and we write

$$X = X_1 \sqcup X_2.$$

The open sets of $X_1 \sqcup X_2$ are then simply the unions $U_1 \cup U_2$ for U_1 open in X_1, U_2 open in X_2.

EXAMPLE Let $X = [0, 2] \setminus \{1\}$ with its usual topology as a subspace of $\mathbf{R}$. Then $X = [0, 1[\sqcup]1, 2]$. On the other hand, $[0, 2]$ itself is not $[0, 1] \sqcup]1, 2]$.

Intuitively, a sum should be thought of as a space which is in two pieces. But one should bear in mind that any X is $X \sqcup \varnothing$.

3.1.1 *Let X_1, X_2 be disjoint subspaces of the topological space X such that $X = X_1 \cup X_2$. The following conditions are equivalent.*

(*a*) $X = X_1 \sqcup X_2$,
(*b*) *X_1, X_2 are both open in X,*
(*c*) *X_1 is both open and closed in X,*
(*d*) $\overline{X}_1 \cap X_2 = \varnothing$ *and* $X_1 \cap \overline{X}_2 = \varnothing$.

Proof That $(c) \Rightarrow (b)$ is immediate from the definition of $X_1 \sqcup X_2$, while $(a) \Leftrightarrow (c)$ follows from the fact that $X_1 = X \setminus X_2$.

If X_1 is open and closed in X, then so also is X_2. Hence, $\overline{X}_1 = X_1$, $\overline{X}_2 = X_2$, and so $(c) \Rightarrow (d)$. Conversely, $\overline{X}_1 \cap X_2 = \varnothing$ implies X_2 is open, so that X_1 is closed, while $X_1 \cap \overline{X}_2 = \varnothing$ implies X_1 is open. Thus $(d) \Rightarrow (c)$.

Finally we prove that (*b*) ⇒ (*c*). Let U_α be open in X_α, $\alpha = 1, 2$. Then U_α is open in X and so $U_1 \cup U_2$ is open in X. □

The most useful property of the sum $X_1 \sqcup X_2$ is concerned with functions $X_1 \sqcup X_2 \to Y$. Let $i_1 : X_1 \to X_1 \sqcup X_2$, $i_2 : X_2 \to X_1 \sqcup X_2$ be the two inclusion functions.

3.1.2 *If $f_1 : X_1 \to Y$, $f_2 : X_2 \to Y$ are maps, then there is a unique map $f : X_1 \sqcup X_2 \to Y$ such that $f i_1 = f_1$, $f i_2 = f_2$.*

Proof We suppose f_1, f_2 given. Then $f : X_1 \sqcup X_2 \to Y$ given by $x \rightsquigarrow f_\alpha(x)$ for $x \in X_\alpha$ ($\alpha = 1, 2$) is the only function $X_1 \sqcup X_2 \to Y$ such that $f i_1 = f_1$, $f i_2 = f_2$. We prove that f is continuous.

Let U be open in Y. Then

$$\begin{aligned} f^{-1}[U] &= (f^{-1}[U] \cap X_1) \cup (f^{-1}[U] \cap X_2) \\ &= f_1^{-1}[U] \cup f_2^{-1}[U]. \end{aligned}$$

Therefore, $f^{-1}[U]$ is open in $X_1 \sqcup X_2$. □

The situation of 3.1.2 is summed up in the diagram

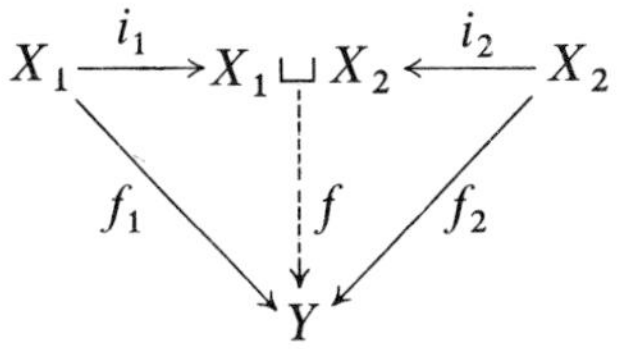

in which the dotted arrow indicates a function to be constructed.

A consequence of 3.1.2 is that a function $f : X_1 \sqcup X_2 \to Y$ is continuous if $f i_1$, $f i_2$ are continuous. This fact is used in the proof of the next result, in which $\{1, 2\}$ is a discrete space.

3.1.3 *Let X_1, X_2 be subspaces of X. Then $X = X_1 \sqcup X_2$ if and only if there is a map $f : X \to \{1, 2\}$ such that $X_\alpha = f^{-1}[\alpha]$, $\alpha = 1, 2$.*

Proof Suppose $X = X_1 \sqcup X_2$. Define $f : X \to \{1, 2\}$ by $f[X_\alpha] = \{\alpha\}$, $\alpha = 1, 2$. Then for $\alpha = 1, 2$ $f i_\alpha : X_\alpha \to \{1, 2\}$ is a constant function, and hence continuous. Therefore f is continuous.

Now suppose that $f : X \to \{1, 2\}$ is continuous, and $X_\alpha = f^{-1}[\alpha]$, $\alpha = 1, 2$. Then X_1, X_2 are disjoint and have union X. Also $\{1\}$ is open and closed in $\{1, 2\}$ and so X_1 is open and closed in X. □

3.1.4 *Let Y be any space with underlying set $X_1 \cup X_2$ such that the inclusions $j_\alpha : X_\alpha \to Y$, $\alpha = 1, 2$ are continuous. Then the topology of $X_1 \sqcup X_2$ is finer than that of Y.*

Proof Let $f : X_1 \sqcup X_2 \to Y$ be the identity function, so that $f\,i_\alpha = j_\alpha$, $\alpha = 1, 2$. By 3.1.2, f is continuous. □

This result is expressed roughly by: $X_1 \sqcup X_2$ has the finest topology such that the inclusions $i_\alpha : X_\alpha \to X_1 \sqcup X_2$ are continuous.

We shall need in chapter 4 to form a sum of spaces X_1, X_2 which are not disjoint. For this we take the universal property 3.1.2 as a definition.

Definition A *sum* of topological spaces X_1, X_2 is a pair $i_1 : X_1 \to X$, $i_2 : X_2 \to X$ which is φ-universal: that is, if $f_1 : X_1 \to Y, f_2 : X_2 \to Y$ are any maps, then there is a unique map $f : X \to Y$ such that $f\,i_1 = f_1$, $f\,i_2 = f_2$.

The usual universal argument [Appendix §4] shows that the space X is then uniquely defined up to a homeomorphism. For this reason, we denote a sum simply by $X_1 \sqcup X_2$.

A sum $X_1 \sqcup X_2$ can always be constructed: Its underlying set is to be a sum of the underlying sets of X_1, X_2, so that we have injections $i_1 : X_1 \to X_1 \sqcup X_2$, $i_2 : X_2 \to X_1 \sqcup X_2$. The open sets of $X_1 \sqcup X_2$ are to be the sets

$$i_1[U_1] \cup i_2[U_2] = U_1 \sqcup U_2$$

for U_1 open in X_1, U_2 open in X_2. We leave the reader to verify that this does define a sum.

If $f_\alpha : X_\alpha \to Y$ are maps ($\alpha = 1, 2$) then the map $X_1 \sqcup X_2 \to Y$ defined by f_1, f_2 is written $(f_1, f_2)'$.

EXERCISES

1. Prove that if X_1, X_2 are metrizable, then $X_1 \sqcup X_2$ is metrizable.
2. Let X_1, X_2 be subspaces of X such that $X = X_1 \sqcup X_2$. Let $f : Y \to X$ be a map. Prove that Y is the sum of $f^{-1}[X_1]$ and $f^{-1}[X_2]$.
3. Let X_1, X_2, X_3 be topological spaces. Prove that there are homeomorphisms,

$$X_1 \sqcup X_2 \to X_2 \sqcup X_1$$
$$X_1 \sqcup (X_2 \sqcup X_3) \to (X_1 \sqcup X_2) \sqcup X_3$$
$$X_1 \times (X_2 \sqcup X_3) \to (X_1 \times X_2) \sqcup (X_1 \times X_3).$$

4. Prove that the following properties hold for $X_1 \sqcup X_2$ if and only if they hold for both X_1 and X_2: separable, first axiom of countability, Hausdorff.
5. Let X_0, X_1, X_2 be subspaces of X such that $X = X_1 \cup X_2$, $X_0 = X_1 \cap X_2$ and

$$X \setminus X_0 = (X_1 \setminus X_2) \sqcup (X_2 \setminus X_1).$$

Prove that a function $f : X \to Y$ is continuous if $f \mid X_1, f \mid X_2$ are continuous.

3.2 Connected spaces

Let X be a topological space. A pair $\{X_1, X_2\}$ of subspaces of X is called a *partition* of X if X_1, X_2 are non-empty, disjoint, and $X = X_1 \sqcup X_2$. Intuitively, X has a partition if it falls into two bits. This leads to the definition; X is *connected* if it has no partition, and otherwise is *disconnected*. We have immediately from 3.1.1 and 3.1.3:

3.2.1 *Let X be a topological space. The following conditions are equivalent.*

(*a*) *X is connected.*
(*b*) *If a subset A of X is open and closed in X, then $A = \varnothing$ or $A = X$.*
(*c*) *If $X = A \cup B$ where $\bar{A} \cap B = \varnothing$, $A \cap \bar{B} = \varnothing$ then $A = \varnothing$ or $B = \varnothing$.*
(*d*) *If $f : X \to \{1, 2\}$ is continuous, then f is constant.*

The last condition is probably the most useful.

A subset Y of a topological space X is connected if Y with its induced topology is a connected space. The connectedness of Y can be described in terms of the closure operator in X: Let $A, B \subset Y$, and let $\bar{B}$, $\mathrm{Cl}_Y B$ be the closures of B in X, Y respectively. Then $\mathrm{Cl}_Y B = Y \cap \bar{B}$ [2.4.2], so that $A \cap \bar{B} = A \cap \mathrm{Cl}_Y B$; and, similarly, $\bar{A} \cap B = \mathrm{Cl}_Y A \cap B$. So Y is connected if and only if the conditions $Y = A \cup B$, $\bar{A} \cap B = A \cap \bar{B} = \varnothing$ imply $A = \varnothing$ or $B = \varnothing$.

The connectedness of Y can also be described in terms of the open sets of X [cf. Exercises 6, 7, 8].

It is to be expected from the fact that connectedness involves the open sets of X that connectedness is a *topological invariant:* that is, if X is homeomorphic to Y, then X is connected if and only if Y is connected. In fact, we prove a stronger result.

3.2.2 *If X is connected, and $f : X \to Y$ is continuous, then* $\mathrm{Im}\, f$ *is connected.*

Proof Let $f' = f \mid X, \mathrm{Im}\, f$. If $\mathrm{Im}\, f$ is disconnected, then there is continuous surjection $g : \mathrm{Im}\, f \to \{1, 2\}$, and $gf' : X \to \{1, 2\}$ is a continuous surjection. This implies that X is disconnected. □

A discrete space with more than one point is disconnected, while $\varnothing$ and $\{a\}$ are connected spaces. By 1.3.4, the real line $\mathbf{R}$ is connected. Any open interval of $\mathbf{R}$ is homeomorphic to $\mathbf{R}$ and hence is connected by 3.2.2.

3.2.3 *If A is dense in a topological space X, and A is connected, then X is connected.*

Proof Let A be connected and dense in X. Let $f : X \to \{1, 2\}$ be continuous. Then $f \mid A$ is continuous and hence constant, say with value 1. By 2.6.2(*d*)

$$f[\bar{A}] \subset \overline{f[A]} = \overline{\{1\}} = \{1\}.$$

Since $\bar{A} = X$, this implies that f is constant. □

A corollary of 3.2.3 is that if $A \subset B \subset \bar{A} \subset X$ and A is connected, then B is connected—for the proof replace X in 3.2.3 by B with its relative topology.

3.2.4 *A subset X of* **R** *is connected if and only if X is an interval.*

Proof If X is an interval of **R**, then X is contained in the closure of an open interval, and so is connected by the remark following 3.2.3.

Suppose, conversely, that X is connected. If X contains at most one point, then it is an interval. If X contains points a, b with $a < b$, let $x \in]a, b[$ and suppose $x \notin X$. Then the set

$$X \cap]\leftarrow, x] = X \cap]\leftarrow, x[$$

is a non-empty proper subset of X both open and closed in X. This contradicts the assumption that X is connected. Hence, $x \in X$, and so X is an interval. □

A very direct way of proving that an interval X of **R** is connected is to show that any map $f : X \to \{0, 1\}$ is constant—an outline proof is as follows. Let $x, y \in X$ and suppose for example $x < y$ and $fx = 0$. Let $s = \sup \{z : x \leqslant z \leqslant y \text{ and } fz = 0\}$. It is easy to prove that $fs = 0$ and to derive a contradiction from the assumption $s < y$. Hence $fy = 0$, and so $fx = fy$.

EXAMPLES 1. The interval $\mathbf{I} = [0, 1]$ is connected. The function $t \rightsquigarrow e^{2\pi i t}$ is a continuous surjection $\mathbf{I} \to \mathbf{S}^1$. Hence $\mathbf{S}^1$ is connected.

2. Let $f : A \to \mathbf{R}$ be a map where A is an interval of **R**. By the last two results, $\operatorname{Im} f$ is an interval. Hence f takes any value between two given values.

3. A space is *totally disconnected* if its only (non-empty) connected subsets consist of single points. Examples of totally disconnected spaces are discrete spaces, **Q**, $\mathbf{R} \setminus \mathbf{Q}$ and also

$$\mathbf{L} = \{0\} \cup \{n^{-1} : n \text{ a positive integer}\}.$$

(The notation **L** will be standard for this space.) The proof that the last three spaces are totally disconnected is easy using 3.2.4.

3.2.5 *Let $(A_\lambda)_{\lambda \in \Lambda}$ be a family of connected subspaces of X, whose intersection is non-empty. Then*

$$A = \bigcup_{\lambda \in \Lambda} A_\lambda$$

is connected.

Proof Let $f : A \to \{1, 2\}$ be continuous. Then $f \mid A_\lambda$ is constant (since A_λ is connected) and so f is constant (since $\bigcap_{\lambda \in \Lambda} A_\lambda \neq \varnothing$). Therefore, A is connected. □

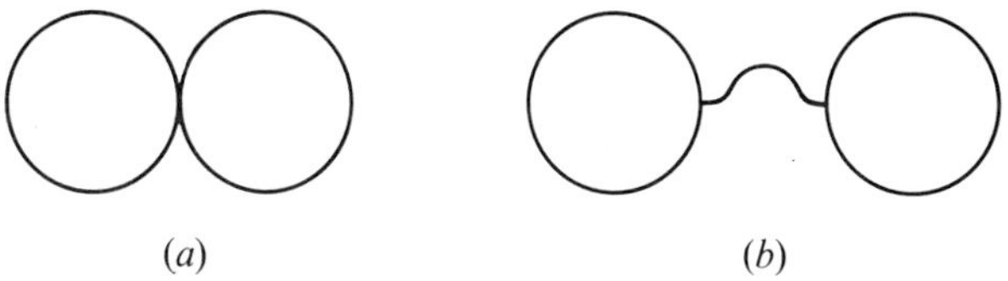

Fig. 3.1

For example, the space illustrated in (*a*) of Fig. 3.1 is connected, being the union of two spaces, homeomorphic to $\mathbf{S}^1$ and meeting in a single point. By two applications of 3.2.5, the space illustrated in (*b*) of Fig. 3.1 is connected.

3.2.6 *If X, Y are connected, then so also is $X \times Y$.*
Proof Let X, Y be connected and let $f : X \times Y \to \{1, 2\}$ be continuous. We prove that f is constant.

Let $(x, y), (x', y') \in X \times Y$. The space $\{x\} \times Y$ is homeomorphic to Y and hence is connected. Therefore, f is constant on $\{x\} \times Y$ and, in particular, $f(x, y) = f(x, y')$. Similarly, $f(x, y') = f(x', y')$. Therefore, $f(x, y) = f(x', y')$, and f is constant. □

EXAMPLES 4. Since $\mathbf{R}$ is connected, so also is $\mathbf{R}^n$.
5. $\mathbf{I}^n$, the n-fold product of $\mathbf{I} = [0, 1]$, is connected.
6. Let $X = \{(x, \sin \pi/x) : 0 \neq x \in \mathbf{R}\}$, let $J = \{0\} \times [-1, 1]$ and let $Y = X \cup J$ (cf. Fig. 2.5, p. 37). We prove that Y is connected.

Let $X_+ = \{(x, y) \in X : x > 0\}$, $X_- = \{(x, y) \in X : x < 0\}$. The function $\mathbf{R}^{>0} \to X_+$ sending $x \rightsquigarrow (x, \sin \pi/x)$ is continuous and surjective. Hence X_+ is connected. By 3.2.3, $X_+ \cup J = \overline{X_+}$ is connected. Similarly, $X_- \cup J$ is connected. By 3.2.5, $X \cup J$ is connected.

EXERCISES

1. Let X be a connected metric space with unbounded metric. Prove that every sphere $S(a, r)$ in X is non-empty. Is this true for X disconnected?

2. Let $X \subset \mathbf{R}^2$ be the subspace of points (x, y) such that either (i) x is irrational and $0 \leqslant y \leqslant 1$, or (ii) x is rational and $-1 \leqslant y < 0$. Prove that X is connected. Prove also that if $f : [0, 1] \to X$ is continuous, and $p_1 : X \to \mathbf{R}$ is the projection on the first coordinate, then $p_1 f$ is constant.

3. Let A be a connected subset of the connected space X. Let B be open and closed in $X \setminus A$. Prove that $A \cup B$ is connected. [Use Exercise 5 of 3.1 to extend a map $A \cup B \to \{1, 2\}$ over X.]

4. Prove 3.2.3, 3.2.5 by using directly condition 3.2.1 (b) for connectedness.

5. Let A be a connected subset of the topological space X. Is Int A necessarily connected?

6. Let A be a subset of the metric space X. Prove that A is disconnected if and only if there are sets U, V open in X such that (i) $U \cap V = \varnothing$, (ii) $A \subset U \cup V$, and (iii) $A \cap U$, $A \cap V$ are non-empty.

7. Show that Exercise 6 is false for arbitrary spaces by considering the space $X = \{0, 1, 2\}$ in which $\varnothing$, X, $\{0\}$, $\{0, 1\}$, $\{0, 2\}$ are the only open sets.

8. Prove that a subset A of a topological space X is disconnected if and only if there are sets U, V open in X such that (i) $U \cap V \subset X \setminus A$, (ii) $A \subset U \cup V$, and (iii) $A \cap U$, $A \cap V$ are non-empty.

9. A *domain* in a topological space X is a non-empty, connected, open subset of X. Prove that if X is connected and A, B are domains in X such that $\bar{A} \subset B$ and Fr A is connected, then $B \setminus A$ is connected.

10. Let $\leqslant$ be an order relation on X. This order is *without gaps* if $]x, y[$ is non-empty for each x, y in X such that $x < y$. Prove that X with its order topology is connected if and only if the order is complete and without gaps.

3.3 Components and locally connected spaces

Let X be a topological space and let $x \in X$. The *component of x in X* is $C(x)$, the union of all connected sets containing x. By 3.2.5, $C(x)$ is connected; therefore, $C(x)$ is the largest connected set containing x. But $\overline{C(x)}$ is also connected [3.2.3]. Therefore, $C(x) = \overline{C(x)}$. This proves:

3.3.1 *The component of x in X is a closed subset of X.*

A component need not be open. For example, in the space $\mathbf{L}$ the component of 0 is $\{0\}$, which is closed but not open in $\mathbf{L}$. Again, the components of points of $\mathbf{Q}$ are not open in $\mathbf{Q}$. An obvious question is therefore: under what conditions are the components always open?

Definition A space X is *locally connected at a point x* in X if the connected neighbourhoods of x form a base for the neighbourhoods at x. (This is sometimes expressed as: x has a base of connected neighbourhoods.) The space X is *locally connected* if it is locally connected at each x in X.

Thus X is locally connected if, for each x in X, each neighbourhood of x contains a connected neighbourhood of x.

One more definition: if $A \subset X$, the *components of* A are the components of the points of the subspace A. So the components of A are subsets of A, except that the empty set $\varnothing$ has no components.

3.3.2 *X is locally connected if and only if the components of each open set of X are open sets of X.*

Proof Suppose X is locally connected, V is open in X, C is a component of V, and $x \in C$. Since V is a neighbourhood of x, and X is locally connected, there is a connected neighbourhood U of x such that $U \subset V$. Therefore, $U \subset C$, and C is a neighbourhood of x. Therefore, C is open.

For the converse, we start with a neighbourhood V of x which we may suppose to be open (otherwise we replace V by Int V). The component of V which contains x is open in X (by assumption), and so is a connected neighbourhood of x contained in V. □

A special case of 3.3.2 is that, if X is locally connected, then each component of X is open.

EXAMPLES 1. Neither **L**, **Q** nor **R** \ **Q** are locally connected.
2. A connected set need not be locally connected: the space $Y = X \cup J$ of Example 6, p. 65, is not locally connected, since points of J have no 'small' connected neighbourhoods [cf. Fig. 3.2]. In fact we leave as an

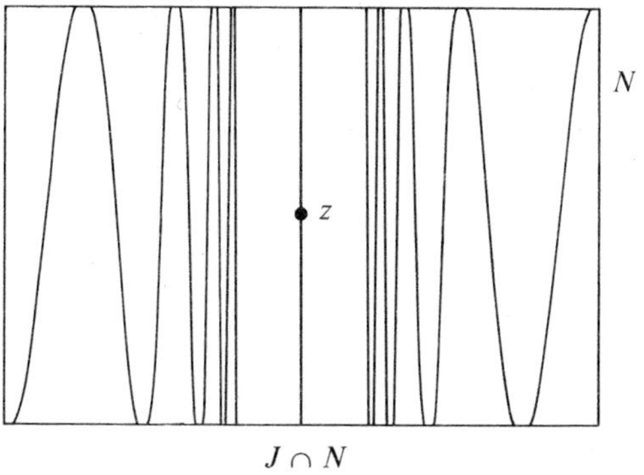

Fig. 3.2

exercise for the reader the detailed proof of the following statement: if $z \in J$, there is in Y a neighbourhood N of z such that the component of z in N is $N \cap J$.

3. Let X consist of the line segment joining $(1, 1)$ in $\mathbf{R}^2$ to the points of $\mathbf{L} \times \{0\}$ [Fig. 3.3]. Then X is connected, but not locally connected, since $(0, 0)$ has no 'small' connected neighbourhoods. (Here also, we leave a detailed proof to the reader.) Is $X \setminus \{(1, 1)\}$ connected?

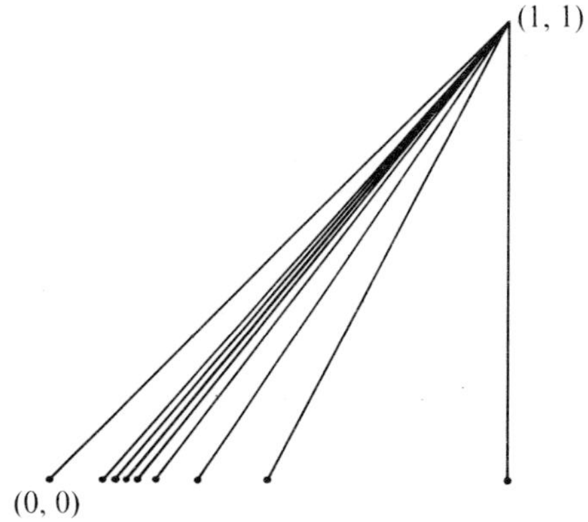

Fig. 3.3

4. A discrete space is locally connected and also totally disconnected.

Cut points

Let $f : X \to Y$ be a homeomorphism. If C is a component of a point x of X, then $f[C]$ is a component of $f(x)$ in Y: this follows from the fact that $A \subset X$ is connected if and only if $f[A]$ is connected. Consequently, f induces a bijection of the components of X to the components of Y. So the number of components of X is a topological invariant of X. However, this is not a very subtle invariant, since it fails to distinguish different connected spaces.

Definition Let X be connected space, and k a natural number or $\mathbf{N}$. A point x in X is a *cut point of order* k if $X \setminus \{x\}$ has k components.

Let X be connected, and $f : X \to Y$ a homeomorphism. A point x in X is a cut point of order k in X if and only if $f(x)$ is a cut point of order k in Y. Therefore, the number of cut points of order k is a topological invariant of X.

EXAMPLES 5. The closed interval $[0, 1]$ has two cut points of order 1; the half-open interval $[0, 1[$ has one cut point of order 1; the open interval $]0, 1[$ has no cut points of order 1. Therefore, no two of the spaces $[0, 1]$, $[0, 1[$ and $]0, 1[$ are homeomorphic.

6. The following 1-dimensional† spaces can be distinguished by the numbers of cut points of various orders:

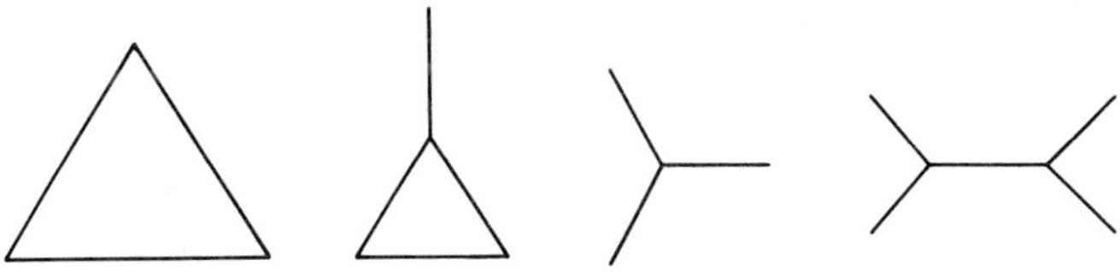

Fig. 3.4

7. These methods however fail to distinguish between the following spaces:

Fig. 3.5

However, if we remove two points from the first space, two components are left, while with the second space we can get one, two, or three components by removing different pairs of points. Hence, the two spaces are not homeomorphic.

8. The space $\mathbf{R}^n$ has cut points of order 1 only if $n > 1$. That is, if $a \in \mathbf{R}^n$, then $\mathbf{R}^n \setminus \{a\}$ is connected if $n > 1$. For let $x, y \in \mathbf{R}^n \setminus \{a\}$, and let $z \in \mathbf{R}^n \setminus \{a\}$ be such that the lines L, M joining z to x, y respectively do not pass through a (this is possible since $n > 1$). Then $L \cup M$ is a connected set containing x and y, so that x and y belongs to the same component of $\mathbf{R}^n \setminus \{a\}$.

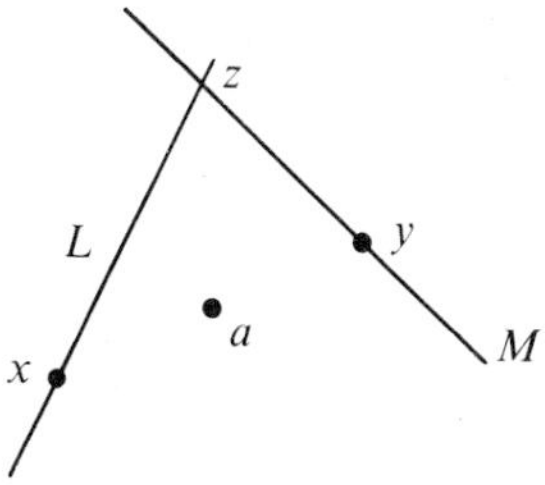

Fig. 3.6

† The term 1-dimensional is to be understood at this stage only intuitively.

This proves that $\mathbf{R}^n$ is not homeomorphic to $\mathbf{R}^1$ if $n \neq 1$. It is much more difficult to prove the *Invariance of Dimension:* if $\mathbf{R}^m$ is homeomorphic to $\mathbf{R}^n$, then $m = n$. All the present proofs of this theorem, as of the Invariance of Domain, use techniques of homology theory or of subdivisions of simplicial complexes.

Actually, a stronger result than the Invariance of Dimension is true; if $f : \mathbf{R}^m \to \mathbf{R}^n$ is a continuous bijection, then $m = n$ and f is a homeomorphism.

9. Another problem is to distinguish between surfaces, for example the 2-sphere $\mathbf{S}^2$ and the torus $T^2 = \mathbf{S}^1 \times \mathbf{S}^1$. This latter space is homeomorphic to the anchor ring [Fig. 3.7(*b*)].

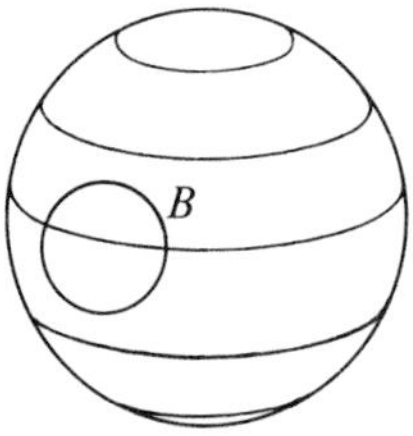

(i) $\mathbf{S}^2$

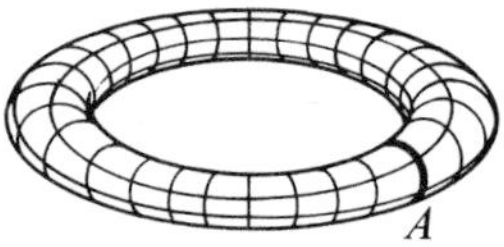

(ii) T^2

Fig. 3.7

Since removing points is a useful method for distinguishing 1-dimensional spaces, it seems reasonable that to distinguish 2-dimensional spaces we need to remove 1-dimensional subspaces (supposing also that we know what the words 1-dimensional and 2-dimensional mean). For example, let A be the meridian circle on T^2; then $T^2 \setminus A$ has only one component. On the other hand, it seems likely that if B is any subspace of $\mathbf{S}^2$ homeomorphic to $\mathbf{S}^1$, then $\mathbf{S}^2 \setminus B$ has two components. This is in fact true, but is far from easy to prove, being equivalent to the *Jordan Curve Theorem:* if B is a subspace of $\mathbf{R}^2$ homeomorphic to $\mathbf{S}^1$ (i.e., B is a simple closed curve in $\mathbf{R}^2$) then $\mathbf{R}^2 \setminus B$ has two components.

As we shall see in chapter 8, $\mathbf{S}^2$ and T^2 can be distinguished by their fundamental groups.

10. If two spaces are homeomorphic, they must have the same 'local' properties. Let X be the graph of $x \rightsquigarrow \sin \pi/x$, and let $Y = X \cup \{(0, 0)\}$. Then Y is connected, and any point of Y is a cut point of order 2. But Y is not homeomorphic to $\mathbf{R}$, since $\mathbf{R}$ is locally connected, and Y is not locally connected at (0, 0).

11. We can also define *local cut points.* A point x in X is a local cut point of order k if each neighbourhood V of x contains a connected neighbour-

hood U of x such that $U \setminus \{x\}$ has k components. If X is homeomorphic to Y, then X and Y must have the same number of local cut points of order k for each $k = 1, 2, \ldots$. The spaces of Fig. 3.8 are distinguished by the

Fig. 3.8

fact that one has a local cut point of order 4, and the other does not.

By the use of homology theory these methods can be generalized to higher dimensional spaces.

EXERCISES

1. Prove that if X has a finite number of components, then each component is open.
2. Prove that the space X of Exercise 2 of 3.2 is not locally connected.
3. Decide whether or not the following 1-dimensional spaces are homeomorphic

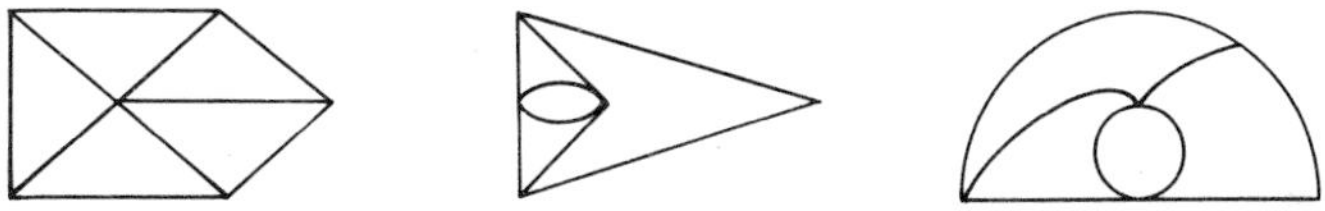

Fig. 3.9

4. Construct a locally connected subspace X of $\mathbf{R}^2$ in which for each $r > 0$ there is an x in X such that $B(x, r)$ (the open ball in X) is not connected.
5. Prove that $\mathbf{R}^n$ is locally connected.
6. Let $X = (\mathbf{L} \times \mathbf{I}) \setminus (\{0\} \times]0, 1[)$. Prove that the components of $(0, 0)$ and $(0, 1)$ in X are single points. Let $f : X \to \{1, 2\}$ be continuous. Prove that $f(0, 0) = f(0, 1)$.
7. Let X be the subspace of $\mathbf{R}^2$

$$\{0\} \cup \{(x, x \sin \pi/x) : x \neq 0\}$$

with the Euclidean metric. Prove that if $r > 0$ is sufficiently small then $B(0, r)$ (the open ball in X) is not connected.

*8. Let A be a non-empty subset of a metric space X. We say A is *bounded* if $\sup \{d(x, y) : x, y \in A\}$ exists, and then this number is called the *diameter* of A. Let $x, y \in X$. If there is a connected set containing x, y and of diameter < 1, let $\sigma(x, y)$ be the infinum of the diameters of such sets. If no such set exists, let $\sigma(x, y) = 1$. Prove that $(x, y) \rightsquigarrow \sigma(x, y)$ is a metric on X.

Let $L \subset X$ be the subset of X of points at which X is locally connected. Prove that σ induces the discrete topology on $X \setminus L$ (if $X \setminus L \neq \varnothing$) and that σ induces the same topology on L as does d.

Prove that in L, each open ball for the metric σ is connected if of radius < 1.

*9. Let X be connected and locally connected, and $f : X \to Y$ continuous. Prove that Im f is locally connected if f maps closed sets of X to closed sets of Y, but not in general.

*10. Let C be a component of the open set U of the locally connected space X. Prove that Fr $B \subset$ Fr $U \subset X \setminus U$.

3.4 Path-connectedness

In this section, we discuss a type of connectedness which is stronger (in the precise sense given by 3.4.4 below) than that of 3.2, and which is to some extent more intuitive. For 'nice' spaces, for example, the cell-complexes of chapter 4, the two notions of connectedness are equivalent.

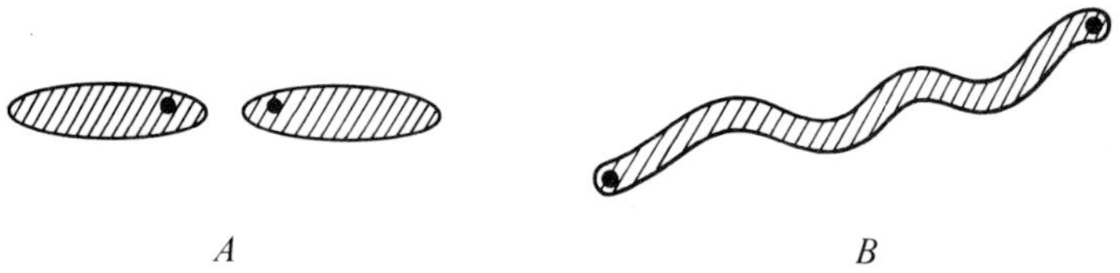

Fig. 3.10

Consider the spaces A and B of Fig. 3.10. It is intuitively clear that any two points of B (such as those shown by dots) can be joined by a continuous curve lying wholly in B. But this is false for the space A. The best general expression of these ideas is in terms of paths and path-connectedness.

Let X be a topological space and let $r \in \mathbf{R}^{\geqslant 0}$.

Definition A *path in X of length r* is a continuous function $a : [0, r] \to X$. We write $|a|$ for r. Then $a(0)$, the *initial point* of a is written a_ι, and $a(r)$ the *final point* of a is written a_φ; we say *a joins* a_ι to a_φ. We call a_ι and a_φ the *end points* of a.

A point x in X determines a unique constant path of length r with value x. If $r = 0$, this path is called the *zero path at x*.

It is important to note that a path in X is not just a set of points, but is a function. For example, the two paths $[0, 1] \to \mathbf{R}$ given by $t \rightsquigarrow t$ and $t \rightsquigarrow t^2$ are distinct paths in $\mathbf{R}$ joining 0 to 1. Our illustrative figures should then show the graph of a path—but it is usually more convenient to illustrate the image of the path.

We now consider two simple operations on paths. The *reverse* of a path $a : [0, r] \to X$ is the path

$$\begin{aligned} -a : [0, r] &\to X \\ t &\leadsto a(r - t). \end{aligned}$$

Thus $|-a| = |a|$ and $-a$ joins a_φ to a_ι.

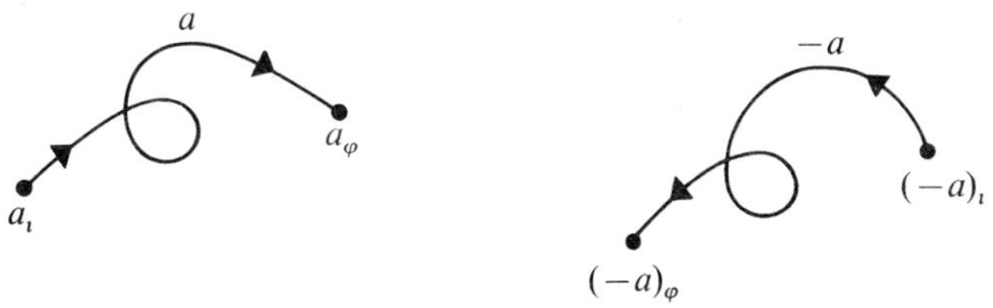

Fig. 3.11

The reverse of a path is always defined. On the other hand, the *sum* $b + a$ of two paths is defined if and only if the final point of a coincides with the initial point of b (that is, if and only if $a_\varphi = b_\iota$). In such case, $b + a$ is the path

$$\begin{aligned} b + a : [0, |b| + |a|] &\to X \\ t &\leadsto \begin{cases} a(t), & 0 \leqslant t \leqslant |a| \\ b(t - |a|), & |a| \leqslant t \leqslant |a| + |b|. \end{cases} \end{aligned}$$

Clearly the condition $a_\varphi = b_\iota$ is essential for $b + a$ to be defined; and with this condition, $b + a$ is continuous by the Glueing Rule [2.5.12].

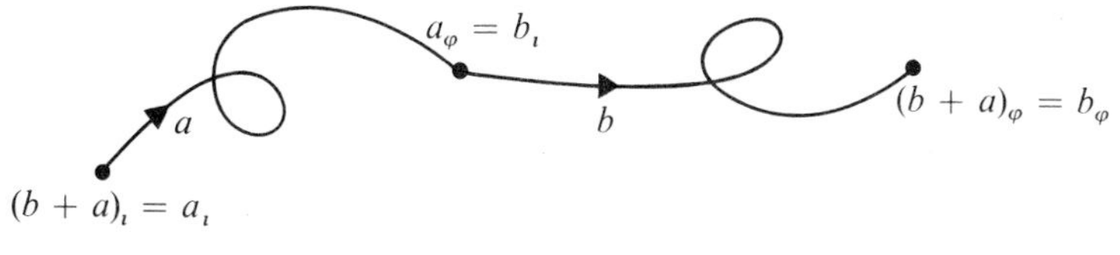

Fig. 3.12

We use the additive notation because of its convenience in dealing with differences ($b - c$ is a good abbreviation of $b + (-c)$). Our convention that $b + a$ means first a then b [Fig. 3.12] is related to the convention that the composition gf of functions means first apply f, then apply g.

The reader should be warned that addition of paths is not commutative: if $b + a$ is defined, then $a + b$ need not be defined. And even if both $b + a$ and $a + b$ are defined, they are usually unequal.

Definition A topological space X is *path-connected* if for any x, y in X, there is a path in X joining x to y.

EXAMPLE 1. We recall that a subset A of a normed vector space V is *convex* if, for any x, y in A, the *line segment*

$$[x, y] = \{(1 - t)x + ty : 0 \leqslant t \leqslant 1\}$$

is contained in A. A convex set A is path-connected, since if $x, y \in A$ then $t \rightsquigarrow (1 - t)x + ty$ is a path in A from x to y of length 1. Examples of convex sets are $B(x, r)$, $E(x, r)$. For example, if $y, z \in B(x, r)$ and $0 \leqslant t \leqslant 1$, then

$$\begin{aligned}\|(1 - t)y + tz - x\| &\leqslant \|(1 - t)(y - x)\| + \|t(z - x)\| \\ &< (1 - t)r + tr = r\end{aligned}$$

whence $(1 - t)y + tz \in B(x, r)$. Again, any interval of $\mathbf{R}$ is convex.

Other examples of path-connected sets may be constructed from the following results.

3.4.1 *Let $f : X \to Y$ be continuous and surjective. If X is path-connected, then so also is Y.*

Proof Let $y, y' \in Y$; then there are points x, x' in X such that $f(x) = y$, $f(x') = y'$. Since X is path-connected, there is a path a joining x to x'. Then fa joins y to y'. □

3.4.1 (*Corollary 1*) *Let X be homeomorphic to Y. Then X is path-connected if and only if Y is path-connected.*

3.4.2 *Let $(A_\lambda)_{\lambda \in \Lambda}$ be a family of path-connected subspaces of X such that $\bigcap_{\lambda \in \Lambda} A_\lambda$ is non-empty. Then $A = \bigcup_{\lambda \in \Lambda} A_\lambda$ is path-connected.*

Proof Let $x, y \in A$, $z \in \bigcap_{\lambda \in \Lambda} A_\lambda$, and suppose $x \in A_\lambda$, $y \in A_\mu$. Since A_λ is path-connected, there is a path a joining x to z. Since A_μ is path-connected, there is a path b joining z to y. Then $b + a$ is a path joining x to y. □

3.4.3 *If X, Y are path-connected, then $X \times Y$ is path-connected.*

Proof Let $(x, y), (x', y') \in X \times Y$. Let a be a path in X joining x to x', b be a path in Y joining y to t'. Then $t \rightsquigarrow (a(t), y)$ joins (x, y) to (x', y), and $t \rightsquigarrow (x', b(t))$ joins (x', y) to (x', y'). The sum of these paths joins (x, y) to (x', y'). □

We can use 3.4.2 to define path-components of X: if $x \in X$, then the *path-component of* x is the union of all path-connected subsets P of X such that P contains x. This union will be path-connected by 3.4.2, and will contain x. Hence it is the largest path-connected subset of X containing x.

A path-component need not be closed. To show this, we consider a little the relationship between the two kinds of connectedness.

3.4.4 *If X is path-connected, then X is connected.*

Proof Let $x, y \in X$, and let a be a path joining x to y. Then $\operatorname{Im} a$ is by 3.2.2 a connected set containing x and y. Hence x and y belong to the same component of X. Thus X has only one component, and must be connected. □

EXAMPLE 2. Let $Y = X \cup J$ be the connected space of Example 2, p. 67 and Example 6, p. 65. Then Y is not path-connected.

Proof Let $g : Y \to \{0, 1\}$ be the function which sends points of J to 0 and points of X to 1. Of course, g is not continuous, but we prove that for any path $f : [0, r] \to Y$, the composite gf is continuous.

Let $x \in [0, r]$. If $fx \in X$, then fx has a neighbourhood N which does not meet J; hence there is a neighbourhood M of x such that $f[M] \subset N$, whence $gf[M] = \{1\}$; continuity of gf at x follows easily.

On the other hand, suppose $fx \in J$. As stated on p. 67, there is a neighbourhood (in Y) of fx such that the component of fx in N is $N \cap J$. Therefore, there is a connected neighbourhood M of x such that $f[M] \subset N \cap J$. Hence, $gf[M] = \{0\}$, and continuity of gf at x follows easily.

We have now proved that gf is continuous. Since $[0, r]$ is connected it follows that gf is constant. So $\operatorname{Im} f$ is contained either in X or in J, and Y is not path-connected. □

The philosophy of local path-connectedness is different from that of local connectedness. The definition of the former concept that first suggests itself is that a space X is locally path-connected if each point x in X has a base of path-connected neighbourhoods. However, because of extensions of this property to higher-dimensional kinds of connectedness, we take a different definition which, it turns out, is equivalent to the above.

Definition A space X is *locally path-connected* if, for each point x in X, any neighbourhood of U of x contains a neighbourhood V of x such that any two points of V can be joined by a path in U.

3.4.5 *A space X is locally path-connected $\Leftrightarrow$ each point of X has a base of open path-connected neighbourhoods.*

Proof The implication $\Leftarrow$ is trivial, and so we prove the implication $\Rightarrow$. Let $x \in X$ and let U be an open neighbourhood of x. Let U' be the path-component of U containing x.

Let $y \in U'$. Then U is a neighbourhood of y, and so U contains a neighbourhood V of y such that any two points of V can be joined by a path in U. This implies that V is contained in U'. Hence U' is a neighbourhood of y. Therefore (since $x \in U'$) U' is an open, path-connected neighbourhood of x. □

EXERCISES

1. In the following, X is a subspace of $\mathbf{R}^2$ and x_0, x_1 are points of X. Write down, if possible, explicit paths in X joining x_0 to x_1.

(i) $X = \{(x, y) \in \mathbf{R}^2 : |x| \geqslant 1, |y| \geqslant 1\}$, $x_0 = (-2, 0)$, $x_1 = (2, 0)$.
(ii) $X = \{(x, y) \in \mathbf{R}^2 : x + y \neq 1\}$, $x_0 = (4, -5)$, $x_1 = (-6, 7)$.
(iii) $X = \{(x, y) \in \mathbf{R}^2 : [x] = [y]\}$, $x_0 = (-\frac{1}{2}, -1)$, $x_1 = (1, \frac{3}{2})$.
(iv) $X = \{(x, y) \in \mathbf{R}^2 : [x] + [y] = 1\}$, $x_0 = (\frac{3}{2}, 0)$, $x_1 = (0, \frac{3}{2})$.

2. Let V be a normed vector space over $\mathbf{R}$. A 'bent-line' in V is the union of a finite number of line segments $[u_i, u_{i+1}]$, $i = 1, \ldots, n - 1$, and such a bent line is said to *join* u_1 to u_n. A subset A of V is 'polygonally connected' if, for any two points u, v of A, there is a bent line joining u to v and lying wholly in A.

Prove that any open, connected subset of V is polygonally connected. Give examples of subsets of $\mathbf{R}^2$ which are path-connected but not polygonally connected.

3. Let $X = \{a, b, c\}$ with the topology whose open sets are $\varnothing$, $\{c\}$, $\{a, c\}$, $\{b, c\}$, X. Prove that X is path-connected.

4. Prove that X is locally path-connected if and only if the path-components of each open set of X are open. Give an example of a space which is path-connected but not locally path-connected.

5. Prove that two points x, y in X lie in the same path-component if and only if they can be joined by a path of length 1.

6. Let V be a normed vector space over $\mathbf{R}$ of dimension > 1. Prove that $S(V)$ is path-connected.

7. Prove that if X is a countable subset of $\mathbf{R}^n$ ($n > 1$), then $\mathbf{R}^n \setminus X$ is path-connected.

8. Let X_1, X_2 be subsets of X such that $X = \operatorname{Int} X_1 \cup \operatorname{Int} X_2$. Prove that, if X is path-connected, then each path-component of X_1 meets X_2.

3.5 Compactness

The reader will certainly be aware of the importance for mathematics of the distinction between finite and infinite sets. As examples consider the statements (*a*) the sum of the elements of a set of A of real numbers is well-defined, (*b*) a set A of real numbers has a least element, (*c*) the intersection of the elements of a set A of open sets is open. Each of these is true if A is finite but may be false if A is infinite. This wide range of techniques applicable to finite sets but not infinite ones is the reason for the importance of the notion of a compact space in topology.

In order to define compactness we need some preliminary definitions.

Let X be a topological space. A *cover* of X is a set $\mathscr{A}$ of sets such that the union of the elements of $\mathscr{A}$ contains X. A *subcover* of $\mathscr{A}$ is a subset $\mathscr{B}$ of $\mathscr{A}$ such that $\mathscr{B}$ covers X. A cover $\mathscr{A}$ of X is *open* if each set of $\mathscr{A}$ is open in X.

EXAMPLES 1. Let $k \in \mathbf{R}^{>0}$, and let $\mathscr{A}$ be the set of intervals $]x - k, x + k[$ for each $x \in \mathbf{R}$. Then $\mathscr{A}$ is an open cover of $\mathbf{R}$. Similarly, in any metric space X, the set of all open balls $B(x, k)$, $x \in X$, is an open cover of X.
2. Let X be a metric space and let $x \in X$. The set of open balls $B(x, n)$ for all positive n in $\mathbf{N}$ is an open cover of X.
3. The set of intervals $]1/n, 1]$ for n, a positive integer, is an open cover of $]0, 1]$.
4. If $\mathscr{A}$ is an open cover of Y, and $f : X \to Y$ is continuous, then the set of $f^{-1}[A]$ for all A in $\mathscr{A}$ is an open cover of X.
5. For any topological space X, the set $\{X\}$ is an open cover of X as is $\{X, \varnothing\}$.

Definition A topological space X is *compact* if every open cover of X has a finite subcover.

This means of course, that to prove a space X is compact we have to start with any open cover $\mathscr{A}$ of X and construct a finite subcover of $\mathscr{A}$. To prove X non-compact, we have to produce an open cover $\mathscr{A}$ of X without finite subcover.

EXAMPLES 6. An infinite discrete space X is not compact, since the set of singletons $\{x\}$ for each x in X is an open cover of X without finite subcover.
7. Any finite space X is compact, since any open cover of X is a finite set.
8. The interval $]0, 1]$ is not compact—the open cover of Example 3 has no finite subcover.
9. If X is a metric space with unbounded metric, then X is not compact. To prove this, let $x \in X$ and consider the open cover of Example 2. This cover has a finite subcover if and only if $X = B(x, n)$ for some n, in which case X has bounded metric.
10. The previous example has a converse, namely, that if X is a non-compact metric space, then X admits an equivalent unbounded metric. The proof of this theorem is not as simple as that of its converse.

Compactness is a topological invariant. In fact, we have the stronger result.

3.5.1 *Let X be compact and $f : X \to Y$ continuous and surjective. Then Y is compact.*

Proof Let $\mathscr{A}$ be an open cover of Y. Since f is continuous the set $\mathscr{B}$ of sets $f^{-1}[A]$ for all A in $\mathscr{A}$ is an open cover of X. Since X is compact, $\mathscr{B}$ has a finite subcover $\mathscr{C}$. The set of A in $\mathscr{A}$ for which $f^{-1}[A]$ is in $\mathscr{C}$ is a finite subcover of $\mathscr{A}$. □

3.5.2 *Remark*

Let C be a subspace of the topological space X. Then we have covers of C by sets open in X, and by sets open in C. We distinguish these by calling them open covers of the *set* C, and of the *space* C, respectively. An open cover of the space C clearly consists of sets $A \cap C$ for A in an open cover of the set C. So the statements (a) every open cover of the space C has a finite subcover, and (b) every open cover of the set C has a finite subcover, are equivalent, and either may be used as a criterion for compactness of C.

3.5.3 *Remark*

Let $\mathscr{A}$ be a cover of the space X. A *refinement* of $\mathscr{A}$ is a cover $\mathscr{B}$ of X such that each set of $\mathscr{B}$ is contained in some set of $\mathscr{A}$. Suppose $\mathscr{B}$ is an open cover which refines an open cover $\mathscr{A}$. Then, if $\mathscr{B}$ has a finite subcover, so also does $\mathscr{A}$. Thus, when trying to construct finite subcovers of an open cover $\mathscr{A}$ we may at will replace $\mathscr{A}$ by an open refinement.

The next theorem gives the simplest non-trivial example of a compact space. (The proof given here is due to R. M. F. Moss and G. Roberts.)

3.5.4 *The unit interval* **I** *is compact*

Proof Let $\mathscr{A}$ be an open cover of **I**. For each x in **I** we choose an interval U_x, open in **I**, such that U_x contains x and is contained in some set of $\mathscr{A}$. The set $\mathscr{B}$ of these intervals U_x is an open cover of **I** which refines $\mathscr{A}$. By Remark 3.5.3, we may assume from the start that each element of $\mathscr{A}$ is an interval.

Let $f : \mathbf{I} \to \{0, 1\}$ be the function defined by $fx = 0$ if $[0, x]$ can be covered by a finite number of sets of $\mathscr{A}$, and $fx = 1$ otherwise. We shall prove that f is constant on each set of $\mathscr{A}$.

Let $U \in \mathscr{A}$, let $x \in U$ and suppose $fx = 0$. Then $[0, x]$ is covered by a finite subset $\mathscr{B}$ of $\mathscr{A}$ and so for any y in U, since U is an interval, $[0, y]$ is covered by the finite set $\mathscr{B} \cup \{U\}$. Thus we have shown, as required, that f is 0 either on all or on none of U.

It follows immediately that f is continuous; therefore, f is constant (since **I** is connected) and the image of f is $\{0\}$ (since $f0 = 0$). Hence $[0, 1]$ can be covered by a finite number of sets of $\mathscr{A}$. □

It is instructive to examine the failure of similar attempts to prove that the intervals $]0, 1]$ and $[0, 1[$ are compact (they are non-compact by Example 8 and a similar example for $[0, 1[$). The proof for $]0, 1]$ breaks down because we cannot prove that the unique value of f is 0. The proof for [0, 1[breaks down because $f1$ is not defined.

3.5.5 *A closed subset of a compact space is compact*

Proof Let C be a closed subset of the compact space X and, applying 3.5.2, let $\mathscr{A}$ be an open cover of the set C. Since $X \setminus C$ is open,

$$\mathscr{A}' = \mathscr{A} \cup \{X \setminus C\}$$

is an open cover of X. By compactness of X, $\mathscr{A}'$ has a finite subcover $\mathscr{B}$ say.

Here $\mathscr{B}$ is an open cover of X, and so an open cover of the set C. If $\mathscr{B}$ does not contain $X \setminus C$, then $\mathscr{B}$ is a finite subcover of $\mathscr{A}$. In any case, $\mathscr{B} \setminus \{X \setminus C\}$ is certainly a finite subcover of $\mathscr{A}$. □

The next theorem has a slightly complicated formulation—this is due to the fact that we wish to include in one theorem (and one proof) a number of highly important special cases.

3.5.6 *Let B, C be compact subsets of X, Y respectively, and let $\mathscr{W}$ be a cover of $B \times C$ by sets open in $X \times Y$. Then B, C have open neighbourhoods U, V respectively such that $U \times V$ is covered by a finite number of sets of $\mathscr{W}$.*

Proof The proof is carried out in two steps, first when B has a single point b, and next for B arbitrary.

Step 1 – $B = \{b\}$

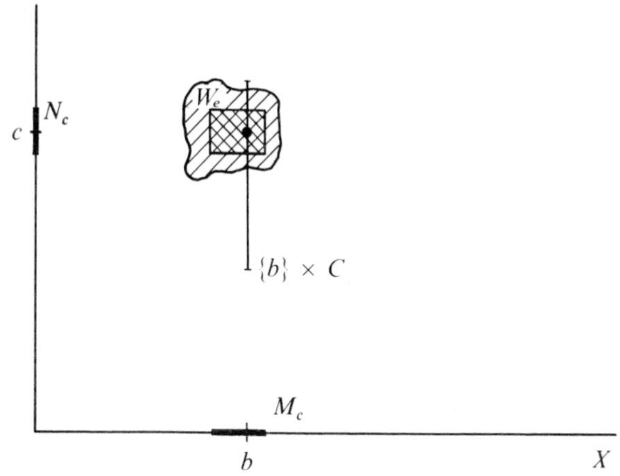

Fig. 3.13

For each c in C there are open neighbourhoods of M_c of b (in X), N_c of c (in Y) such that $M_c \times N_c$ is contained in some set W_c of $\mathscr{W}$. The set $\{N_c : c \in C\}$ is an open cover of the set C which, by compactness, has a finite subcover $\{N_f : f \in F\}$. Let

$$U = \bigcap_{f \in F} M_f, \qquad V = \bigcup_{f \in F} N_f.$$

Since A is finite, U is an open neighbourhood of b. Clearly,

$$\{b\} \times C \subset U \times V \subset \bigcup_{f \in F} W_f.$$

Step 2—B arbitrary

By step 1, for each b in B there are open neighbourhoods U_b of b (in X), V_b of C (in Y), such that $U_b \times V_b$ is contained in a finite union, say $\bigcup \{W_f : f \in F_b\}$, of sets of $\mathscr{W}$. The set $\{U_b : b \in B\}$ is an open cover of B which, by compactness, has a finite subcover $\{U_g : g \in G\}$. Let

$$U = \bigcup_{g \in G} U_g, \qquad V = \bigcap_{g \in G} V_g.$$

Then U, V are open neighbourhoods of B, C respectively and $U \times V$ is contained in the union of the finite number of sets $W_f, f \in F_g, g \in G$. □

3.5.6 (*Corollary 1*) *The product of compact spaces is compact.*

Proof This follows from 3.5.6 by taking $B = X$, $C = Y$ (and so $U = X$, $V = Y$). □

3.5.6 (*Corollary 2*) *Let B, C be compact subsets of X, Y respectively, and let W be an open subset of $X \times Y$ containing $B \times C$. Then B, C have open neighbourhoods U, V respectively such that $U \times V \subset W$.*

Proof This is the case of 3.5.6 when $\mathscr{W} = \{W\}$. □

3.5.6 (*Corollary 3*) *If B, C are disjoint compact subsets of the Hausdorff space X, then B, C have disjoint open neighbourhoods.*

Proof In 3.5.6 (Corollary 2), let $X = Y$ and $W = (X \times X) \setminus \Delta[X]$ where Δ is the diagonal of $X \times X$; W is open since X is Hausdorff. Since B and C are disjoint, $B \times C$ is contained in W. By 3.5.6 (Corollary 2), B, C have open neighbourhoods U, V such that $U \times V \subset W$. Hence $U \cap V = \varnothing$. □

3.5.6 (*Corollary 4*) *A compact subset of a Hausdorff space is closed.*

Proof Let C be a compact subset of the Hausdorff space X, and let $x \in X \setminus C$. By 3.5.6 (Corollary 3), x and C have disjoint neighbourhoods. Hence $X \setminus C$ is open. □

As an application of these rules we prove

3.5.7 *A subset of Euclidean n-space* $\mathbf{R}^n$ *is compact if and only if it is closed and bounded.*

Proof Let A be a closed, bounded subset of $\mathbf{R}^n$. Let J_r be the subspace of $\mathbf{R}^n$ of points $(x_1, \ldots, x_n)$ such that $|x_i| \leqslant r$—thus J_r is the product of the interval $[-r, r]$ with itself n times [cf. Exercise 8 of 2.3]. The interval $[-r, r]$ is homeomorphic to $[0, 1]$ (if $r > 0$) and so is compact. By 3.5.6 (Corollary 1), J_r is compact. Since A is bounded, it is contained in J_r for some r. Thus A is a closed subset of a compact space and so is compact.

The converse follows from Example 9 and 3.5.6 (Corollary 4). □

Examples 11. The Cantor set K [Example 1 of 1.3] is a closed, bounded subset of $\mathbf{R}$ and so is compact.

12. The subsets of $\mathbf{R}^n$ (with the Euclidean norm)

$$\begin{aligned} \mathbf{S}^{n-1} &= \{x \in \mathbf{R}^n : \|x\| = 1\} \\ \mathbf{E}^n &= \{x \in \mathbf{R}^n : \|x\| \leqslant 1\} \end{aligned}$$

are both closed, bounded subsets of $\mathbf{R}^n$ and so are compact.

13. If X is indiscrete, then any subset A of X is compact, but A will be closed in X only if $A = \varnothing$ or $A = X$.

14. Let W be the subspace of $\mathbf{R}^2$ of points (x, y) such that $y > |x|$. Let $B = \{0\}$, $C =]0, 1]$. Then W is an open set of $\mathbf{R}^2$ containing $B \times C$ but, if U is a neighbourhood of 0, then W does not contain even $U \times C$.

A map $f : X \to Y$ of spaces is called *closed* if $f[C]$ is closed in Y for each closed set C of X. For example, a continuous bijection $f : X \to Y$ is closed if and only if f is a homeomorphism.

3.5.8 *Any map from a compact space to a Hausdorff space is closed.*

Proof Let $f : X \to Y$ be a map where X is compact and Y is Hausdorff. Let C be closed in X. Then C is compact [3.5.5] whence $f[C]$ is compact [3.5.1] and so $f[C]$ is closed [3.5.6 (Corollary 4)]. □

3.5.8 (*Corollary 1*) *A continuous bijection from a compact space to a Hausdorff space is a homeomorphism.*

Proof If f is a closed bijection, then f^{-1} is continuous. □

The following proposition is required later; the simple proof is left as an exercise.

3.5.9 *A topological space which is a finite union of compact spaces is itself compact.*

There is a characterization of compactness by means of closed sets which is of great importance in some contexts but not, as it turns out, in this book. For this reason it has been left as an exercise [Exercise 10].

There is a generalization of 3.5.6 which uses the product topology defined in 5.7:

3.5.10 (*Tychonoff's theorem*) *The topological product of any family of compact spaces is compact.*

This theorem is of great importance in functional analysis and in a further study of some of the topics in this book. However the theorem is not essential to our present purposes, and so we refer the reader to Kelley [1] or Simmons [1] for a proof.

EXERCISES

1. Give an example of a space with two points in which not all compact sets are closed.
2. Prove that **R** with the Zariski topology is compact.
3. Prove that a discrete space is compact if and only if it is finite.
4. Use the result of 3.2 and 3.5 to prove that if $a, b \in \mathbf{R}$ $(a < b)$ and $f : [a, b] \to \mathbf{R}$ is continuous, then $\operatorname{Im} f$ is a closed, bounded interval.
5. Let X be a compact topological space, and $f : X \to \mathbf{R}$ a continuous function. Prove that there are elements a, b in X such that $f(a) = \inf(\operatorname{Im} f), f(b) = \sup(\operatorname{Im} f)$. Deduce that if $f(x) > 0$ for all x in X, then there is a positive real number r such that $f(x) > r$ for all x in X.
6. Let $X = \mathbf{R} \times Y$ where Y is an indiscrete space with two elements a, b. Prove that in X the sets $A = [0, 1[\times \{a\} \cup [1, 2] \times \{b\}$, $B = [0, 1] \times \{a\} \cup]1, 2] \times \{b\}$ are both compact, but $A \cap B$ is non-compact.
7. Let $(C_i)_{i \in I}$ be a family of closed, compact subsets of X. Prove that $\cap_{i \in I} C_i$ is compact.
8. Let A, B be non-empty subsets of the metric space X. Let $\operatorname{dist}(A, B) = \inf\{(d(a, b) : a \in A, b \in B\}$. Prove that if A, B are disjoint and closed, and A is compact, then there is an element a of A such that

$$\operatorname{dist}(A, B) = \operatorname{dist}(a, B) > 0.$$

9. Prove that $\mathbf{I}^2$ with the television topology is compact, connected and Hausdorff.
10. Prove that the following conditions on a topological space X are equivalent. (i) X is compact. (ii) If $(C_i)_{i \in I}$ is a family of closed subsets of X such that $\cap_{i \in I} C_i = \varnothing$, then $\cap_{a \in A} C_a = \varnothing$ for some finite subset A of I. (iii) If $(C_i)_{i \in I}$ is a family of closed subsets of X such that $\cap_{a \in A} C_a \neq \varnothing$ for all finite subsets A of I, then $\cap_{i \in I} C_i \neq \varnothing$.
11. Let $f : \mathbf{R}^n \to V$ be a linear isomorphism, where $\mathbf{R}^n$ is Euclidean space and V is a normed vector space over **R**. Prove that the function $h : \mathbf{S}^{n-1} \to \mathbf{R}$ which sends $x \rightsquigarrow \|f(x)\|$ is continuous, and that there is a positive real number δ such that $h(x) \geqslant \delta$ for all x in $\mathbf{S}^{n-1}$. Show that for all y in V, $|f^{-1}(y)| \leqslant \delta^{-1} \|y\|$. Finally, prove that f is a homeomorphism.

12. Prove that n-dimensional complex space $\mathbf{C}^n$ is linearly homeomorphic to $\mathbf{R}^{2n}$. Prove that, if V is a finite dimensional normed vector space over $\mathbf{R}$ or $\mathbf{C}$, then all norms on V are equivalent. [This theorem is proved in Dieudonné [1] by completeness methods. An advantage of such methods is that they also prove that any finite dimensional subspace of a normed vector space V is closed in V.]
13. Let X_n be the subset $\{n^{-1}\} \times [-n, n]$ of $\mathbf{R}^2$ and let $Y = \mathbf{R}^2 \setminus \bigcup_{n \geq 1} X_n$. Prove that Y is connected but not path-connected.
*14. Let $\varphi : X \times C \to \mathbf{R}$ be a map where C is compact. Prove that the function $x \rightsquigarrow \sup_{c \in C} \varphi(x, c)$ is a continuous function $X \to \mathbf{R}$.

3.6 Further properties of compactness

The basic results on compactness are given in the last section. The more technical results of this section will be used in later chapters, but the study of these results can be omitted till they are needed.

Locally compact spaces, normal spaces

Definition A topological space X is *locally compact* if each x in X has a base of compact neighbourhoods.

EXAMPLES 1. Euclidean n-space $\mathbf{R}^n$ is locally compact since, if $x \in \mathbf{R}^n$, then the closed balls $E(x, r)$ for $r > 0$ are compact and form a base for the neighbourhoods of x.
2. The space $\mathbf{Q}$ of rational numbers is not locally compact since a neighbourhood of 0 in $\mathbf{Q}$ cannot be closed in $\mathbf{R}$ and so cannot be compact.

In the literature, it is common to define a space X to be locally compact if each point of X has a compact neighbourhood. We have not adopted this definition for two reasons:
(i) It would be contrary to the general spirit of local properties. If P is a property of topological spaces, it is usual to say X is locally P if each point of X has a base of neighbourhoods with property P.
(ii) The property of locally compact spaces needed later is exactly the one we have taken for a definition.

For Hausdorff spaces the two definitions are equivalent. This is an easy consequence of the following result.

3.6.1 *A compact Hausdorff space is locally compact.*

Proof Let X be compact and Hausdorff, let $x \in X$ and let W be an open neighbourhood of x. We must find a compact neighbourhood of x contained in W.

Let $C = X \setminus W$. Then C is closed in X and so is compact. By 3.5.6 (Corollary 3), x and C have disjoint open neighbourhoods M, N say. The closure $\overline{M}$ of M is contained in $X \setminus N$ which is itself contained in W. Also $\overline{M}$ is compact (since it is closed in X) and is a neighbourhood of x. □

Definition A topological space X is *normal* if disjoint closed sets of X have disjoint neighbourhoods.

It is immediate from 3.5.5 and 3.5.6 (Corollary 3) that any compact, Hausdorff space is normal. We showed in 2.9 that any metric space is normal.

Normal spaces have another property important in many parts of topology, for example in metrization theorems and in the theory of ANRs. For a proof of the following theorem see, for example, Kelley [1] or Simmons [1].

3.6.2 (*Tietze extension theorem*) *A space X is normal if and only if for any closed subspace C of X any map $f : C \to \mathbf{I}$ extends over X (i.e., is the restriction of a map $X \to \mathbf{I}$).*

The proof that the extension condition implies normality is easy since let $C = C_1 \cup C_2$ where C_1, C_2 are disjoint, non-empty closed subsets of X and let $f : C \to \mathbf{I}$ be 0 on C_1 and 1 on C_2. Let $g : X \to \mathbf{I}$ be an extension of f over X. Then, for each r in $]0, 1[$, the sets

$$g^{-1}[0, r[, \qquad g^{-1}]r, 1]$$

are disjoint open sets containing C_1, C_2 respectively; in fact for various r these sets form a kind of 'continuous family' of open sets between C_1 and C_2. One word of warning—it is not always possible to find g such that $C_1 = g^{-1}[0]$, $C_2 = g^{-1}[1]$; conditions for this will be mentioned in the Exercises.

Proper maps

Definition Let $f : X \to Y$ be a map of topological spaces. Then f is *proper* if, for all spaces Z,

$$f \times 1 : X \times Z \to Y \times Z$$

is a closed map.

By taking Z to consist of a single point, we see that a proper map is always closed. Similarly, to say that a constant map $X \to \{y\}$ is proper is equivalent to saying that for all Z the projection $X \times Z \to Z$ is closed. The result on proper maps that we shall need (in 5.8) is the following and its corollary.

3.6.3 *If $f : X \to Y$ is a closed map such that $f^{-1}[y]$ is compact for each y in Y, then f is proper.*

Proof Let $h = f \times 1 : X \times Z \to Y \times Z$, let C be a closed subset of $X \times Z$ and let $D = h[C]$—we must prove that D is closed, i.e., that the set $D' = (Y \times Z) \setminus D$ is open.

Let $(y, z) \in D'$ and let $C' = (X \times Z) \setminus C$, so that C' is open. It is easily verified that

$$f^{-1}[y] \times \{z\} \subset C'$$

and so, by our assumptions and 3.5.6 (Corollary 2), there are open sets U, V such that

$$f^{-1}[y] \times \{z\} \subset U \times V \subset C'.$$

Let $U' = X \setminus U$, $V' = Y \setminus V$. Then $C \subset (U' \times Z) \cup (X \times V')$ and so

$$D = h[C] \subset (f[U'] \times Z) \cup (f[X] \times V') = Q \text{ say.}$$

Since f is a closed map, Q is a closed set. Therefore, $Q' = (Y \times Z) \setminus Q$ is an open set contained in D'. But $y \notin f[U']$, nor does $z \in V'$, so it follows that $(y, z) \in Q'$. Hence D' is open. □

The converse of 3.6.3 is true [see the Exercises], but will not be needed here.

3.6.3 (*Corollary 1*) *Any map from a compact space to a Hausdorff space is proper.*

Proof Let $f : X \to Y$ be a map where X is compact and Y is Hausdorff. Then f is closed by 3.5.8. If $y \in Y$, then $\{y\}$ is closed in Y; hence $f^{-1}[y]$ is closed in X and so $f^{-1}[y]$ is compact. □

Lebesgue covering lemma

Let X be a metric space and $\mathscr{A}$ an open cover of X. We consider the following question: is there a real number $r > 0$ such that the open cover $\mathscr{B}_r = \{B(x, r) : x \in X\}$ refines $\mathscr{A}$? Clearly the set of all r for which this is so is an interval L of $\mathbf{R}$, and L may be empty. If L is non-empty, then the real number $l = \sup L$ is called the *Lebesgue number* of the cover, $\mathscr{A}$ (we allow the rather boring case $l = \infty$, which, intuitively, means $\mathscr{A}$ has lots of large sets).

EXAMPLES 3. If $\mathscr{A} = \mathscr{B}_r$ then the Lebesgue number of $\mathscr{A}$ is r.
4. Let $X = \mathbf{R}$, and let $\mathscr{A}$ consist of the open intervals $]n, n + 2[$ for each $n \in \mathbf{Z}$. Then the Lebesgue number of $\mathscr{A}$ is $\frac{1}{2}$.
5. Let $X =]0, 1[$ and let $\mathscr{A}$ consist of the open intervals $]n^{-1}, 1[$ for all positive integral n. Then $\mathscr{A}$ has no Lebesgue number.
6. Let $X = [0, 2] \setminus \{1\}$, and let $\mathscr{A}$ consist of the intervals $[0, 1[$ and $]1, 2]$. Then $\mathscr{A}$ has no Lebesgue number.

3.6.4 (*Lebesgue covering lemma*) *If X is a compact metric space, then any open cover of X has a Lebesgue number.*

Proof Let $\mathscr{A}$ be an open cover of X. Since X is compact, $\mathscr{A}$ has a finite subcover and any refinement of this is a refinement of $\mathscr{A}$. So we may assume $\mathscr{A}$ finite.

For each A in $\mathscr{A}$ and x in X let

$$f_A(x) = \text{dist}\,(x, X \setminus A)$$
$$f(x) = \max\,\{f_A(x) : A \in \mathscr{A}\}.$$

Each f_A is continuous and hence f is continuous.

If $x \in X$ then $x \in A$ for some A in $\mathscr{A}$ and for that A $f_A(x) > 0$ (since $X \setminus A$ is closed). Hence $f(x) > 0$ for all x in X.

The set $f[X]$ is a compact and hence closed subset of **R**. Therefore, $r = \inf f[X]$ belongs to $f[X]$, and so $r > 0$.

If $x \in X$ then $f(x) \geqslant r$ whence $f_A(x) \geqslant r$ for some A in $\mathscr{A}$ and so, by definition of f_A, $B(x, r) \subset A$. This proves that the cover $\mathscr{B}_r$ refines $\mathscr{A}$. □

We now illustrate the main type of application of 3.6.4. Let $a : [0, r] \to Z$ be a path in a topological space Z. A *subdivision* of a is a sequence $a_1, \ldots, a_n$ (for some n) of paths in Z such that

$$a = a_n + \cdots + a_1. \qquad (*)$$

Such a subdivision is usually denoted by the expression (*).

3.6.4 (*Corollary 1*) *Let $\mathscr{U}$ be an open cover of Z and $a : [0, r] \to Z$ a path in Z. Then there is a subdivision $a = a_n + \cdots + a_1$ such that for each $i = 1, \ldots, n$, $\operatorname{Im} a_i$ is contained in some set of $\mathscr{U}$.*

Proof Let δ be the Lebesgue number of the covering $\{a^{-1}[U] : U \in \mathscr{U}\}$ of the compact metric space $[0, r]$. Let n be an integer such that $0 < r/n < \frac{1}{2}\delta$, and let a_{i+1} be the path $t \rightsquigarrow a(t + ir/n)$, $i = 0, \ldots, n - 1$ of length r/n. Then clearly $a = a_n + \cdots + a_1$. Further, for each $i = 0, \ldots, n - 1$, the interval $[ir/n, (i + 1)r/n]$ is contained in some set $a^{-1}[U]$, $U \in \mathscr{U}$; hence

$$\operatorname{Im} a_i = a[ir/n, \quad (i + 1)r/n] \subset U. \quad \square$$

EXERCISES

1. Prove that the product of two locally compact spaces is locally compact.
2. Prove that a closed subspace of a locally compact space is locally compact, but that an arbitrary subspace of a locally compact space need not be locally compact.
3. Let A be a compact subspace of a locally compact space, and let W be a neighbourhood of A. Prove that there is a compact neighbourhood M of A such that $M \subset W$.

4. A space X is a *k-space* if a subset A of X is closed in X if and only if $A \cap C$ is closed in C for every compact subset C of X. Prove that a locally compact space is a k-space, as is any sequential space [Exercise 5 of 2.10]. Prove that the space X of Exercise 4 of 2.9 is not a k-space.

5. Show that the following statement is true if f is open but not in general: if X is locally compact and $f : X \to Y$ is a continuous surjection, then Y is locally compact.

*6. *The Alexandroff 1-point compactification.* Let X be a topological space, let ω be a point not in X and let $X^* = X \cup \{\omega\}$. Define a neighbourhood topology on X^* by (i) if $x \in X$ and M is a neighbourhood in X of x, then M and $M \cup \{\omega\}$ are neighbourhoods in X^* of x; (ii) if A is a closed compact subset of X, then $X^* \setminus A$ is a neighbourhood of ω. Prove (*a*) X^* is compact and X is a subspace of X^*, (*b*) X is locally compact and Hausdorff $\Leftrightarrow$ X^* is Hausdorff, (*c*) if X^* is Hausdorff and $i_1 : X \to X_1^*$ is any homeomorphism into a compact Hausdorff space X_1^* such that the image of i_1 is the complement of a single point of X_1^*, then there is a unique homeomorphism $g : X^* \to X_1^*$ such that $gi = i_1$, where $i : X \to X^*$ is the inclusion. Prove also that, if X is compact, then $X^* = X \sqcup \{\omega\}$. [The point ω is called the *point at infinity* of X^*.]

7. Given an open cover $\mathscr{U}$ of X, prove that the following prescription defines a topological space X'. (i) $X' = X \cup \{\omega\}$ where ω is a point not belonging to X, (ii) if $x \in X$ then any subset N of X' such that $x \in N$ is a neighbourhood of x, (iii) a subset M of X' is a neighbourhood of ω if and only if $\omega \in M$ and $X' \setminus M$ is contained in a finite union of sets of $\mathscr{U}$. Prove that with this topology on X', the set $\{\omega\}$ is open in X' if and only if $\mathscr{U}$ has a finite subcover.

8. Let X' be the space defined by an open cover of X as in the previous exercise. Prove that the projection $X \times X' \to X'$ is closed if and only if $\{\omega\}$ is open in X'. Deduce that if a constant map $X \to \{y\}$ is proper, then X is compact.

9. Let $f : X \to Y$ be continuous and injective. Then the following are equivalent: (*a*) f is proper; (*b*) f is closed; (*c*) f is a homeomorphism onto a closed subspace of Y.

10. Let $f : X \to Y$ be continuous, let $B \subset Y$ and let $A = f^{-1}[B]$. Prove that, if f is proper, then so also is $f \mid A, B$.

11. Prove that, if $f : X \to Y$ is proper, then f is closed and $f^{-1}[y]$ is compact for each y in Y.

12. Prove that if $f : X \to X', g : Y \to Y'$ are proper then so also is $f \times g$. Deduce that if $f : X \to X'$ is proper and X is Hausdorff, then $\operatorname{Im} f$ is Hausdorff.

13. Let $f : X \to Y, g : Y \to Z$ be continuous. Prove that (*a*) if f, g are proper and $\operatorname{Im} f$ is closed, then gf is proper; (*b*) if gf is proper and f is surjective, then g is proper; (*c*) if gf is proper and g is injective then f is proper; (*d*) if gf is proper and Y is Hausdorff then f is proper. Deduce that (*e*) if f is proper, so also is $f \mid A$ for any closed subset A of X, (*f*) if X is Hausdorff and $f : X \to Y$, $g : X \to Z$ are proper, then $(f, g) : X \to Y \times Z$ is proper.

14. Prove that if $f : X \to Y$ is continuous, $\operatorname{Im} f$ is a Hausdorff k-space and $f^{-1}[K]$ is compact for every compact subset K of $\operatorname{Im} f$, then f is proper.

15. Let X, Y be locally compact, Hausdorff spaces and X^*, Y^* their Alexandroff compactifications. Let $f : X \to Y$ be continuous and let $f^* : X^* \to Y^*$ be the

extension of f which sends the point at infinity to the point at infinity. Prove that f is proper and has closed image if and only if f^* is continuous.

16. Prove the following (i) if C is closed in the normal space X then there is a map $f : X \to \mathbf{I}$ such that $C = f^{-1}[0]$ if and only if C is a G_δ-set; (ii) if C_1, C_2 are disjoint, closed G_δ-sets in the normal space X, then there is a map $f : X \to \mathbf{I}$ such that $C_1 = f^{-1}[0]$, $C_2 = f^{-1}[1]$. [You should assume the Tietze extension theorem.]

17. A *continuum* is a compact, connected space. Read the proof of the following theorem in Hocking–Young [1] (Theorem 2.9): If a, b are two points of a compact Hausdorff space X, and if X is not the union of two disjoint open sets one containing a and the other containing b, then X contains a continuum containing a and b.

18. A subset Q of X is a *quasicomponent* of X if for any partition $\{X_1, X_2\}$ of X, Q is contained in X_1 or in X_2, and Q is maximal with respect to this property. Prove that (i) the quasicomponents of X cover X, (ii) each quasicomponent of X is closed, (iii) every component of X is contained in a quasicomponent, (iv) in a compact, Hausdorff space the components and quasicomponents coincide.

NOTES

The first three chapters of Hocking–Young [1] form excellent supplementary reading to the topics we have discussed, particularly for the results on continua (i.e., compact, connected spaces). Most other books on general topology are biased more towards analysis. For more results on proper maps and, in particular, for solutions of some of the exercises, see Bourbaki [1]; however, the proofs there use filters in an essential way. The Tietze extension theorem has led to a large theory of retracts which is surveyed in Hu [3]. This theorem is also used in the metrization problem—the problem of finding necessary and sufficient conditions of a topological character for a space to be metrizable. An account of the solution of this is given in Kelley [1]. In this context, an important role is played by the paracompact spaces, although the use of these spaces in topology is beginning to be taken over by the use of partitions of unity, cf. Mokobodzki [1], Dold [1].

4. Identification spaces and cell complexes

4.1 Introduction

In chapter 1, we considered briefly some examples of topological spaces obtained by identifications. In this chapter, we shall discuss this process in full generality. But first we shall consider some examples in order to clarify the set-theoretic processes involved.

EXAMPLES 1. The interval [0, 2] can be thought of as obtained by joining two intervals of length 1. The circle $\mathbf{S}^1$ is obtained from [0, 1] by identifying 0 and 1 [Fig. 4.1].

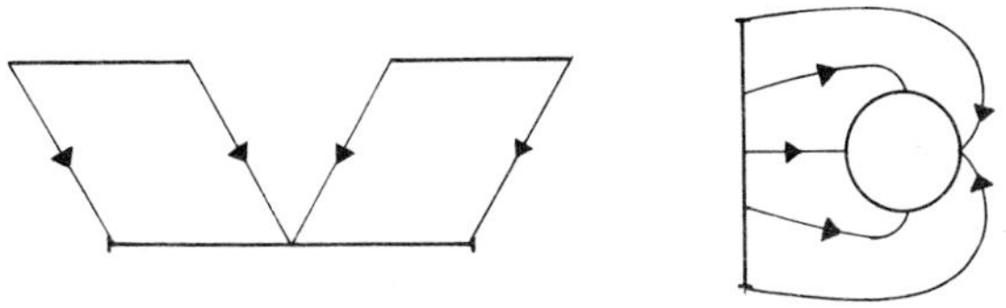

Fig. 4.1

2. From the square $\mathbf{I}^2 = \mathbf{I} \times \mathbf{I}$ we can obtain two spaces by identifying two opposite sides according to the schemes shown in Fig. 4.2 (the

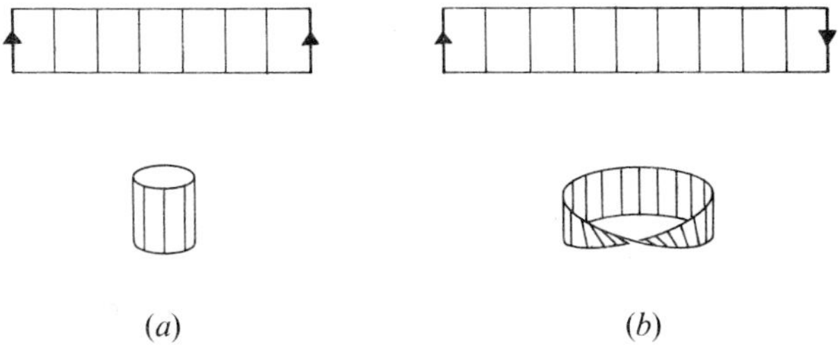

Fig. 4.2

figures have used topological licence in stretching and bending). The arrows on the sides indicate whether the sides should be stuck together in the same or in opposite senses. In (*a*) the sides are stuck together in the same sense and the result is the cylinder $\mathbf{S}^1 \times \mathbf{I}$. In (*b*), the sides are stuck together in the opposite sense, and the result is the Möbius band.

3. By identifying two pairs of sides of $\mathbf{I}^2$ we can obtain the *torus* (or *anchor ring*), which is simply $\mathbf{S}^1 \times \mathbf{S}^1$ [cf. Exercise 9 of 2.7].

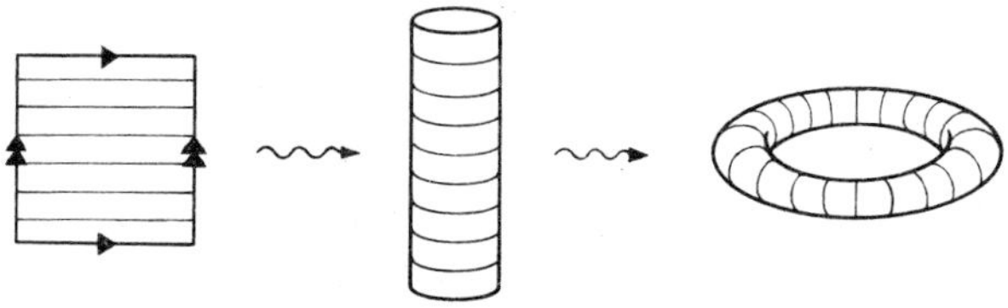

Fig. 4.3

4. If we try and identify the sides of $\mathbf{I}^2$ according to the scheme of Fig. 4.4 then, by making one identification, we obtain a cylinder but the final identification cannot be represented properly in three dimensions. We do in fact obtain a space as we shall show in detail later; this space is called the *Klein bottle* and is homeomorphic to the space constructed at

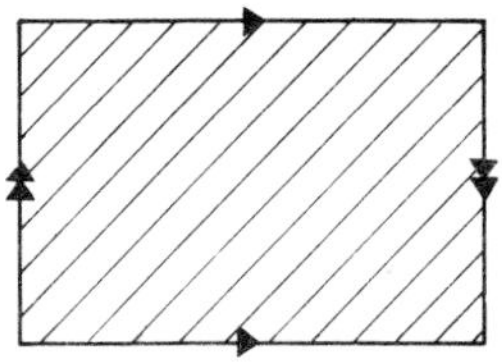

Fig. 4.4

the end of chapter 1. There is a 3-dimensional model of the Klein bottle [Fig. 4.5], but this model has 'self-intersections' which the 'real' Klein bottle does not have.

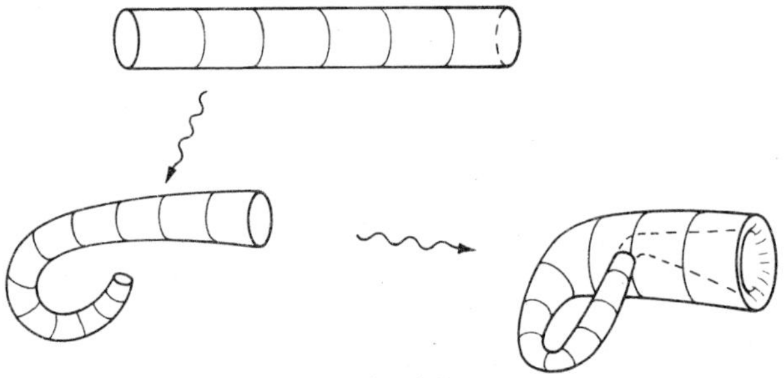

Fig. 4.5

This last example illustrates a general principle, namely, that it may be possible to describe a space Y as obtained by identifications from a simple space X, and this description may be adequate to show us the properties of Y even though Y cannot be visualized.

We have to formulate in general the notion of constructing a space Y by identifications from a given space X. For the moment, let us forget about the topologies on X and Y.

In X, we suppose, given a relation R, where the interpretation of $a\ R\ b$ is that a shall be identified with b. But if a is to be identified with b then b must be identified with a; and if, further, b is to be identified with c then a must also be identified with c. Also a is 'identified' with itself. This shows that we must consider not only R but also the equivalence relation E generated by R [Appendix A.4], and that the totality of points of X identified with a given point a of X are the points b such that $b\ E\ a$; the set of such b is, of course, the equivalence class cls a.

The equivalence classes of E form a set of disjoint, non-empty, subsets of X whose union is X. This set X / E of equivalence classes shall be the set Y. There is a function $f : X \to Y$ which sends $a \rightsquigarrow$ cls a. Clearly, f is a surjection. So we have passed from the original relation R to a surjection $f : X \to Y$, characterized by the universal property of 4.6 of the Appendix.

Conversely, suppose given a surjection $f : X \to Y$, where Y is now any set. Then the relation $a\ E_f\ b \Leftrightarrow f(a) = f(b)$ is an equivalence relation in X whose equivalence classes are the sets $f^{-1}[y]$ for each y in Y. The function f identifies all the elements of $f^{-1}[y]$ to the point y. This shows that the notions of *set with identifications* and *surjective function* are closely related.

We shall need a generalization of the above identifications. Suppose there is given a family $(X_\alpha)_{\alpha \in A}$ of sets and for each α, β in A a relation $R_{\alpha\beta}$ from X_α to X_β; that is $R_{\alpha\beta}$ is a subset of $X_\alpha \times X_\beta$, and $a\ R_{\alpha\beta}\ b$ means $a \in X_\alpha$, and $(a, b) \in R_{\alpha\beta}$. The following result shows that there is a set Y obtained from the family (X_α) by identifying a with b whenever $a \in X_\alpha$, $b \in X_\beta$, and $a\ R_{\alpha\beta}\ b$. The set Y is characterized, as in the case of identifications in a single set, by a φ-universal property.

4.1.1 *There is a set Y and a family $(f_\alpha : X_\alpha \to Y)_{\alpha \in A}$ of functions such that*

(a) for all a, b, $a\ R_{\alpha\beta}\ b \Rightarrow f_\alpha a = f_\beta b$,
(b) if $(g_\alpha : X_\alpha \to Z)_{\alpha \in A}$ *is a family such that for all a, b*

$$a\ R_{\alpha\beta}\ b \Rightarrow g_\alpha\, a = g_\beta b,$$

then there is a unique function $g^* : Y \to Z$ *such that*

$$g^* f_\alpha = g_\alpha \quad \text{all } \alpha \text{ in } A.$$

Proof Let $X = \bigsqcup_{\alpha \in A} X_\alpha$ be the sum of the family, and let $i_\alpha : X_\alpha \to X$ be the injections. In X we define a relation R by

$$i_\alpha(a) \; R \; i_\beta(b) \Leftrightarrow a \; R_{\alpha\beta} \; b.$$

Let E be the equivalence relation generated by R, let $Y = X / E$ and let $f : X \to Y$ be the projection. Let $f_\alpha = f i_\alpha : X_\alpha \to Y$.

The functions g_α define a function $g : X \to Z$ such that $g \, i_\alpha = g_\alpha$, α in A. By (b), for any a, b in X, $a \; R \; b \Rightarrow ga = gb$. So g defines $g^* : Y \to Z$ such that $g^*f = g$, whence $g^*f_\alpha = g^*f \, i_\alpha = g \, i_\alpha = g_\alpha$, α in A.

If $g' : Y \to Z$ satisfies $g'f_\alpha = g_\alpha$, then $g'f \, i_\alpha = g \, i_\alpha$. Hence $g'f = g$ and so $g' = g^*$. □

These considerations may, initially, seem abstract. However we shall see that the main point of spaces with identifications is that they give a means of constructing functions, and this is both a necessary and a sufficient condition for their utility. Put in another way, we have moved from a local consideration—what happens in a given space—to a global consideration—the relation of this space to other space. This widening of the point of view has proved very fruitful in mathematics.

4.2 Final topologies, identification topologies

We have now treated the set theoretic part of the notion of identification space, and shown that it can be subsumed under the general notion of a family $(X_\alpha)_{\alpha \in A}$ of sets and a family

$$(f_\alpha : X_\alpha \to Y)_{\alpha \in A}$$

of functions. In specific situations, (f_α) is often the universal family constructed in 4.1.1, but for the moment this fact is irrelevant.

We now suppose that each X_α is a topological space and we construct from (f_α) a reasonable topology on Y. The problem centres of course on the word reasonable. In the case considered in 4.1.1, the specific virtue of Y was that we could construct functions from Y. So the topology on Y which we choose is that which enables us to decide whether or not functions from Y are continuous.

Definition A topology $\mathscr{F}$ on Y is said to be *final* with respect to the functions (f_α) if, for any topological space Z and function $g : Y_{\mathscr{F}} \to Z$, we have g is continuous if and only if $g f_\alpha : X_\alpha \to Z$ is continuous for each α in A.

EXAMPLE 1 Suppose, for example, that we are in the situation of 4.1.1 and $(f_\alpha : X_\alpha \to Y)$ is a family satisfying the φ-universal property considered there. Let each X_α be a topological space and let Y have the final topology. Then we have: *if* $(g_\alpha : X_\alpha \to Z)_{\alpha \in A}$ *is any family of continuous functions such that*

$$a \, R_{\alpha\beta} \, b \Rightarrow g_\alpha a = g_\beta b$$

then there is a unique continuous function $g^* : Y \to Z$ *such that* $g^* f_\alpha = g_\alpha$, $\alpha \in A$. In fact, the universal property ensures that there exists $g^* : Y \to Z$, such that $g^* f_\alpha = g_\alpha$, $\alpha \in A$. The continuity of g^* follows from the fact that Y has the final topology.

We shall show that a final topology always exists, but we first point out some simple consequences.

4.2.1 *If* $\mathscr{F}$ *is the final topology on* Y *with respect to* (f_α) *then*

(*a*) *each* $f_\alpha : X_\alpha \to Y_{\mathscr{F}}$ *is continuous,*
(*b*) *if* $\mathscr{T}$ *is any topology on* Y *such that each* $f_\alpha : X_\alpha \to Y_{\mathscr{T}}$ *is continuous, then* $\mathscr{F}$ *is finer than* $\mathscr{T}$.

Proof (*a*) The identity $1 : Y_{\mathscr{F}} \to Y_{\mathscr{F}}$ is continuous and hence $1 f_\alpha : X_x \to Y$ is continuous for each α in A. Since $1 f_\alpha = f_\alpha$, it follows that f_α is continuous.
(*b*) Let $g : Y_{\mathscr{F}} \to Y_{\mathscr{T}}$ be the identity function. Then $g f_\alpha = f_\alpha : X_\alpha \to Y_{\mathscr{T}}$ is continuous for each α in A. Hence g is continuous, and so $\mathscr{F}$ is finer than $\mathscr{T}$. □

We now show that the final topology exists. We suppose given (X_α) and (f_α) as before.

4.2.2 *The final topology* $\mathscr{F}$ *on* Y *with respect to* (f_α) *exists and is characterized by either of the following conditions:*

(*a*) *If* $U \subset Y$, *then* U *is open in* $\mathscr{F}$ *if and only if* $f_\alpha^{-1}[U]$ *is open in* X_α *for each* α *in* A.
(*b*) *The same as* (*a*), *but with 'open' replaced by 'closed'.*

Proof We first show that (*a*) does define a topology. Clearly $\emptyset$, Y are open in $\mathscr{F}$. If U, V are open in $\mathscr{F}$, then

$$f_\alpha^{-1}[U \cap V] = f_\alpha^{-1}[U] \cap f_\alpha^{-1}[V]$$

which is open in X_α, and so $U \cap V$ belongs to $\mathscr{F}$. Similarly, the formula $f_\alpha^{-1}[\bigcup U_i] = \bigcup f_\alpha^{-1}[U_i]$, shows that the union of any family of sets open in $\mathscr{F}$ is again open in $\mathscr{F}$. Therefore (*a*) does define a topology. The proof that (*b*) also defines a topology in terms of closed sets is similar, and the relation $f_\alpha^{-1}[Y \setminus U] = X \setminus f_\alpha^{-1}[U]$ shows that these topologies are the same.

We now prove that this topology is the final topology. Clearly, each

$f_\alpha : X_\alpha \to Y_{\mathscr{F}}$ is continuous. Suppose $g : Y_{\mathscr{F}} \to Z$ is a function where Z is a topological space.

If g is continuous, then so also is each composite gf_α. Suppose, conversely, that each gf_α is continuous. Let U be open in Z. Then

$$f_\alpha^{-1}g^{-1}[U] = (gf_\alpha)^{-1}[U]$$

and, therefore, $f_\alpha^{-1}g^{-1}[U]$ is open in X_α. It follows that $g^{-1}[U]$ is open in $Y_{\mathscr{F}}$. Hence g is continuous, as we were required to prove. □

EXAMPLE 2. Let $X = \bigsqcup_{\alpha \in A} X_\alpha$ be the sum of the underlying sets of the family $(X_\alpha)_{\alpha \in A}$ of topological space, and let $i_\alpha : X_\alpha \to X$ be the injections. The final topology on X with respect to (i_α) is called the *sum topology*. Such a topology was defined in chapter 3 when the indexing set was finite—clearly the definitions coincide in this case.

By means of the topological sum we can reduce final topologies with respect to a family (f_α) to final topologies with respect to a single function f. In fact, with the assumptions as for 4.2.1, 4.2.2 we have:

4.2.3 *Let X be the sum of spaces (X_α) and let $f : X \to Y$ be the function determined by (f_α). Then the final topologies on Y with respect to f and with respect to (f_α) coincide.*

Proof Let $i_\alpha : X_\alpha \to X$ be the injection. Let $g : Y \to Z$ be a function where Z is a topological space. Then

$$gf_\alpha = gfi_\alpha$$

since $fi_\alpha = f_\alpha$. Also gf is continuous if and only if each gfi_α is continuous. Thus the conditions (a) g is continuous if and only if gf is continuous, and (b) g is continuous if and only if each gf_α is continuous, are equivalent. □

We now concentrate attention on the case of a single function $f : X \to Y$. Let Y have the final topology with respect to f. Let $Y_1 = Y \setminus f[X]$. If $y \in Y_1$ then $f^{-1}[y]$ is empty, and so $\{y\}$ is both open and closed in Y. Also, $f[X]$ is open and closed in Y. Therefore, Y is the topological sum of $f[X]$ and the discrete space Y_1. This shows that the case of major interest is when f is a surjection.

Let X, Y be topological spaces and $f : X \to Y$ a function. We say f is an *identification map* if f is a surjection and Y has the final topology with respect to f. This topology on Y is also called the *identification topology* with respect to f, and we say Y is an *identification space* of f.

There is a useful characterization of identification topologies in addition to those given by the definition and by 4.2.2. A subset A of X is *saturated* (more precisely, *saturated with respect to f*, or *f-saturated*) if

$f^{-1}f[A] = A$. For example, any set $f^{-1}[B]$ is saturated.

Let $f : X \to Y$ be an identification map. If B is a subset of Y then $ff^{-1}[B] = B$ (since f is surjective). Also, if A is saturated, then $f[A] = ff^{-1}f[A]$. Hence, the *open sets of Y are the sets $f[V]$ for all saturated open sets V of X*; and the same statement holds with open replaced by closed.

There is a difficulty in the description of neighbourhoods in Y. Let $A \subset Y$ and suppose N is a neighbourhood A. Then there is an open set U such that $A \subset U \subset N$, whence

$$f^{-1}[A] \subset f^{-1}[U] \subset f^{-1}[N].$$

So N is a neighbourhood of A implies $f^{-1}[N]$ is a neighbourhood of $f^{-1}[A]$. The converse of this last implication is false: for suppose $f^{-1}[N]$ is a neighbourhood of $f^{-1}[A]$. Then we can find an open set V such that $f^{-}[A] \subset V \subset f^{-1}[N]$, but it may be impossible to find such a V which is saturated with respect to f [Exercise 1 of 4.3].

The following result gives a useful class of identification maps.

4.2.4 *Let $f : X \to Y$ be a continuous surjection. If f is an open map, or a closed map, then f is an identification map.*

Proof Suppose that f is an open map. Let U be a subset of Y. By continuity, if U is open in Y then $f^{-1}[U]$ is open in X. On the other hand f is a surjection, so $ff^{-1}[U] = U$. Hence $f^{-1}[U]$ is open if and only if U is open.

A similar proof applies with open replaced by closed. □

4.2.4 (*Corollary 1*) *A continuous surjection from a compact space to a Hausdorff space is an identification map.*

Proof Such a function is a closed map, by 3.5.8 (Corollary 1). □

A consequence of 4.2.4 (Corollary 1) is that if X is compact and $f : X \to Y$ is a surjection, then there is at most one Hausdorff topology on Y such that f is continuous. If such a Hausdorff topology exists, it is clearly the most 'reasonable' topology on Y. Unfortunately, the identification topology need not be Hausdorff, nor need it be, when X is non-compact, the most 'reasonable' Hausdorff topology.

Since identification topologies are special cases of final topologies, we can apply to identification topologies the results of 4.2.2. We illustrate this in the proof of our next result, which can also be proved directly from the definition.

4.2.5 *The composite of identification maps is an identification map.*

Proof Let $f : X \to Y$, $g : Y \to Z$ be identification maps. Then $gf : X \to Z$ is certainly continuous and surjective. We show that if $h : Z \to W$ and

hgf is continuous, then h is continuous—the implication $(c) \Rightarrow (a)$ of 4.2.2 then shows that gf is an identification map.

Suppose then $hgf : X \to W$ is continuous. The implication $(a) \Rightarrow (c)$ of 4.2.2 shows first that hg is continuous (since f is an identification map) and second that h is continuous (since g is an identification map). □

EXERCISES

1. Let X, Y be topological spaces and $f : X \to Y$ an injection. Prove that f is an identification map if and only if f is a homeomorphism.
2. Let $f : X \to Y$ be an identification map. What is the topology of Y if X is discrete? indiscrete?
3. Prove that the following maps are identification maps
 (i) The projections $X \times Y \to X$, $X \times Y \to Y$.
 (ii) $\{(x, y) \in \mathbf{R}^2 : xy = 0\} \to \mathbf{R}, \quad (x, y) \rightsquigarrow x$.
 (iii) $\{(x, y) \in \mathbf{R}^2 : x^2y^2 = 1\} \to \mathbf{R}^{\neq 0}, \quad (x, y) \rightsquigarrow x$.
 (iv) $\mathbf{I} \to \mathbf{S}^1, \quad t \rightsquigarrow e^{2\pi i t}$.
4. Let Y have the final topology with respect to $f : X \to Y$. Prove that $f \mid X, f[X]$ is an identification map.
5. Let X, Y be topological spaces and $f : X \to Y$ a continuous surjection. Suppose that each point y in Y has a neighbourhood N such that $f \mid f^{-1}[N], N$ is an identification map. Prove that f is an identification map. Deduce that the *covering map*

$$p : \mathbf{R} \to \mathbf{S}^1, t \rightsquigarrow e^{2\pi i t}$$

is an identification map.
6. Let A be a subspace of X. A *retraction of X onto A* is a map $r: X \to A$ such that $r \mid A$ is the identity. Prove that a retraction of X onto A is an identification map. Deduce that the map

$$\mathbf{R}^{n+1} \setminus \{0\} \to \mathbf{S}^n, \quad x \rightsquigarrow x/|x|$$

ia an identification map.
7. Prove that if $f: X \to Y, f' : X' \to Y', g : Y \to Z$ are open surjections, then so also are $gf : X \to Z$ and $f \times f' : X \times X' \to Y \times Y'$.
8. Let $f : X \to Y$ be an identification map. For each $A \subset X$, let $f\dagger[A] = \{a \in A : f^{-1}f[a] \subset A\}$. Prove that the following conditions are equivalent. (i) f is a closed map. (ii) If A is closed in X, then also is $f^{-1}f[A]$. (iii) If A is open in X, then so also is $f\dagger[A]$. (iv) For each y in Y, every neighbourhood N of $f^{-1}[y]$ contains a saturated neighbourhood of $f^{-1}[y]$.
9. Let $f : X \to Y$ be an identification map, and let $R_f = \{(x, x') \in X \times X : fx = fx'\}$. Prove that, if Y is Hausdorff, then R_f is closed in $X \times X$. Prove that if f is an open map, and R_f is closed in $X \times X$, then Y is Hausdorff.
10. Prove the following 'transitive law' for final topologies. Suppose there are given functions $f_{\alpha\lambda} : X_{\alpha\lambda} \to Y_\lambda$ for each λ in Λ and α in A_λ, and functions $g_\lambda : Y_\lambda \to Z$ for each λ in Λ, and a topology for each $X_{\alpha\lambda}$. Prove that if Y_λ has the final topology with respect to $(f_{\alpha\lambda})_{\alpha \in A_\lambda}$, then the final topology on Z with respect to

$(g_\lambda)_{\lambda\in\Lambda}$ coincides with the final topology with respect to the family of composites $(g_\lambda f_{\alpha\lambda})_{\lambda\in\Lambda,\,\alpha\in A_\lambda}$. Show that 4.2.3 and 4.2.5 are corollaries of this transitive law.

11. Let Σ be a set of subspaces of X and for each $S \in \Sigma$ let $i_s : S \to X$ be the inclusion. The final topology on X with respect to $(i_s)_{s\in\Sigma}$ is called the *fine topology with respect to* Σ (or the *weak topology* with respect to Σ); and the set X with this topology is written X_Σ. Prove that: (i) Each inclusion $i_s : S \to X_\Sigma$ is a homeomorphism into. (ii) If $X \in \Sigma$ then $X = X_\Sigma$. (iii) Σ is a set of subspaces of X_Σ, and $(X_\Sigma)_\Sigma = X_\Sigma$. (iv) The identity of $X_\Sigma \to X$ is continuous.

12. Let Σ, Σ' be sets of subspaces of X, X' respectively. Let $f : X \to X'$ be a function such that for each S of Σ (i) there is an S' of Σ' such that $f[S] \subset S'$, (ii) $f \mid S$ is continuous. Prove that $f : X_\Sigma \to X'_{\Sigma'}$ is continuous. Suppose now that $X = X'$: prove that $X_\Sigma = X_{\Sigma'}$ if both Σ is a refinement of Σ' and Σ' is a refinement of Σ.

4.3 Subspaces, products, and identification maps

The following result is important in itself and will also help the discussion of examples of identification spaces.

Let $f : X \to Y$ be an identification map, and let $A \subset X$. Then $A, f[A]$ are subspaces of X, Y respectively, and we ask: is $g = f \mid A, f[A]$ an identification map?

4.3.1 *The following conditions are equivalent.*
(a) $g = f \mid A, f[A]$ is an identification map,
(b) each g-saturated set which is open in A is the intersection of A with an f-saturated set open in X,
(c) as for (b) but with 'open' replaced by 'closed'.

Proof We state an elementary exercise in set theory [Exercise 1.5 of the Appendix]: if $U \subset f[A]$ and $V \subset Y$, then

$$U = f[A] \cap V \Leftrightarrow g^{-1}[U] = A \cap f^{-1}[V].$$

$(a) \Rightarrow (b)$ Let U' be open in A and g-saturated. Then $U = g[U']$ is open in $f[A]$, whence $U = f[A] \cap V$ where V is open in Y. So

$$U' = g^{-1}g[U'] = g^{-1}[U] = A \cap f^{-1}[V].$$

Clearly, $f^{-1}[V]$ is open in X and f-saturated.

$(b) \Rightarrow (a)$ Let U be a subset of $f[A]$ such that $g^{-1}[U]$ is open in A. By condition (b), $g^{-1}[U] = A \cap f^{-1}[V]$ for some V open in Y. Hence $U = f[A] \cap V$, and so U is open in $f[A]$.

The proof of $(a) \Leftrightarrow (c)$ is the same as the above but with the word 'open' replaced by 'closed'. □

4.3.1 *(Corollary 1) Each of the following conditions implies that* $g = f \mid A, f[A]$ *is an identification map.*

(*a*) *For all* U, *if* U *is* g*-saturated and open in* A, *then* $f^{-1}f[U]$ *is open in* X.
(*b*) *As for* (*a*), *but with 'open' replaced by 'closed'.*
(*c*) *The set* A *is* f*-saturated and open in* X.
(*d*) *As for* (*c*), *but with 'open' replaced by 'closed'.*

Proof We give the proof only for cases (*a*) and (*c*).
(*a*) For any subset U of A

$$g^{-1}g[U] = A \cap f^{-1}f[U].$$

So condition (*b*) of 4.3.1 is satisfied.
(*c*) We reduce this to case (*a*). Let U be g-saturated and open in A. Then

$$f^{-1}f[U] \subset f^{-1}f[A] = A$$

whence

$$f^{-1}f[U] = g^{-1}g[U] = U.$$

Also U is open in X since A is open in X. □

EXAMPLES 1. Consider the subspace of $\mathbf{R}^2$

$$X = \mathbf{I} \times \{0\} \cup \mathbf{I} \times \{1\}.$$

Let $Y = [0, 1[\cup \{y_0, y_1\}$ where y_0, y_1 are distinct points not in $[0, 1[$.

Fig. 4.6

Let $f : X \to Y$ be the function

$$(t, \varepsilon) \rightsquigarrow \begin{cases} t, & t \neq 1 \\ y_0, & t = 1, \varepsilon = 0 \\ y_1, & t = 1, \varepsilon = 1. \end{cases}$$

Thus f identifies $(t, 0)$ with $(t, 1)$ for each t in $[0, 1[$. Let

$$I_0 = [0, 1[\cup \{y_0\}, \qquad I_1 = [0, 1[\cup \{y_1\}.$$

Let $f_0 : \mathbf{I} \times \{0\} \to I_0$ be the restriction of f. It is an easy deduction from 4.3.1 (Corollary 1*a*) that f_0 is an identification map. Since f_0 is a bijection, it is therefore a homeomorphism [Exercise 1 of 4.2]. Thus I_0 is (both as a set and topologically) essentially $\mathbf{I}$ with 1 renamed y_0. Similar remarks apply to I_1.

From this we have that the sets

$$]t, 1[\cup \{y_0\}, \quad t \in [0, 1[$$

form a base for the neighbourhood of y_0, and there is a similar base for the neighbourhood of y_1. Hence every neighbourhood of y_0 meets every neighbourhood of y_1; that is, Y is not Hausdorff.

2. Let A be a subset of the topological space X. Then *X with A shrunk to a point* is a topological space, written X/A, which is obtained from X by identifying all of A to a single point. More precisely, the elements of X/A are the equivalence classes in X under the equivalence relation generated by

$$x \sim y \Leftrightarrow x \in A \text{ and } y \in A.$$

These equivalence classes are therefore the sets $\{x\}$ for x in $X \setminus A$ and also, when A is non-empty, the set A.

It is convenient to identify the point x of $X \setminus A$ with $\{x\}$ of X/A, and so regard $X \setminus A$ as a subset of X/A. This causes confusion only when $A \in X \setminus A$, in which case we can always replace the equivalence class A by some point not in $X \setminus A$. Let $f : X \to X/A$ be the projection

$$x \rightsquigarrow \begin{cases} x, & x \in X \setminus A \\ A, & x \in A. \end{cases}$$

We give X/A the final topology with respect to f.

Now f is surjective and so an identification map. If A is empty, or consists of a single point then X/A can be identified with X.

The set $X \setminus A$ is f-saturated and f is the identity on $X \setminus A$. By 4.3.1 (Corollary 1), $X \setminus A$ is a subspace of X/A if A is open or is closed in X.

The application of Example 1 of 4.2 is important: *if $g : X \to Z$ is any map such that $g[A]$ consists of a single point of Z, then there is a unique map $g^* : X/A \to Z$ such that $g^*f = g$.*

The following terminology is convenient. A function $g : X \to Z$ is *constant on A* if $g \mid A$ is a constant map; also *g shrinks A to a point* if g is constant on A, is surjective, and $g \mid X \setminus A$ is injective. Thus, in the latter case, the function $g^* : X/A \to Z$ defined by g is a bijection. The universal property of 4.1.1 and Example 1 above shows that if $g : X \to Z$, $g' : X \to Z'$ are two identification maps which shrink A to a point, then there is a unique homeomorphism $h : Z \to Z'$ such that $hg = g'$.

3. We consider a special case of the last example. Let Y be any space. The *cone on Y* is

$$CY = Y \times \mathbf{I}/Y \times \{0\}.$$

The point v of CY which is the set $Y \times \{0\}$ is called the *vertex* of CY.

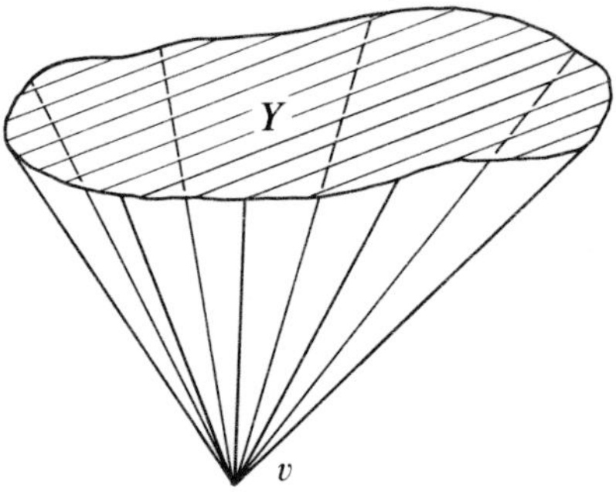

Fig. 4.7

Let B be a subspace of Y, let $i : B \times \mathbf{I} \to Y \times \mathbf{I}$ be the inclusion function, and let $p' : B \times \mathbf{I} \to CB$, $p : Y \times \mathbf{I} \to CY$ be the identification maps. Then $pi[B \times \{0\}]$ is the vertex of CY and so there is a unique map $g : CB \to CY$ such that $gp' = pi$. Also, g is injective. We now show that g need not be a homeomorphism into.

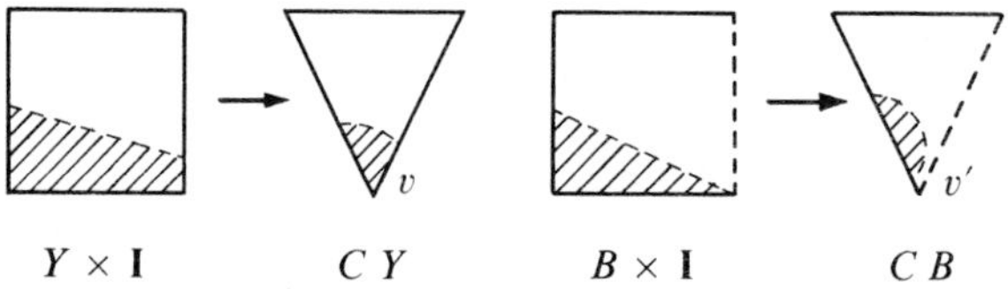

Fig. 4.8

Let $Y = \mathbf{I}$, $B = [0, 1[$, and let v, v' be the vertices of CY, CB respectively. The difference between the topologies of CY and CB is illustrated in Fig. 4.8. The neighbourhoods in CY of v look as expected, but the shaded set in CB is a neighbourhood of v' since it is the image of a saturated open neighbourhood of $B \times \{1\}$. (A detailed justification of these pictures is left as an exercise.)

In this case, it might be considered more reasonable to give CB the topology which makes $g : CB \to CY$ a homeomorphism into. Actually, in chapter 5 we shall discuss the *coarse topology* which behaves better with regard to subspaces than does the identification topology. Until chapter 5 we shall need a topology on CY for which we can construct continuous functions *from* CY *to* some other space. For these purposes the identification topology is the best.

We shall see in 4.6, that when Y is compact Hausdorff so also is CY; in this case, the coarse and identification topologies on CY coincide.

Products of identification maps†

Let $f: X \to Y$, $f': X' \to Y'$ be identification maps. It is not true in general that $f \times f': X \times X' \to Y \times Y'$ is an identification map—an example is given below. However we can prove:

4.3.2 *Let $f: X \to Y$ be an identification map and let B be locally compact. Then*

$$f \times 1 : X \times B \to Y \times B$$

is an identification map.

Proof

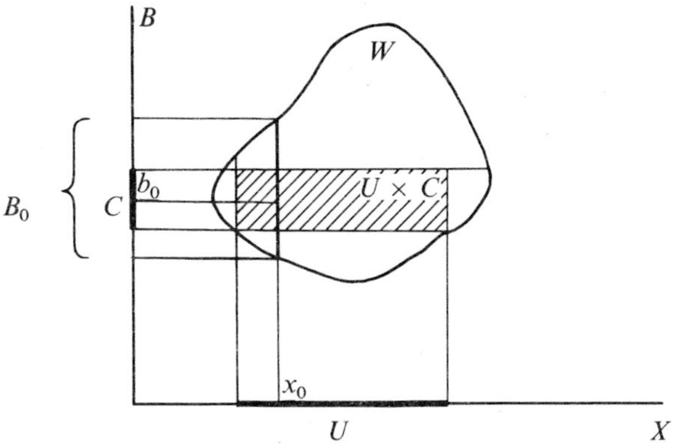

Fig. 4.9

Let $W \subset X \times B$ be open and saturated with respect to

$$h = f \times 1.$$

We must prove that $h[W]$ is open in $Y \times B$.

To this end, let $(y_0, b_0) \in h[W]$ and suppose $y_0 = fx_0$. Then $(x_0, b_0) \in W$ since W is saturated. Let B_0 be the subset of B such that

$$\{x_0\} \times B_0 = (\{x_0\} \times B) \cap W;$$

this is the 'x_0-section' of W. Since W is open, B_0 is a neighbourhood of b_0. Since B is locally compact, B_0 contains a compact neighbourhood C of b_0.

Let $U \subset X$ be the largest set such that $U \times C \subset W$, that is,

$$U = \{x \in X : \{x\} \times C \subset W\}.$$

† The main result here and its application in 4.6.6 are not needed until chapter 7.

Then

$$(y_0, b_0) \in f[U] \times C \subset h[W].$$

So to prove that $h[W]$ is a neighbourhood of (y_0, b_0), it is sufficient to prove that $f[U]$ is a neighbourhood of y_0.

U is open in X Let $x \in U$ so that $\{x\} \times C \subset W$. Since C is compact and W is open, there is a neighbourhood M of x such that $M \times C \subset W$ [3.5.6 (Corollary 2)]. This implies that $M \subset U$, by definition of U. So U is open.

U is f-saturated We have $U \subset f^{-1}f[U]$ and

$$f^{-1}f[U] \times C = h^{-1}h[U \times C] \subset h^{-1}h[W] = W.$$

So $f^{-1}f[U]$, by definition of U. Hence $f^{-1}f[U] = U$.

It follows that $f[U]$ is a neighbourhood of y_0. □

EXAMPLE 4 We now show that 4.3.2 is false without some assumptions on B. Let $f : \mathbf{Q} \to \mathbf{Q}/\mathbf{Z}$ be the identification map and let $h = f \times 1 : \mathbf{Q} \times \mathbf{Q} \to (\mathbf{Q}/\mathbf{Z}) \times \mathbf{Q}$. Then h is not an identification map.

Proof Let $r_0 = 1$ and for each non-zero n in $\mathbf{Z}$ let $r_n = \sqrt{2}/|n|$—thus r_n is irrational and $r_n \to 0$ as $n \to \infty$. Let A_n be any open subset of $[n, n + 1] \times \mathbf{R}$ such that the closure of A_n meets $\{n, n + 1\} \times \mathbf{R}$ in the two points (n, r_n) and (n, r_{n+1}) (such a set A_n is shaded in Fig. 4.10.) Let A be the union of these sets A_n, $n \in \mathbf{Z}$, and let $B = \bar{A} \cap (\mathbf{Q} \times \mathbf{Q})$. Then B is closed in $\mathbf{Q} \times \mathbf{Q}$ and saturated with respect to h. We leave it as an exercise to the reader to prove that in $(\mathbf{Q}/\mathbf{Z}) \times \mathbf{Q}$ the point $(f0, 0)$ belongs to the closure of $h[B]$.

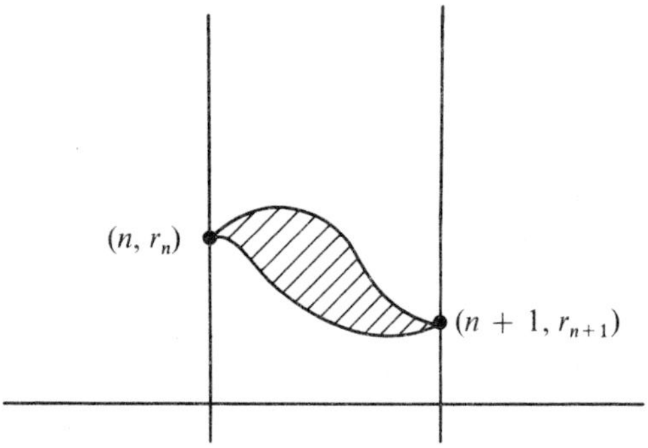

Fig. 4.10

The fact that 4.3.2 is false in general has led to suggestions for changing the maps used in topology, or for changing the product topology [cf. Brown [1], [2], Spanier [2] and Steenrod [2]].

EXERCISES

1. Let A be the subset $\mathbf{L} \setminus \{0\}$ of $\mathbf{I}$ and let $p : \mathbf{I} \to \mathbf{I}/A$ be the identification map. Let $N = [0, \frac{1}{2}] \cup \{1\}$. Prove that N is a saturated neighbourhood of 0, but $p[N]$ is not a neighbourhood of $p0$.

2. Let $f : \mathbf{R} \to [-1, 1]$ be the function

$$x \rightsquigarrow \begin{cases} \sin 1/x & x \neq 0 \\ 0 & x = 0 \end{cases}$$

and let $[-1, 1]$ have the identification topology with respect to f. Prove that the subspace $[-1, 1] \setminus \{0\}$ has its usual topology but the only neighbourhood of 0 is $[-1, 1]$.

3. Let A, B be subsets of X such that A is closed and $A \subset B$. Prove that B/A is a subspace of X/A.

4. Let $A \subset X$. Prove that $X \setminus A$ is a subspace of X/A if and only if the following condition holds: a subset U of $X \setminus A$ is open in $X \setminus A$ if and only if $U = V \cap (X \setminus A)$ where V is open in X and V either contains A or does not meet A. Give an example of X, A for which this condition holds, yet A is neither open nor closed in X. Give an example of X, A for which this does not hold.

5. Let A be a subspace of X. Prove that X/A is Hausdorff if (i) $X \setminus A$ is Hausdorff, (ii) $X \setminus A$ is a subspace of X/A, and (iii) if $x \in X \setminus A$ then x and A have disjoint neighbourhoods in X. Prove that if $X = [0, 2]$, $A =]1, 2]$, then X/A is not Hausdorff.

6. Let A be a closed subset of X and let B be a proper subset of A. Let $X' = X \setminus B$, $A' = A \setminus B$. Prove that X'/A' is homeomorphic to X/A.

7. Let $f : X \to Y$ be an identification map and let $A \subset X$. Suppose that there is a map $u : X \to A$ such that for all $x \in X$, $fux = fx$. Prove that $f[A] = Y$ and $g \mid A$ is an identification map. Apply this result to the case $X = \mathbf{R}$, Y is the set of real numbers mod 1, and $A = [0, 1]$.

8. [This and the following exercises use the notation of Exercises 12, 13 of 4.2]. Let Σ be a set of subspaces of X and let $A \subset X$. The *restriction* of Σ to A is the set $\Sigma \mid A = \{S \cap A : S \in \Sigma\}$. So we can form the space $A_{\Sigma \mid A}$. On the other hand, A with its relative topology as a subset of X_Σ determines a space A_Σ say. Prove that the identity function $A_{\Sigma \mid A} \to A_\Sigma$ is continuous. Prove also that if A and each S of Σ is closed in X, then $A_{\Sigma \mid A} = A_\Sigma$ and A_Σ is a closed subspace of X_Σ.

9. For any space X, let $\mathscr{C}X$ denote the set of compact subspaces of X, and write kX for $X_{\mathscr{C}X}$. Refer to the definition of a k-space in Exercise 4 of 3.6 in order to prove that X is a k-space if and only if $X = kX$. Prove also that (i) X, kX have the same compact subspaces, (ii) kX is a k-space, (iii) the topology of kX is the finest topology $\mathscr{T}$ on the underlying set of X such that X and $X_{\mathscr{T}}$ have the same compact subspaces.

10. Prove that a closed subspace of a k-space is a k-space.

11. Prove that an identification space of a k-space is a k-space.

12. The *weak product* of spaces X, Y is the space $X \times_W Y = k(X \times Y)$. Prove that $X \times_W Y$ has the fine topology with respect to the sets $A \times B$ for A compact in X, B compact in Y. Prove also that (i) $X \times_W Y = kX \times_W kY$, (ii) the weak product is associative and commutative, (iii) the projections from $X \times_W Y$ to X

and Y are continuous, (iv) if $X \times Y = X \times_W Y$, then X and Y are k-spaces, (v) the diagonal map $X \to X \times_W X$ is continuous if and only if X is a k-space, (vi) $X \times_W Y = X \times Y$ if X and Y satisfy the first axiom of countability.

13. Prove that if X is a k-space and Y is locally compact, then $X \times Y$ is a k-space. [Use 4.2.3 to represent X as an identification space of a topological sum of compact spaces, and then apply 4.3.2.]

14. A function $f : X \to Y$ of spaces is called *k-continuous* if $f \mid C$ is continuous for each compact subspace C of X. Prove that the following are equivalent. (i) $f : X \to Y$ is k-continuous, (ii) $f : kX \to Y$ is continuous, (iii) $f : kX \to kY$ is continuous. Prove also that the identity $X \to X$ is k-continuous and that the composite of k-continuous functions is again k-continuous. Prove that if $f : X \to Y$, $g : X \to Z$ are k-continuous, then so also is $(f, g) : X \to Y \times Z$.

15. A function $f : X \to Y$ is called a *k-identification map* if f is k-continuous and for all Z and all functions $g : Y \to Z$, g is k-continuous if gf is k-continuous. Prove that $f : X \to Y$ is a k-identification map if and only if $f : kX \to kY$ is an identification map.

16. Let $f : X \to Y$ be a k-identification map and let Q satisfy: each compact subspaces of Q is locally compact. Prove that $f \times 1 : X \times Q \to Y \times Q$ is a k-identification map.

17. Let $f : X \to Y$ be an identification map. Suppose also that X, Q and $X \times Q$ are k-spaces. Prove that $Y \times Q$ is a k-space if and only if $f \times 1 : X \times Q \to Y \times Q$ is an identification map. Hence give an example of k-spaces Y, Q such that $Y \times Q$ is not a k-space.

4.4 Cells and spheres

In this section, we shall be considering real n-space $\mathbf{R}^n$ with its Euclidean norm, written $|\ |$, and its Cartesian norm, written $\|\ \|$. The map

$$\begin{aligned} i : \mathbf{R}^n &\to \mathbf{R}^{n+1} \\ x &\rightsquigarrow (x, 0) \end{aligned}$$

is called the *natural inclusion*—it is linear homeomorphism onto the closed subspace $\mathbf{R}^n \times \{0\}$ of $\mathbf{R}^{n+1}$. Also i is norm preserving, that is

$$\left.\begin{aligned} |i(x)| &= |x|, \\ \|i(x)\| &= \|x\|, \end{aligned}\right. \quad x \in \mathbf{R}^n. \tag{4.4.1}$$

We have already defined the standard n-cell, n-ball and $(n - 1)$-sphere as

$$\begin{aligned} \mathbf{E}^n &= \{x \in \mathbf{R}^n : |x| \leqslant 1\} \\ \mathbf{B}^n &= \{x \in \mathbf{R}^n : |x| < 1\} \\ \mathbf{S}^{n-1} &= \{x \in \mathbf{R}^n : |x| = 1\}. \end{aligned}$$

For $n = 0$, this gives $\mathbf{E}^0 = \mathbf{B}^0 = \{0\}$, $\mathbf{S}^{-1} = \varnothing$. Notice that $\mathbf{S}^{n-1}$, $\mathbf{E}^n$ are closed, bounded subsets of $\mathbf{R}^n$, and so are compact.

It is clear from equation (4.4.1) that

$$i[\mathbf{E}^n] = \mathbf{E}^{n+1} \cap (\mathbf{R}^n \times \{0\})$$
$$i[\mathbf{S}^{n-1}] = \mathbf{S}^n \cap (\mathbf{R}^n \times \{0\}).$$

We call $i[\mathbf{S}^{n-1}]$ the *equatorial* $(n-1)$-*sphere* of $\mathbf{S}^n$. This $(n-1)$-sphere divides $\mathbf{S}^n$ into two parts called the northern and southern hemispheres

$$E^n_+ = \{(x, t) \in \mathbf{S}^n : x \in \mathbf{R}^n, t \geqslant 0\}$$
$$E^n_- = \{(x, t) \in \mathbf{S}^n : x \in \mathbf{R}^n, t \leqslant 0\}.$$

Thus $\mathbf{S}^n = E^n_+ \cup E^n_-$ and $E^n_+ \cap E^n_- = i[\mathbf{S}^{n-1}]$.

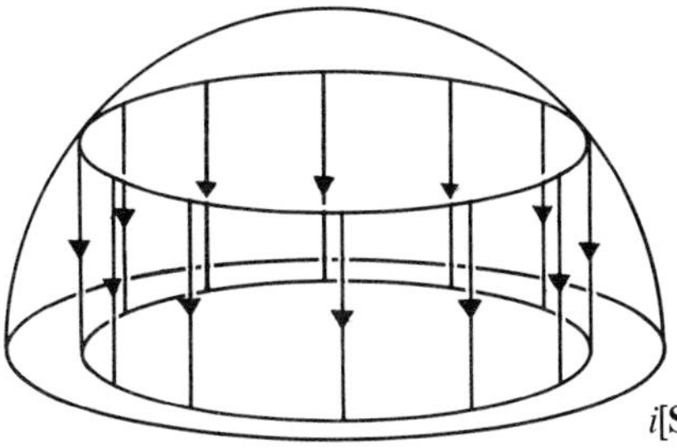

Fig. 4.11

We shall continue to regard $\mathbf{R}^{n+1}$ as $\mathbf{R}^n \times \mathbf{R}$, so that the projection $p : \mathbf{R}^{n+1} \to \mathbf{R}^n$ simply omits the last coordinate. If $x \in \mathbf{S}^n$, then $|x| = 1$ whence $|p(x)| \leqslant 1$, i.e., $p[\mathbf{S}^n] \subset \mathbf{E}^n$.

4.4.2 *The projection* $p : \mathbf{R}^{n+1} \to \mathbf{R}^n$ *maps both* E^n_+ *and* E^n_- *homeomorphically onto* $\mathbf{E}^n$.

Proof The function

$$\mathbf{E}^n \to E^n_+$$
$$x \rightsquigarrow (x, \sqrt{(1 - |x|^2)})$$

is well-defined and is a continuous inverse to $p \mid E^n_+, \mathbf{E}^n$ [Fig. 4.11]. This proves the result for E^n_+, and the proof for E^n_- is similar. □

The sets $\mathbf{B}^n$, $\mathbf{E}^n$ are convex and hence path-connected.

4.4.3 $\mathbf{S}^n$ *is path-connected for* $n \geqslant 1$.

Proof From 4.4.2, E^n_+, E^n_- are path-connected. But $E^n_+ \cap E^n_-$ is non-empty for $n \geqslant 1$, and so $\mathbf{S}^n$ is path-connected by 3.4.2. □

There is an important connection between cells, spheres, and the cone construction. We recall that for any topological space X, the cone on X is

$$CX = X \times \mathbf{I}/X \times \{0\}.$$

4.4.4 *There is a homeomorphism*

$$h : C\mathbf{S}^{n-1} \to \mathbf{E}^n.$$

Proof The function

$$k : \mathbf{S}^{n-1} \times \mathbf{I} \to \mathbf{E}^n$$
$$(x, t) \rightsquigarrow tx$$

is continuous and surjective. Since all the spaces concerned are compact and Hausdorff, k is an identification map. Also k shrinks $\mathbf{S}^{n-1} \times \{0\}$ to the point 0 of $\mathbf{E}^n$. It follows that k defines a homeomorphism $h : C\mathbf{S}^{n-1} \to \mathbf{E}^n$. □

For any space X, let the *suspension* of X be

$$SX = CX/X \times \{1\}.$$

Thus SX is obtained from $X \times \mathbf{I}$ by shrinking to a point first $X \times \{0\}$ and then $X \times \{1\}$. Let l be the composite of the identification maps $X \times \mathbf{I} \to CX \to SX$; by 4.2.5, l is an identification map.

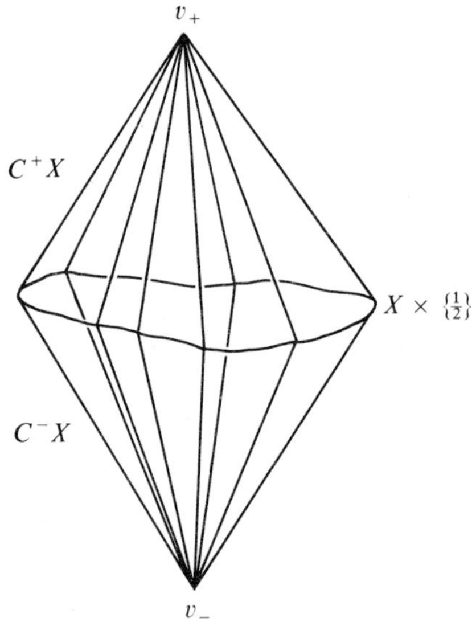

Fig. 4.12

The picture for SX is Fig. 4.12 in which each section $X \times \{t\}$, $t \neq 0, 1$ is homeomorphic to X under the projection $X \times \{t\} \to X$. The subspaces $X \times [\frac{1}{2}, 1]$, $X \times [0, \frac{1}{2}]$ are mapped by l to subspaces C^+X, C^-X. By an easy application of (*d*) of 4.3.1 (Corollary 1), l restricts to an identification map $X \times [0, \frac{1}{2}] \to C^- X$ and so we have an identification map

$$\begin{aligned} X \times \mathbf{I} &\to X \times [0, \tfrac{1}{2}] \to C^- X \\ (x, t) &\rightsquigarrow (x, t/2) \rightsquigarrow l(x, t/2) \end{aligned}$$

which shrinks to a point $X \times \{0\}$. It follows that there is a homeomorphism $CX \to C^- X$. A similar argument (using the homeomorphism $\mathbf{I} \to [\frac{1}{2}, 1]$ given by $t \rightsquigarrow 1 - t/2$) shows that there is a homeomorphism $CX \to C^+ X$; on $X \times \{1\}$ both of these homeomorphisms send $(x, 1) \rightsquigarrow (x, \frac{1}{2})$.

The argument of the last paragraph generalizes the representation of $\mathbf{S}^n$ as the union of its Northern and Southern hemispheres. In fact, by the last paragraph, 4.4.2, and 4.4.4, we can construct homeomorphisms

$$E^n_+ \to C^+ \mathbf{S}^{n-1}, \qquad E^n_- \to C^- \mathbf{S}^{n-1}$$

which on $i[\mathbf{S}^{n-1}]$ are given by $(x, 0) \rightsquigarrow (x, \frac{1}{2})$. By the glueing rule for homeomorphisms [2.7.3] these homeomorphisms define a homeomorphism

$$\mathbf{S}^n \to S\mathbf{S}^{n-1}.$$

The composite

$$\mathbf{E}^n \xrightarrow{h^{-1}} C\mathbf{S}^{n-1} \xrightarrow{p} S\mathbf{S}^{n-1}$$

(where p is the identification map) is injective on $\mathbf{B}^n$ and shrinks the boundary $\mathbf{S}^{n-1}$ of $\mathbf{E}^n$ to a point. It follows that $\mathbf{E}^n/\mathbf{S}^{n-1}$ is homeomorphic to $S\mathbf{S}^{n-1}$, and hence there is a homeomorphism

$$k : \mathbf{E}^n/\mathbf{S}^{n-1} \to \mathbf{S}^n.$$

This homeomorphism sends the shrunken $\mathbf{S}^{n-1}$ to the South Pole $P = (0, \ldots, 0, -1)$ of $\mathbf{S}^n$. Further, $\mathbf{B}^n$ is a subspace of $\mathbf{E}^n/\mathbf{S}^{n-1}$ and k maps $\mathbf{B}^n$ homeomorphically onto $\mathbf{S}^n \setminus \{P\}$. For this reason, we write

$$\mathbf{S}^n = e^0 \cup e^n$$

where e^r denotes an *open r-cell*, or *r-ball*, that is, a space homeomorphic to $\mathbf{B}^r$. In the case of $\mathbf{S}^n$, $e^0 = \{P\}$, $e^n = \mathbf{S}^n \setminus \{P\}$.

It might be expected that $\mathbf{S}^n \setminus \{x\}$ is homeomorphic to $\mathbf{B}^n$ for any point x of $\mathbf{S}^n$. This is in fact true, and is a consequence of the *homogeneity* of $\mathbf{S}^n$ by which is meant that for any points x, y of $\mathbf{S}^n$ there is a homeomorphism $\sigma : \mathbf{S}^n \to \mathbf{S}^n$ such that $\sigma(x) = y$ [cf. Exercise 5 of 5.4].

We conclude this section by proving a simple result which will have a more general analysis in chapter 5.

4.4.5 *There is a homeomorphism* $\mathbf{E}^m \times \mathbf{E}^n \to \mathbf{E}^{m+n}$.

Proof Let $\| \ \|$ denote the Cartesian norm on $\mathbf{R}^n$ and let

$$\mathbf{J}^n = \{x \in \mathbf{R}^n : \|x\| \leqslant 1\}.$$

Thus $\mathbf{J}^n$ is the n-fold product of the interval $[-1, 1]$ with itself. The associativity of the product shows that $\mathbf{J}^m \times \mathbf{J}^n$ is homeomorphic to $\mathbf{J}^{m+n}$. So our result is proved if we exhibit a homeomorphism from $\mathbf{E}^m$ to $\mathbf{J}^m$.

Consider the function

$$f : \mathbf{E}^m \to \mathbf{J}^m$$
$$x \rightsquigarrow \begin{cases} \|x\|^{-1}|x|\,x, & x \neq 0 \\ 0, & x = 0. \end{cases}$$

Then f is a bijection, and is continuous certainly at all $x \neq 0$. We prove f is continuous at 0.

The closed ball about 0 of radius r in the Euclidean norm is $r\mathbf{E}^m$, and in the Cartesian norm is $r\mathbf{J}^m$. Also

$$f^{-1}[r\mathbf{E}^m] = r\mathbf{J}^m$$

But the sets $r\mathbf{E}^m$, and also the sets $r\mathbf{J}^m$, taken for all $r > 0$ form bases for the neighbourhoods of 0. Therefore, f is continuous at 0.

A similar argument shows that f^{-1} is continuous. □

Now $\mathbf{E}^i$ is a subset of $\mathbf{R}^i$, $i = m, n, m + n$, so that $\mathbf{E}^m \times \mathbf{E}^n$ is a subset of $\mathbf{R}^{m+n}$. Let 'interior' mean the interior operator on these Euclidean spaces. Then the homeomorphism $h : \mathbf{E}^m \times \mathbf{E}^n \to \mathbf{E}^{m+n}$ constructed above maps the interior of $\mathbf{E}^m \times \mathbf{E}^n$ bijectively onto the interior of $\mathbf{E}^{m+n}$, that is, it maps $\mathbf{B}^m \times \mathbf{B}^n$ bijectively onto $\mathbf{B}^{m+n}$. Therefore, h maps the boundary of $\mathbf{E}^m \times \mathbf{E}^n$ bijectively onto the boundary of $\mathbf{E}^{m+n}$; hence h restricts to a homeomorphism

$$\mathbf{S}^{m-1} \times \mathbf{E}^n \cup \mathbf{E}^m \times \mathbf{S}^{n-1} \to \mathbf{S}^{m+n-1}. \qquad (4.4.6)$$

EXERCISES

1. Let $x_0 \in \mathbf{S}^n$ and let Π be the hyperplane in $\mathbf{R}^{n+1}$ which is perpendicular to x_0 and which passes through the origin. Let $s : \mathbf{S}^n \setminus \{x_0\} \to \Pi$ be the *stereographic projection* defined as follows: if $x \in \mathbf{S}^n \setminus \{x_0\}$ then $s(x)$ is the unique point of Π such that $s(x)$, x, and x_0 are collinear. Let $p : \mathbf{R}^{n+1} \to \Pi$ be the perpendicular projection onto Π. Prove that for each $x \in \mathbf{S}^n \setminus \{x_0\}$

$$s(x) = p(x) \,/\, (1 - \cos\theta)$$

where θ is the angle between x and x_0. Prove also that s is a homeomorphism.

2. Prove that E^n_+ is a retract of $\mathbf{S}^n$. Prove also that each point x of $\mathbf{S}^n$ has a base $\mathscr{B}$ for the neighbourhoods (in $\mathbf{S}^n$) of x such that each element of $\mathscr{B}$ is a retract of $\mathbf{S}^n$.

3. Let points of $\mathbf{R}^{p+q+2}$ be denoted by (x, y) where $x \in \mathbf{R}^{p+1}$, $y \in \mathbf{R}^{q+1}$. Define subsets A_+, A_- of $\mathbf{S}^{p+q+1}$ by

$$\begin{aligned} A_+ &= \{(x, y) \in \mathbf{S}^{p+q+1} : |x| \geqslant |y|\} \\ A_- &= \{(x, y) \in \mathbf{S}^{p+q+1} : |x| \leqslant |y|\}. \end{aligned}$$

Prove that A_+ is homeomorphic to $\mathbf{S}^p \times \mathbf{E}^{q+1}$, A_- is homeomorphic to $\mathbf{E}^{p+1} \times \mathbf{S}^q$, and $A_+ \cap A_-$ is homeomorphic to $\mathbf{S}^p \times \mathbf{S}^q$.

4. Let X be a topological space and let $x_0 \in X$. Define the *reduced cone* and *reduced suspension* respectively by

$$\begin{aligned} \Gamma X &= X \times \mathbf{I} \,/\, (X \times \{0\} \cup \{x_0\} \times \mathbf{I}) \\ \Sigma X &= X \times \mathbf{I} \,/\, (X \times \{0, 1\} \cup \{x_0\} \times \mathbf{I}). \end{aligned}$$

Prove that $\Gamma \mathbf{S}^{n-1}$ is homeomorphic to $\mathbf{E}^n$, and that $\Sigma \mathbf{S}^{n-1}$ is homeomorphic to $\mathbf{S}^n$.

4.5 Adjunction spaces

Arbitrary identification spaces are so general that it is difficult to say anything about them. We therefore concentrate attention on a special kind of identification space—the adjunction spaces—which occur commonly and for which we can prove useful theorems.

Suppose there is given a topological space X, a closed subspace A of X and a map $f : A \to B$. The *adjunction space*

$$B \,{}_f\!\sqcup X$$

is obtained by 'glueing' X to B by means of f. More precisely, we have a relation from X to B given by

$$x \sim b \Leftrightarrow x \in A \text{ and } fx = b.$$

The elements of $B \,{}_f\!\sqcup X$ are then the equivalence classes of $B \sqcup X$ under the equivalence relation generated by $\sim$.

Let $F : B \sqcup X \to B \,{}_f\!\sqcup X$ be the projection. We have a diagram

$$\begin{array}{ccc} A & \xrightarrow{\quad f \quad} & B \\ {\scriptstyle i}\downarrow & & \downarrow{\scriptstyle \bar{i}} \\ X & \xrightarrow[\quad \bar{f} \quad]{} & B \,{}_f\!\sqcup X \end{array} \qquad (4.5.1)$$

in which i is the inclusion and $\bar{i}$, $\bar{f}$ are the composites

$$B \xrightarrow{i_1} B \sqcup X \xrightarrow{F} B \,{}_f\!\sqcup X, \qquad X \xrightarrow{i_2} B \sqcup X \xrightarrow{F} B \,{}_f\!\sqcup X.$$

Clearly, diagram (4.5.1) is commutative, that is, $\bar{f}i = \bar{i}f$, since if $a \in A$

$$i_2(a) \sim i_1 f(a) \quad \text{in} \quad B \sqcup X.$$

An equivalence class in $B \sqcup X$ either contains a single element $i_2(x)$ for x in $X \setminus A$ or is $\{i_1(b)\} \cup i_2 f^{-1}[b]$ for b in B. It is usual, therefore, to identify x in $X \setminus A$ and b in B with their corresponding equivalence classes in $B \sqcup X$, and so to regard $B\, {}_f\!\sqcup X$ as the union of the sets B and $X \setminus A$. This convention causes confusion only when B meets $X \setminus A$.

The topology on $B\, {}_f\!\sqcup X$ is to be the final topology with respect to $\bar{i}, \bar{f}$ or, equivalently, the identification topology with respect to

$$F : B \sqcup X \to B\, {}_f\!\sqcup X.$$

If $g : B\, {}_f\!\sqcup X \to Y$ is any map, then we can construct a commutative diagram of maps

$$\begin{array}{ccc} A & \xrightarrow{f} & B \\ {\scriptstyle i}\downarrow & & \downarrow{\scriptstyle i'} \\ X & \xrightarrow[f']{} & Y \end{array} \tag{4.5.2}$$

where $i' = g\bar{i}, f' = g\bar{f}$. The converse of this statement gives a method of constructing maps from $B\, {}_f\!\sqcup X$ to Y.

4.5.3 *If $i' : B \to Y, f' : X \to Y$ are maps such that $f'i = i'f$, then there is a unique map $g : B\, {}_f\!\sqcup X \to Y$ such that*

$$g\bar{i} = i', \qquad g\bar{f} = f'.$$

Proof Let $x \in X, b \in B$. The condition $f'i = i'f$ ensures that

$$x \sim b \Rightarrow f'(x) = i'(b).$$

The result is immediate from Example 1 of 4.2. □

In the usual way, the 'universal' property 4.5.3 characterizes $\bar{i}, \bar{f}$ up to a homeomorphism—if the maps f', i' of diagram (4.5.2) are universal then the map $g : B\, {}_f\!\sqcup X \to Y$ of 4.5.3 is a homeomorphism. This type of universal property occurs in diverse situations and so deserves a name. Suppose we are given a commutative diagram of maps of topological spaces

$$\begin{array}{ccc} X_0 & \xrightarrow{i_1} & X_1 \\ {\scriptstyle i_2}\downarrow & & \downarrow{\scriptstyle u_1} \\ X_2 & \xrightarrow[u_2]{} & X \end{array} \tag{4.5.4}$$

Then we say (u_1, u_2) is a *pushout* of (i_1, i_2), and also that the square (4.5.4) is a *pushout*, if the following property holds: *if* $u_1' : X_1 \to X', u_2' : X_2 \to X'$ *are maps such that* $u_1' i_1 = u_2' i_2$, *then there is a unique map* $u : X \to X'$ *such that* $uu_1 = u_1'$, $uu_2 = u_2'$. As usual, this property characterizes the pair (u_1, u_2) up to a homeomorphism of X. For this reason, it is common to make an abuse of language and refer to X as *the* pushout of (i_1, i_2). In such case, we write

$$X = X_1 \, {}_{i_1}\!\sqcup_{i_2} X_2 .$$

(In chapter 7 we shall need the notion of *weak pushout*: the definition of this is the same as that of pushout except that the word *unique* is omitted. Of course, the property that (4.5.4) is a weak pushout does not characterize (u_1, u_2) up to a homeomorphism of X.)

We shall, as above, restrict the term adjunction space to a pushout of a pair $(f; i)$ in which i is an inclusion of a closed subspace; and, of course, we are abbreviating $B \, {}_f\!\sqcup_i X$ to $B \, {}_f\!\sqcup X$.

One of the good features of adjunction spaces is shown in the following result.

4.5.5 *(a) $\bar{i}$ is a closed map. (b) $\bar{f} \mid X \setminus A$ is an open map.*

Proof (*a*) Let C be closed in B and let $C' = \bar{i}[C]$. Then

$$\bar{i}^{-1}[C'] = C, \qquad \bar{f}^{-1}[C'] = f^{-1}[C].$$

The first set is closed in B (trivially) and the second set is closed in X because f is continuous and A is closed in X. Since $B \, {}_f\!\sqcup X$ has the final topology with respect to $\bar{i}, \bar{f}$, it follows from 4.2.2(*b*) that C' is closed in $B \, {}_f\!\sqcup X$.

(*b*) Let U be open in $X \setminus A$ and let $U' = \bar{f}[U]$. Then

$$\bar{i}^{-1}[U'] = \varnothing, \qquad \bar{f}^{-1}[U'] = U.$$

These sets are open in B, X respectively, and so U' is open in $B \, {}_f\!\sqcup X$. □

It is immediate from 4.5.5 that $\bar{i}$ is a homeomorphism onto a closed subspace, and $\bar{f} \mid X \setminus A$ is a homeomorphism onto an open subspace of $B \, {}_f\!\sqcup X$. So if we identify B and $X \setminus A$ with the corresponding subsets of $B \, {}_f\!\sqcup X$, we obtain:

4.5.5 (*Corollary 1*) *B is a closed subspace, and $X \setminus A$ is an open subspace, of $B \, {}_f\!\sqcup X$.*

In some cases, adjunction spaces can be regarded as identification spaces.

4.5.6 *If $f : A \to B$ is an identification map, then so also is $\bar{f} : X \to B \, {}_f\!\sqcup X$.*

Proof Clearly $\bar{f}$ is surjective. We prove that $B\ {}_f\!\sqcup X$ has the final topology with respect to $\bar{f}$. Suppose $g : B\ {}_f\!\sqcup X \to Y$ is such that $g\bar{f} : X \to Y$ is continuous. Then $g\bar{f}i = g\bar{i}f$ is continuous. Since f is an identification map, the continuity of $g\bar{i}f$ implies the continuity of $g\bar{i}$. Finally, the continuity of $g\bar{f}$ and $g\bar{i}$ implies the continuity of g. □

EXAMPLES 1. If A is empty, then $B\ {}_f\!\sqcup X = B \sqcup X$. If $A = X$, then $B\ {}_f\!\sqcup X = B$. If $X = B$, then $B\ {}_1\!\sqcup X = X$.
2. Suppose $X = B \cup C$, $A = B \cap C$ where B, C are closed in X. Let $j : A \to B$ be the inclusion. Then

$$X = B\ {}_j\!\sqcup C.$$

The only problem here is one of topology; the fact that X has the final topology with respect to the inclusions $B \to X$, $C \to X$ is a consequence of the Glueing Rule [2.5.12].
3. Let B be the space consisting of a single point b, let A be non-empty, and let $f : A \to B$ be the unique map. Clearly, f is an identification map, and therefore so also is $\bar{f} : X \to B\ {}_f\!\sqcup X$. Since $\bar{f}$ simply shrinks A to a point, it follows that B ${}_f\!\sqcup X$ is homeomorphic to X/A; the only difference between these spaces is that in $B\ {}_f\!\sqcup X$ the point A of X/A has been replaced by b.
4. Many common identification spaces can be regarded as adjunction spaces. For example, let A_1, A_2 be closed in X and let $\varphi : A_1 \to A_2$ be a homeomorphism which is the identity on $A_1 \cap A_2$. Let $f : A_1 \cup A_2 \to A_2$ be the identity on A_2 and φ on A_1—the given conditions ensure that f is well-defined and continuous. Let $Y = A_2\ {}_f\!\sqcup X$. It is easy to prove that f is an identification map, and it follows that $\bar{f} : X \to Y$ is an identification map. Thus Y is obtained from X by the identifications $a_1 \sim \varphi\, a_1$ for all $a_1 \in A_1$.
5. One of our principal examples of adjunction spaces is when $X = \mathbf{E}^n$, the standard n-cell, and $A = \mathbf{S}^{n-1}$, the standard $(n - 1)$-sphere. Thus, let $n = 1$. Then $\mathbf{E}^1 = [-1, 1]$, $\mathbf{S}^0 = \{-1, +1\}$ and $B\ {}_f\!\sqcup \mathbf{E}^n$ can be pictured as one of the spaces of Fig. 4.13.

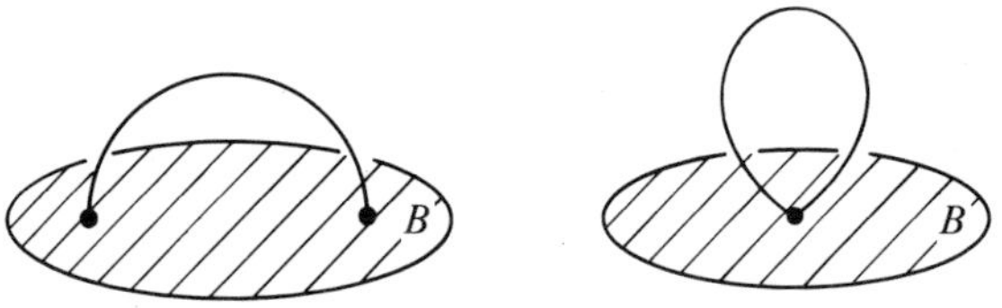

Fig. 4.13

For $n > 1$, and even for $n = 2$, it is usually impossible to visualize $B\ {}_f\!\sqcup \mathbf{E}^n$. But some special cases are of interest. First of all let B consist of a single point: $B\ {}_f\sqcup\ \mathbf{E}^n$ is homeomorphic to $\mathbf{E}^n\ /\ \mathbf{S}^{n-1}$ which, as we have seen in 4.4, is homeomorphic to $\mathbf{S}^n$.

Second, consider the function

$$f : \mathbf{S}^1 \to B$$
$$z \rightsquigarrow \mathrm{Re}\,(z)$$

where $B = [-1, 1]$. Then $B\ {}_f\sqcup\ \mathbf{E}^2$ is homeomorphic to $\mathbf{S}^2$. The picture for this is Fig. 4.14 (which may be thought of as illustrating the operation

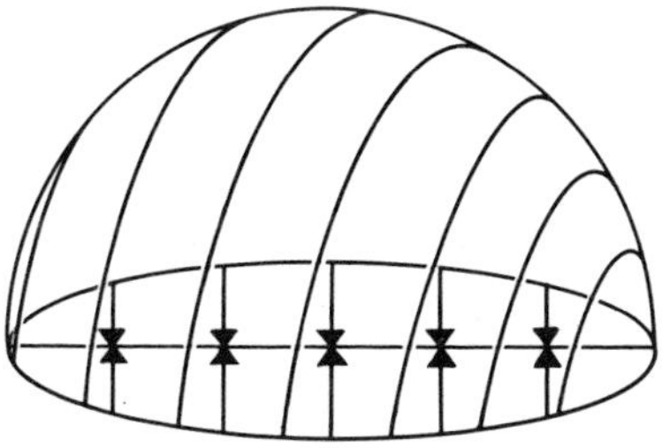

Fig. 4.14

of zipping up a purse). Third, let $f : \mathbf{S}^1 \to \mathbf{S}^1$ be the function $z \rightsquigarrow z^2$. Then f identifies the points $\pm z$ for z in $\mathbf{S}^1$ and f is an identification map. Therefore $\mathbf{S}^1\ {}_f\sqcup\ \mathbf{E}^2$ is homeomorphic to $\mathbf{S}^2$ with antipodal points identified. This space is a model of the real projective plane $P^2(\mathbf{R})$ of chapter 5.

6. A *pointed space* is a topological space X and a point x_0 of X, called the *base point* such that $\{x_0\}$ is closed in X. Let X, Y be pointed spaces. The *wedge* of X and Y is the space $X \vee Y$ obtained from X and Y by identifying x_0 with y_0. Figure 4.15 illustrates $\mathbf{S}^1 \vee \mathbf{S}^2$ (for some choice of base points) and $\mathbf{S}^2 \vee \mathbf{S}^2$ can be thought of as two tangential 2-spheres

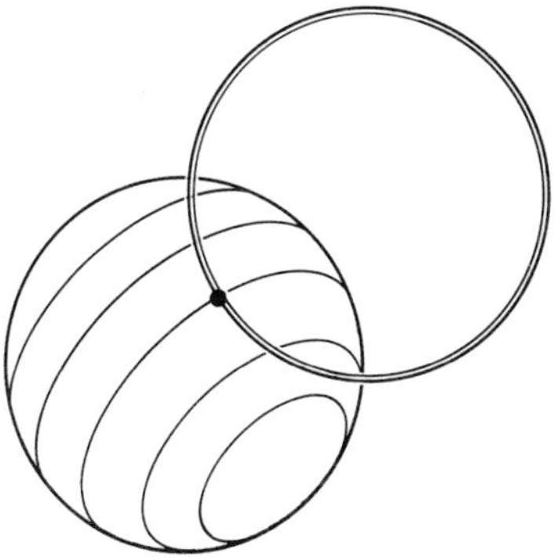

Fig. 4.15

in $\mathbf{R}^3$. If we take the boundary of a square and identify opposite sides according to the schemes of either Fig. 4.3 or Fig. 4.4 on p. 90, then the result is essentially $\mathbf{S}^1 \vee \mathbf{S}^1$.

7. Let T^2 be the torus and let $g : \mathbf{I}^2 \to T^2$ be the identification map by which opposite edges of $\mathbf{I}^2$ are identified. Let $h : \mathbf{E}^2 \to \mathbf{I}^2$ be a homeomorphism which maps $\mathbf{S}^1$ onto the boundary of $\mathbf{I}^2$ (in fact any homeomorphism $\mathbf{E}^2 \to \mathbf{I}^2$ does this), and let $\bar{f} = gh : \mathbf{E}^2 \to T^2$. Then $\bar{f}[\mathbf{S}^1]$ is essentially a subset $\mathbf{S}^1 \vee \mathbf{S}^1$ of T^2 (by the last example). Since $\bar{f}$ is an identification map we can say that T^2 is homeomorphic to $(\mathbf{S}^1 \vee \mathbf{S}^1)\; {}_f\!\sqcup \mathbf{E}^2$ where f is a map $\mathbf{S}^1 \to \mathbf{S}^1 \vee \mathbf{S}^1$.

We have shown how to construct spaces by forming adjunction spaces. On the other hand, it may help us to grasp the structure of a given space Q if we can represent Q as an adjunction space $B\; {}_f\!\sqcup X$, or, at least, a homeomorph of $B\; {}_f\!\sqcup X$.

Suppose there is given then a closed subspace B of Q, a closed subspace A of X and a map $\bar{f} : X \to Q$ such that $\bar{f}[A] \subset B$. Consider the diagram

$$\begin{array}{ccc} A & \xrightarrow{\;f\;} & B \\ {\scriptstyle i}\downarrow & & \downarrow{\scriptstyle \bar{i}} \\ X & \xrightarrow[\;\bar{f}\;]{} & Q \end{array} \qquad (4.5.7)$$

in which i, $\bar{i}$ are inclusions and f is the restriction of $\bar{f}$. We then ask: is (4.5.7) a pushout? This involves questions of set theory and of topology.

If (4.5.7) is a pushout, then $h = \bar{f} \mid X \setminus A, Q \setminus B$ is a bijection. Also, if this map h is a bijection, then (4.5.7) is a pushout of sets—if $f' : X \to Y$, $i' : B \to Y$ are functions such that $f'i = i'f$, then there is a unique function $g : Q \to Y$ such that $g\bar{i} = i'$, $g\bar{f} = f'$, where g is i' on B and $x \rightsquigarrow f'h^{-1}(x)$ on $Q \setminus B$.

Suppose then $h = \bar{f} \mid X \setminus A, Q \setminus B$ is a bijection and, as before, that B is closed in Q, A is closed in X.

4.5.8 *The square* (4.5.7) *is a pushout if* $\bar{f}[X]$ *is closed in* Q *and* $\bar{f} \mid X, \bar{f}[X]$ *is an identification map. These conditions hold if* X *is compact and* Q *is Hausdorff.*

Proof The last statement is clear. For the first statement, we have to prove that Q has the final topology with respect to $\bar{i}, \bar{f}$.

Let C be a subset of Q such that $\bar{i}^{-1}[C]$, $\bar{f}^{-1}[C]$ are closed in B, X respectively. Then $B \cap C = \bar{i}^{-1}[C]$ is closed in B and hence is closed in Q. Also if

$$C' = C \cap f[X] = \bar{f}\bar{f}^{-1}[C],$$

then C' is closed in $\bar{f}[X]$ (since $\bar{f} \mid X, f[X]$ is an identification map) and therefore C' is closed in Q (since $\bar{f}[X]$ is closed in Q). Therefore

$$C = (B \cap C) \cup C'$$

is closed in Q. □

We shall need the notion of attaching a number of spaces; this is a simple extension of previous results, so we sketch the theory, leaving the details as exercises.

Let B be a topological space, let $(X_\lambda)_{\lambda \in \Lambda}$ be a family of spaces, and for each λ in Λ let A_λ be a closed subset of X_λ and let $f_\lambda : A_\lambda \to B$ be a map. The adjunction space

$$Q = B\,_{(f_\lambda)}\sqcup (X_\lambda)$$

is obtained by identifying each a_λ of A_λ with $f_\lambda(a_\lambda)$ in B. Thus Q is an identification space of the sum of B and all the X_λ. Providing the X_λ are disjoint and do not meet B, the underlying set of Q can be identified with the union of B and all the $X_\lambda \setminus A_\lambda$. There is an inclusion $i : B \to Q$, and maps $\bar{f}_\lambda : X_\lambda \to Q$. The topology of Q is the final topology with respect to the family consisting of $\bar{i}$ and all $\bar{f}_\lambda$. The maps $\bar{i}$ and $\bar{f}_\lambda \mid X_\lambda \setminus A_\lambda$ are closed and open maps respectively. The adjunction space Q is characterized by the universal property: if $i_\lambda : A_\lambda \to X_\lambda$ is the inclusion, and $i' : B \to Y$, $g_\lambda : X_\lambda \to Y$ are maps such that $g_\lambda i_\lambda = i' f_\lambda$, $\lambda \in \Lambda$, then there is a unique map $g : Q \to Y$ such that $gi = i'$, $g\bar{f}_\lambda = g_\lambda$, $\lambda \in \Lambda$. In the case Λ is finite, say $\Lambda = \{1, ..., n\}$, we write

$$Q = B\,_{f_1}\sqcup X_1 \cdots {}_{f_n}\sqcup X_n.$$

This notation is unambiguous since f_i is a map to B.

EXERCISES

1. Let Q be the subspace of $\mathbf{R}^2$ which is the union of $\{0\} \times [0, 2]$ and the set $\{(x, y) \in \mathbf{R}^2 : 0 < x \leqslant 1/\pi,\ \sin 1/x \leqslant y \leqslant 2\}$. Let B be the boundary of the subset Q of $\mathbf{R}^2$. Prove that B is closed in Q, that $Q \setminus B$ is homeomorphic to $\mathbf{B}^2$, but that Q is not homeomorphic to $B\,_f\sqcup \mathbf{E}^2$ for any f.
2. Let B be a topological space and let $f_\lambda : A_\lambda \to B$ be maps, where A_λ is a closed subset of X_λ and $\lambda = 1, 2$. Let $\bar{i}_1 : B \to B\,_{f_1}\sqcup X_1$ be the inclusion. Characterize the adjunction space

$$Q' = (B\,_{f_1}\sqcup X_1)_{\bar{i}f_2}\sqcup X_2$$

by a universal property and prove that Q' is homeomorphic to $B\,_{f_1}\sqcup X_1\,_{f_2}\sqcup X_2$. Generalize this result to arbitrary finite indexed families (f_λ), (X_λ).
3. Let B be a closed subspace of Q. For each $\lambda = 1, \ldots, n$, let $\bar{f}_\lambda : X_\lambda \to Q$ be maps, and let A_λ be a closed subspace of X_λ such that (i) $\bar{f}_\lambda[A_\lambda] \subset B$, (ii) $\bar{f}_\lambda \mid X_\lambda \setminus A_\lambda$ is injective, (iii) the sets $\bar{f}_\lambda[X_\lambda \setminus A_\lambda]$ are disjoint and cover $Q \setminus B$, (iv) $\bar{f}_\lambda \mid X_\lambda, \bar{f}_\lambda[X_\lambda]$

is an identification map. Prove that a function $g : Q \to Y$ is continuous if and only if $g\bar{f}_\lambda$ is continuous, $\lambda = 1, \ldots, n$. Prove also that there is a homeomorphism

$$Q \to B\ {}_{f_1}\!\sqcup X_1 \cdots {}_{f_n}\!\sqcup X_n$$

which is the identity on B.

4. Let $B\ {}_f\!\sqcup X$ be an adjunction space, where $f : A \to X$. Prove that B is a retract of $B\ {}_f\!\sqcup X$ if and only if there is a map $g : X \to B$ such that $g \mid A = f$. Deduce that if A is a retract of X then B is a retract of $B\ {}_f\!\sqcup X$.

4.6 Properties of adjunction spaces

Throughout this section we suppose given an adjunction space $B\ {}_f\!\sqcup X$ and the pushout square

$$\begin{array}{ccc} A & \xrightarrow{\ f\ } & B \\ {\scriptstyle i}\downarrow & & \downarrow{\scriptstyle \bar{i}} \\ X & \xrightarrow[\ \bar{f}\ ]{} & B\ {}_f\!\sqcup X \end{array}$$

Let $F : B \sqcup X \to B\ {}_f\!\sqcup X$ be the identification map.

4.6.1 *Let B', X' be subspaces of B, X respectively such that $X' \supset A$ and $B' \supset f[A]$. Let $g = f \mid A, B'$. Then $B'\ {}_g\!\sqcup X'$ is a subspace of $B\ {}_f\!\sqcup X$.*

Proof We have to prove that the restriction of F

$$F' : B' \sqcup X' \to F[B' \sqcup X']$$

is an identification map, and for this we can use 4.3.1.

Notice that F is injective on $(B \setminus B') \sqcup (X \setminus X')$. Suppose that $U \subset B \sqcup X$ and $U' = U \cap (B' \sqcup X')$. Then U is F-saturated $\Leftrightarrow$ U' is F'-saturated (since all the identifications take place in $B' \sqcup X'$). So 4.3.1 applies, and F' is an identification map. □

We now consider the connectivity of $B\ {}_f\!\sqcup X$.

4.6.2 *If A is non-empty and X and B are connected then $B\ {}_f\!\sqcup X$ is connected.*

4.6.3 *If A is non-empty and X and B are path-connected then $B\ {}_f\!\sqcup X$ is path-connected.*

4.6.4 *If $B\ {}_f\!\sqcup X$ and A are connected then B is connected.*

Proof of 4.6.2 The space $B\ {}_f\!\sqcup X$ is the union of the sets B and $\bar{f}[X]$

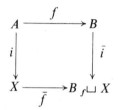

both of which are connected. But B meets $\bar{f}[X]$ (since A is non-empty); therefore $B\ {}_f\sqcup X$ is connected [3.2.5]. □

The proof of 4.6.3 is similar to that of 4.6.2.

Proof of 4.6.4 Suppose that B is not connected and that $i' : B \to Y$ is a map onto the discrete space $Y = \{1, 2\}$. Since A is connected, $i'f : A \to Y$ is constant with value 1 say. Let $f' : X \to Y$ be the constant function with value 1. Then $i'f = f'i$.

By the pushout property 4.5.3, there is a map $g : B\ {}_f\sqcup X \to Y$ such that $g\bar{f} = f'$, $g\bar{i} = i'$. The last condition shows that g is surjective; hence $B\ {}_f\sqcup X$ is not connected. □

If B, X are compact then so also is the sum $B \sqcup X$ and hence $B\ {}_f\sqcup X$ is compact.

The question of whether $B\ {}_f\sqcup X$ is Hausdorff is important but rather delicate. It is true that if B and X are normal then so also is $B\ {}_f\sqcup X$ [cf. Exercise 6 and also Hu [2] 4.3.7]. We shall prove not this but instead a result involving the placing of A in X which is satisfied in many cases [cf. section 7.3].

We recall that a *retraction* of X onto A is a map $r : X \to A$ such that $r \mid A$ is the identity. If such a retraction exists we say A is a *retract* of X. To prove that A is a retract of X it is of course sufficient to produce a retraction $X \to A$—it is rather harder to prove that A is not a retract of X. A retraction $X \to A$ is surjective: so A is not a retract of X if X is connected and A is not connected. For example, $\mathbf{S}^0$ is not a retract of $\mathbf{E}^1$. We shall later be able to prove that $\mathbf{S}^1$ is not a retract of $\mathbf{E}^2$; the proof that $\mathbf{S}^n$ is not a retract of $\mathbf{E}^{n+1}$ for any n is beyond the scope of this book.

We say A is a *neighbourhood retract* of X if A is a retract of some neighbourhood of A in X. This neighbourhood may always be taken to be open since, if $A \subset \operatorname{Int} N$ and $r : N \to A$ is a retraction, then so also is $r \mid \operatorname{Int} N$.

Examples 1. X is a retract of X. If $x \in X$ then $\{x\}$ is a retract of X. The product of retractions is a retraction; hence for any Y, $\{x\} \times Y$ is a retract of $X \times Y$.

2. $\mathbf{S}^n$ is a neighbourhood retract of $\mathbf{E}^{n+1}$ since the map

$$\begin{aligned}\mathbf{E}^{n+1} \setminus \{0\} &\to \mathbf{S}^n\\ x &\rightsquigarrow x/|x|\end{aligned}$$

is a retraction.

3. $\mathbf{L}$ is not a neighbourhood retract of $\mathbf{I}$. For suppose N is a neighbourhood of $\mathbf{L}$ in $\mathbf{I}$ and $r : N \to \mathbf{L}$ is continuous. For some $\varepsilon > 0$, N contains $[0, \varepsilon]$, which is connected. Hence r is constant on $[0, \varepsilon]$ and so cannot be a retraction.

4.6.5 *$B\ {}_f\sqcup X$ is Hausdorff if the following conditions hold:*

(*a*) *B and X are Hausdorff,*
(*b*) *Each x in $X \setminus A$ has a neighbourhood closed in X and not meeting A,*
(*c*) *A is a neighbourhood retract of X.*

Proof Let $Q = B\ {}_f\sqcup X$ and let $y_1, y_2 \in Q$. In order to find disjoint neighbourhoods W_1, W_2 of y_1, y_2, we distinguish three cases.

(i) Suppose $y_1, y_2 \in X \setminus A$. Since $X \setminus A$ is Hausdorff there are disjoint sets W_1, W_2 which are neighbourhoods in $X \setminus A$ of y_1, y_2. But $X \setminus A$ is open in Q [4.5.5 (Corollary 1)]. So W_1, W_2 are also neighbourhoods in Q of y_1, y_2.
(ii) Suppose $y_1 \in X \setminus A$, $y_2 \in B$. Let W_1 be a neighbourhood of y_1 closed in X and not meeting A. Then W_1 is also a neighbourhood in Q of y_1.

Let $W_2 = Q \setminus W_1$. Then

$$\bar{i}^{-1}[W_2] = B, \qquad \bar{f}^{-1}[W_2] = X \setminus W_1$$

which are open in B, X respectively. Hence W_2 is open in Q.
(iii) Suppose $y_1, y_2 \in B$. Since B is Hausdorff there are in B disjoint open neighbourhoods V_1, V_2 of y_1, y_2. Then $f^{-1}[V_1], f^{-1}[V_2]$ are open in A, but not necessarily open in X. We therefore use the neighbourhood retraction to enlarge them into sets open in X.

Let $r : N \to A$ be a retraction where N is open in X. Let

$$V_i' = r^{-1}f^{-1}[V_i], \quad i = 1, 2.$$

Then V_1', V_2' are disjoint sets open in N and so open in X. Notice also that because r is a retraction

$$V_i' \setminus A = V_i' \setminus f^{-1}[V_i], \quad i = 1, 2.$$

Let W_i be the subset of Q

$$W_i = V_i \cup (V_i' \setminus A), \quad i = 1, 2.$$

Then $\bar{i}^{-1}[W_i] = V_i$, and

$$\begin{aligned} \bar{f}^{-1}[W_i] &= \bar{f}^{-1}[V_i] \cup (V_i' \setminus A) \\ &= V_i'. \end{aligned}$$

Therefore W_i is open in Q. Clearly, $y_1 \in W_1$, $y_2 \in W_2$. □

(The existence, for each A closed in X and each x in $X \setminus A$, of a closed neighbourhood of x not meeting A is one of the standard separation axioms. Such a space X is called T_3 by Bourbaki [1], who also calls X *regular* if it is T_3 and Hausdorff. The opposite terminology is used by Kelley [1].)

EXAMPLES 4. If X is a metric space and A is closed in X then condition (b) of 4.6.5 is always satisfied. For suppose $x \in X \setminus A$; then dist $(x, A) = \varepsilon > 0$ [2.9.2] and $E(x, \varepsilon/2)$ is a closed neighbourhood of x not meeting A. The condition is also satisfied if X is Hausdorff and A is compact [3.5.6 (Corollary 3)]. It follows (for either reason!) that, if $f : \mathbf{S}^{n-1} \to B$ and B is Hausdorff, then $B\ {}_f\!\sqcup\ \mathbf{E}^n$ is Hausdorff.

5. Let X, Y be topological spaces and let $f : X \to Y$ be a map. Let $f' : X \times \{0\} \to Y$ be the map $(x, 0) \rightsquigarrow f(x)$. Since $X \times \{0\}$ is a closed subspace of $X \times \mathbf{I}$ we can define the adjunction space [Fig. 4.16]

$$M(f) = Y\ {}_{f'}\!\sqcup\ (X \times \mathbf{I}).$$

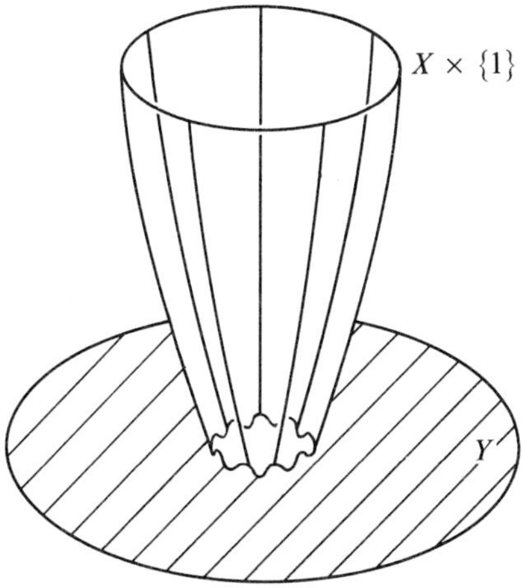

Fig. 4.16

This is the *mapping cylinder* of f. It is a construction of great use in chapter 7. If X is Hausdorff, then so also is $X \times \mathbf{I}$. It is easy to verify the conditions (b) and (c) of 4.6.5 with A, X replaced by $X \times \{0\}$, $X \times \mathbf{I}$. Hence, *if X and Y are Hausdorff, then so also is $M_{(f)}$.*

The *mapping cone* of f is

$$C(f) = M(f)/X \times \{1\}.$$

Again we can prove that if X and Y are Hausdorff, then so also is $C(f)$.

Our next result will become vital when we come to consider homotopies of maps of adjunction spaces in chapters 7, 8.

4.6.6 *Let Y be a locally compact space. In the following diagrams let the first square be a pushout; then the second square is a pushout.*

$$\begin{array}{ccc} A & \xrightarrow{f} & B \\ {\scriptstyle i}\downarrow & & \downarrow{\scriptstyle \bar{i}} \\ X & \xrightarrow[\bar{f}]{} & Q \end{array} \qquad \begin{array}{ccc} A \times Y & \xrightarrow{f \times 1} & B \times Y \\ {\scriptstyle i \times 1}\downarrow & & \downarrow{\scriptstyle \bar{i} \times 1} \\ X \times Y & \xrightarrow[\bar{f} \times 1]{} & Q \times Y \end{array}$$

Proof The maps $\bar{i}, \bar{f}$ define a map $F : B \sqcup X \to Q$, and we know that the first square is a pushout if and only if it is commutative and F is an identification map. If F is an identification map, then so also is

$$F \times 1 : (B \sqcup X) \times Y \to Q \times Y \quad [4.3.2]. \quad \square$$

The main utility of this result is that to specify a map $Q \times Y \to Z$ it is sufficient to give a commutative diagram

$$\begin{array}{ccc} A \times Y & \xrightarrow{f \times 1} & B \times Y \\ {\scriptstyle i \times 1}\downarrow & & \downarrow{\scriptstyle j} \\ X \times Y & \xrightarrow[g]{} & Z \end{array}$$

For further properties of adjunction spaces we refer the reader to §3 of chapter IV of Hu [2] and also to Michael [1].

EXERCISES

1. Prove that the adjunction space $B\ {}_f\!\sqcup X$ is a T_1-space if B and X are T_1-spaces.
2. Let X be a non-normal space. Prove that there are disjoint, closed subsets A, B of X such that $(X/A)/B$ is not Hausdorff.
3. Prove that $\mathbf{R}/\mathbf{Z}$ does not satisfy the first axiom of countability.
4. Prove the following generalization of 4.6.5: an adjunction space $B\ {}_{(f_\lambda)}\!\sqcup (X_\lambda)$ is Hausdorff if for each λ in Λ (i) B and X_λ are Hausdorff, (ii) each x in $X_\lambda \setminus A_\lambda$ has a neighbourhood closed in X_λ and not meeting A_λ, (iii) A_λ is a neighbourhood retract of X_λ.
5. Let X, Y be spaces and let $x_0 \in X$, $y_0 \in Y$. Prove that if P is the subspace $X \times \{y_0\} \cup \{x_0\} \times Y$ of $X \times Y$ then the square of inclusions

$$\begin{array}{ccc} \{(x_0, y_0)\} & \longrightarrow & \{x_0\} \times Y \\ \downarrow & & \downarrow \\ X \times \{y_0\} & \longrightarrow & P \end{array}$$

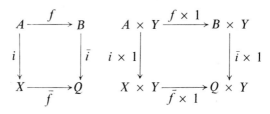

is a pushout square. Deduce that if x_0, y_0 are taken as base points of X, Y respectively, then P is homeomorphic to $X \vee Y$.
6. Prove that an adjunction space $B\ {}_f\!\sqcup X$ is normal if B and X are normal. [Assume the Tietze Extension Theorem.]
7. Let the space X be the union of a family $(X_i)_{i \in \mathbf{N}}$ of subspaces such that (i) for each $i \in \mathbf{N}$, X_i is a closed subset of X_{i+1}, (ii) a set C is closed in X if and only if $C \cap X_i$ is closed in X_i for each i in $\mathbf{N}$. Prove that if each X_i is a normal space, then so also is X. [Assume the Tietze Extension Theorem.]
8. Let A be a closed subspace of X, and let $f : A \to B$, $g : B \to C$ be maps. Prove that the spaces

$$C\ {}_g\!\sqcup (B\ {}_f\!\sqcup X), \quad C\ {}_{gf}\!\sqcup X$$

are homeomorphic.
9. Let $i : A \to X$, $j : X \to Y$ be inclusions of closed subspaces, and let $f : A \to B$ be a map. Prove that the spaces

$$B\ {}_f\!\sqcup Y, \qquad (B\ {}_f\!\sqcup X)\ {}_{\bar{f}}\!\sqcup Y$$

are homeomorphic.
10. Let A be a closed subspace of X and $f : A \to B$ a map. Prove that the subspace $L = (B\ {}_f\!\sqcup X) \times \{0\} \cup B \times \mathbf{I}$ of $M = (B\ {}_f\!\sqcup X) \times \mathbf{I}$ is homeomorphic to $(B \times \mathbf{I})\ {}_{f \times 1}\!\sqcup (X \times \{0\} \cup A \times \mathbf{I})$, where $f \times 1 : A \times \mathbf{I} \to B \times \mathbf{I}$. Hence show that if $X \times \{0\} \cup A \times \mathbf{I}$ is a retract of $X \times \mathbf{I}$, then L is a retract of M.

4.7 Cell complexes

In order to emphasize the intuitive ideas, we shall restrict ourselves to finite cell complexes. Indeed the theory for infinite cell complexes involves, in the main, arranging the topologies so that theorems and proofs for the finite case carry over without change to the infinite case.

There are two useful ways of thinking about cell complexes: (*a*) constructive, (*b*) descriptive. In the first approach, we simply construct cell complexes by starting in dimension -1 with the empty set $\varnothing$, and then attach cells to $\varnothing$ in order of increasing dimension. In the second approach, we suppose there is given a topological space X, and seek to describe X in a useful way as the union of open cells. In both approaches, a vital role is played by the sequence of *skeletons*, the n-skeleton being the union of all cells of dimension $\leqslant n$.

The constructive approach

We construct a sequence of spaces K^n such that $K^n \subset K^{n+1}$, $n \geqslant -1$. The space K^{-1} is to be the empty set. Suppose that K^{n-1} has been constructed. We suppose given maps

$$f_1^n, \ldots, f_{r_n}^n : \mathbf{S}^{n-1} \to K^{n-1}.$$

Then K^n is to be the adjunction space

$$K^{n-1} {}_{f_1^n}\sqcup \mathbf{E}^n \cdots {}_{f_{r_n}^n}\sqcup \mathbf{E}^n.$$

For $n > 0$, we allow the case $r = 0$, when $K^n = K^{n-1}$. In fact, since we are dealing only with finite cell complexes, we shall assume that for some N, $K^n = K^{n-1}$ for $n > N$, and define $K = K^N$. However, we shall always assume K^0 is non-empty.

Let us consider the intuitive picture in low dimensions. K^0 is a non-empty, finite, discrete space. K^1 is formed by attaching to K^0 a finite number (possibly 0) of 1-cells by means of maps $f_1^1, \ldots, f_{r_1}^1 : \mathbf{S}^0 \to K^0$. Now $\mathbf{S}^0$ is the discrete space $\{-1, +1\}$. So for each $i = 1, \ldots, r_1$ the map f_i^1 is either constant or maps to two distinct points of K^0. Thus a representative picture for K^1 is Fig. 4.17, where the dots denote elements of K^0.

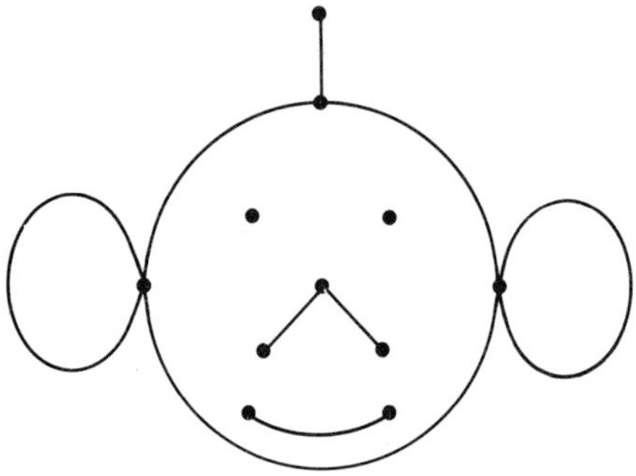

Fig. 4.17

Next we form K^2 by attaching a finite number of 2-cells to K^1 by means of maps $f_1^2, \ldots, f_{r_2}^2 : \mathbf{S}^1 \to K^1$. It can be shown that if K^1 has 1-cells (i.e., if $K^1 \neq K^0$) then there is an uncountable number of maps $\mathbf{S}^1 \to K^1$. It is even possible to construct an uncountable set of spaces of the form $\mathbf{I} {}_f\sqcup \mathbf{E}^2$, no two of which are homeomorphic.

The pictures in Fig. 4.18 are of two simple cell-complexes of the form

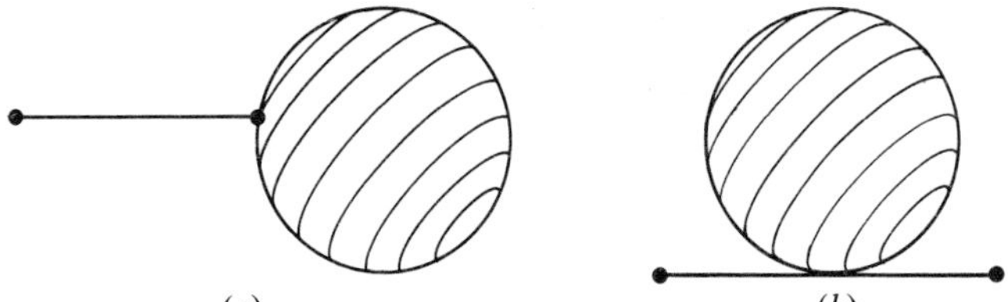

(*a*) (*b*)

Fig. 4.18

$\mathbf{E}^1 {}_f\sqcup \mathbf{E}^2$, where $K^0 = \{-1, +1\}$, $K^1 = \mathbf{E}^1$. In (*a*), $f[\mathbf{S}^1] = \{+1\}$; in (*b*), $f[\mathbf{S}^1] = \{0\}$.

It may be felt that, because of the wide latitude in the attaching maps, cell complexes are bizarre spaces, and that the attaching maps should be restricted in some way, for example to be homeomorphisms. However many important spaces, e.g., the projective spaces, have natural cell structures in which the attaching maps are not homeomorphisms. Also we shall be considering homotopies of attaching maps in chapter 7 and any restriction on these maps would be inconvenient.

The burden of this section is that cell complexes are always 'good' spaces, although additional restrictions on the attaching maps might make them 'better'. In any case, by results of the last section, a cell complex is always a compact, Hausdorff space.

The descriptive approach

We recall that an open n-cell (also called an n-ball) is a space e homeomorphic to the standard n-ball $\mathbf{B}^n$. The *dimension* of e is the natural number $\dim e = n$. However the proof that this dimension is well-defined is non-trivial, and depends on the Invariance of Dimension: if $f : \mathbf{R}^m \to \mathbf{R}^n$ is a homeomorphism, then $m = n$. Since $\mathbf{B}^n$ is homeomorphic to $\mathbf{R}^n$, it follows that if e is homeomorphic to $\mathbf{B}^m$ and to $\mathbf{B}^n$ then $m = n$; hence $\dim e$ is well-defined.

An idea which occurred early in topology is that of a decomposition of a space Q by open cells. In this, Q is given as the union $\bigcup_{\lambda \in \Lambda} e_\lambda$, where e_λ is an open n_λ-cell and the sets e_λ are disjoint.

The difficulty has always been what to do with such a decomposition, since it gives relatively little information on the space Q. For example, the 'bad' space Q of Exercise 1 of 4.5 has a decomposition with two 0-cells, two 1-cells and one 2-cell. The two spaces of Fig. 4.19 have decompositions with one 0-cell, and two 1-cells. Yet by simple local cut point arguments they are not homeomorphic.

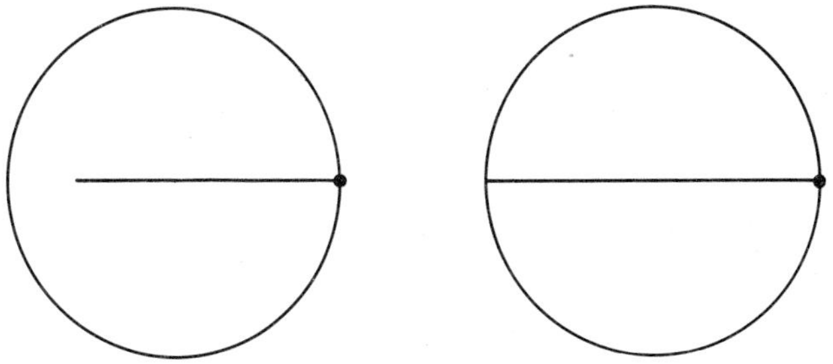

Fig. 4.19

Let us consider what we are given with a decomposition by open cells, $Q = \bigcup_{\lambda \in \Lambda} e_\lambda$. For each λ in Λ, there is a homeomorphism $k_\lambda : \mathbf{B}^{n_\lambda} \to e_\lambda$. Since $\dim e_\lambda = n_\lambda$ is uniquely defined, for each natural number n the *n-skeleton* of Q is well-defined by

$$Q^n = \bigcup_{n_\lambda \leqslant n} e_\lambda.$$

J. H. C. Whitehead realized that a satisfactory theory could be built by assuming the extra condition that the homeomorphism $k_\lambda : \mathbf{B}^{n_\lambda} \to e_\lambda$ extends to a map $h_\lambda : \mathbf{E}^{n_\lambda} \to Q$ such that $h_\lambda[\mathbf{S}^{n_\lambda - 1}] \subset Q^{n_\lambda - 1}$. This ensures that e_λ is obtained from the attaching of a closed n_λ-cell to $Q^{n_\lambda - 1}$; it takes us to the situation considered before and, from the technical point of view, allows proofs by induction on the skeletons.

The last sentence is not quite correct. The reason for this is that if the number of cells is infinite then the cells do not uniquely determine the topology of Q. For example, any Q has a decomposition in which each point of Q is a 0-cell. Therefore we restrict ourselves to the finite case, which retains the geometric ideas without additional topological difficulties.

It is convenient in defining a cell complex to keep the above maps h_λ as part of the structure. It is also necessary to assume that Q is Hausdorff.

Let Q be a non-empty Hausdorff topological space. A *cell structure* on Q is a finite set of maps

$$h_\lambda : \mathbf{E}^{n_\lambda} \to Q, \qquad \lambda \in \Lambda$$

called the *characteristic maps*. Let

$$e_\lambda = h_\lambda[\mathbf{B}^{n_\lambda}].$$

The *n-skeleton* of the cell structure is

$$Q^n = \bigcup_{n_\lambda \leqslant n} e_\lambda.$$

Then we require that the characteristic maps satisfy:

CS1 $h_\lambda \mid \mathbf{B}^{n_\lambda}, e_\lambda$ *is a bijection,*

CS2 *the* e_λ *are disjoint and* $Q = \bigcup_\lambda e_\lambda$,

CS3 $h_\lambda[\mathbf{S}^{n_\lambda - 1}] \subset Q^{n_\lambda - 1}$.

A *cell complex* is a space Q with a cell structure on it. It is usual to denote the cell complex simply by Q.

The space $\mathbf{E}^n$ is compact and Q^{n-1} is Hausdorff. Therefore Exercise 3 of 4.5 implies that Q^n is homeomorphic to an adjunction space,

$$Q^{n-1} {}_{f_1}\sqcup \mathbf{E}^n \dots {}_{f_n}\sqcup \mathbf{E}^n$$

where f_i is a map $\mathbf{S}^{n-1} \to Q^{n-1}$. This shows how the present definition links with the constructive approach considered first.

Another easy consequence of the definition is that

$$\overline{e_\lambda} = h_\lambda[\mathbf{E}^{n\lambda}] :$$

for the continuity of h_λ implies that

$$h_\lambda[\mathbf{E}^{n\lambda}] = h_\lambda[\overline{\mathbf{B}^{n\lambda}}] \subset \overline{h_\lambda[\mathbf{B}^{n\lambda}]};$$

on the other hand $h[\mathbf{E}^{n\lambda}]$ is a compact, and hence closed, set containing e_λ.

4.7.1 *Let Q be a cell complex and $g : Q \to X$ a function to a topological space X. The following conditions are equivalent.*

(*a*) *g is continuous.*
(*b*) *gh_λ is continuous for each characteristic map h_λ.*
(*c*) *$g \mid \overline{e_\lambda}$ is continuous for each cell e_λ.*

Proof Obviously $(a) \Rightarrow (b)$ and $(a) \Rightarrow (c)$. Also $(c) \Rightarrow (b)$ since $h_\lambda[\mathbf{E}^{n\lambda}] = \overline{e_\lambda}$. Since $\mathbf{E}^{n\lambda}$ is compact and Q is Hausdorff, the map $h_\lambda \mid \mathbf{E}^{n\lambda}, \overline{e_\lambda}$ is an identification map.

To complete the proof we show that $(b) \Rightarrow (a)$. The proof is by induction on the skeletons.

Since Q^0 is discrete, $g \mid Q^0$ is certainly continuous. But if $g \mid Q^{n-1}$ is continuous, then by Exercise 3 of 4.5 so also is $g \mid Q^n$. Since $Q = Q^n$ for some n, it follows that g is continuous. □

In order to extend the present theory to infinite cell complexes, it is necessary to take the last result as one of the properties characterizing the topology of Q.

The smallest integer n such that $Q = Q^n$ is called the *dimension* of the cell complex Q. The methods which are used to prove the invariance of dimension can also prove that the dimension of a cell complex is a topological invariant.

Our next result shows one of the ways in which cell-complexes are 'nice' spaces.

4.7.2 *If Q is a cell complex, then the following conditions are equivalent.*

(*a*) *Q^1 is connected,*
(*b*) *Q is path-connected,*
(*c*) *Q is connected.*

Proof Clearly $(b) \Rightarrow (c)$. We prove also $(c) \Rightarrow (a) \Rightarrow (b)$.

$(c) \Rightarrow (a)$ The n-sphere $\mathbf{S}^n$ is connected for $n \geqslant 1$. Further Q^{n+1} is homeomorphic to a space obtained by attaching $(n + 1)$-cells to Q^n.

Therefore by 4.6.4, if Q^{n+1} is connected and $n \geqslant 1$, then Q^n is connected. But $Q^{n+1} = Q$ for some n. So the result follows by downward induction.

(a) ⇒ *(b)* By 4.6.3 and induction on the skeletons, if Q^1 is path-connected then so also is Q. So it is sufficient to prove the more general result that each path-component of Q^1 is both open and closed in Q^1.

Let P be a path-component of Q^1. If P contains no 1-cells of Q, then P consists of an isolated point, which is certainly both open and closed in Q^1. We suppose then that P contains more than one point.

A point x of P must belong to $\bar{e}^1$ for some open cell e^1, for otherwise x would be an isolated point. On the other hand, each $\bar{e}^1$ is path-connected (since it is the continuous image of $\mathbf{E}^1$) and so $\bar{e}^1$ is either contained in or disjoint from P. Therefore P is the union of a finite number of $\bar{e}^1$, and so P is closed.

We now prove P is open. Let $x \in P$. If $x \in Q^0$, then the set N, consisting of x and all open 1-cells e^1 such that $x \in \bar{e}^1$, is a path-connected neighbourhood of x, and so $N \subset P$. If $x \in Q^1 \setminus Q^0$, then x belongs to some open 1-cell e^1; clearly e^1 is contained in P, and therefore P is a neighbourhood of x. □

Let Q be a cell-complex with characteristic maps h_λ, $\lambda \in \Lambda$. Let P be a non-empty subset of Q and let M be the set of λ in Λ such that the image of h_λ is contained in P. For each λ in M, let $g_\lambda : \mathbf{E}^{n_\lambda} \to P$ be the restriction of $h_\lambda : \mathbf{E}^{n_\lambda} \to Q$. We say P is a *subcomplex* of Q if the characteristic maps g_λ, $\lambda \in \mathrm{M}$, form a cell structure on P. In this case P is covered by the open cells e_λ for λ in M and $Q \setminus P$ is covered by the open cells e_λ for λ in $\mathrm{M} \setminus \Lambda$.

EXAMPLES 1. The n-skeleton Q^n of Q is a subcomplex of Q.
2. The intersection and the union of any family of subcomplexes of Q are again subcomplexes of Q.
3. Let X be any subset of Q. Then the intersection of all subcomplexes of Q containing X is the smallest subcomplex of Q containing X.

Let P, R be cell-complexes and let $f : P \to R$ be a continuous function. We say f is a *cellular map* if

$$f[P^n] \subset R^n, \quad n \geqslant 0.$$

4.7.3 *Let Q, R, be cell-complexes, let P be a subcomplex of Q and let $f : P \to R$ be a cellular map. Then the adjunction space $R\, {}_f\!\sqcup Q$ can be given the structure of a cell-complex.*

Proof The open cells of $R\, {}_f\!\sqcup Q$ are to be the open cells of R and of $Q \setminus P$. In order to describe the characteristic maps, let $\bar{f} : Q \to R\, {}_f\!\sqcup Q$, $\bar{\imath} : R \to R\, {}_f\!\sqcup Q$ be the usual maps. Let h_λ, $\lambda \in \Lambda$, be the characteristic maps of Q; g_λ, $\lambda \in \mathrm{M}$, those of P; and k_ν, $\nu \in \mathrm{N}$, those of R. If e_ν is an open cell of R, then its characteristic map in $R\ {}_f\sqcup Q$ is to be $\bar{i}\,k_\nu$. If e_λ is an open cell of $Q \setminus P$, then its characteristic map in $R\ {}_f\sqcup Q$ is to be $\bar{f}\,h_\lambda$.

The conditions CS1 and 2 are obviously satisfied, as is CS3 for the maps $\bar{i}\,k_\nu$, $\nu \in \mathrm{N}$. Since f is cellular, so also is $\bar{f}$. Hence CS3 is satisfied for the maps $\bar{f}h_\lambda$, $\lambda \in \Lambda \setminus \mathrm{M}$. □

We now show that the product of cell complexes is a cell complex. Let P, Q be cell complexes with characteristic maps f_λ, $\lambda \in \Lambda$, g_μ, $\mu \in \mathrm{M}$ respectively. Since P is the union of disjoint open cells e_λ, and Q is the union of disjoint open cells e_μ, $P \times Q$ is the union of the disjoint sets $e_\lambda \times e_\mu$. But $e_\lambda \times e_\mu$ is an open $(n_\lambda + n_\mu)$-cell. So the n-skeleton $(P \times Q)^n$ is well defined and

$$P^m \times Q^n \subset (P \times Q)^{m+n}.$$

Suppose that e_λ is an open m-cell of P, and e_μ is an open n-cell of Q. Let $h_{mn} : \mathbf{E}^{m+n} \to \mathbf{E}^m \times \mathbf{E}^n$ be the homeomorphism constructed in 4.4.5 —we recall that

$$h_{mn}[\mathbf{S}^{m+n-1}] = \mathbf{S}^{m-1} \times \mathbf{E}^n \cup \mathbf{E}^m \times \mathbf{S}^{n-1}.$$

Let $k_{\lambda\mu}$ be the composite

$$\mathbf{E}^{m+n} \xrightarrow{h_{mn}} \mathbf{E}^m \times \mathbf{E}^n \xrightarrow{f_\lambda \times g_\mu} P \times Q.$$

Then $k_{\lambda\mu}$ maps $\mathbf{B}^{m+n}$ bijectively onto $e_\lambda \times e_\mu$ and $k_{\lambda\mu}[\mathbf{S}^{m+n-1}]$ is contained in $P^{m-1} \times Q^n \cup P^m \times Q^{n-1}$ which is itself contained in $(P \times Q)^{m+n-1}$. This shows that the $k_{\lambda\mu}$ are characteristic maps for a cell structure on $P \times Q$.

When exhibiting a space as a cell complex it is common practice to write it simply as the union of open cells, and say afterwards what is the attaching map. For example, the m-sphere has a cell structure with one m-cell and one 0-cell, and we therefore write

$$\mathbf{S}^m = e^0 \cup e^m.$$

Also, if e^m is an open m-cell of a cell complex Q with characteristic map $h : \mathbf{E}^m \to Q$, we make an abuse of language and call $h \mid \mathbf{S}^{m-1}, Q^{m-1}$ the attaching map of e^m. In particular, we say that the cell e^m of $\mathbf{S}^m$ is attached by the constant map.

Consider now $\mathbf{S}^m \times \mathbf{S}^n$. If we take $\mathbf{S}^m = e^0 \cup e^m$, $\mathbf{S}^n = e^0 \cup e^n$, then we can write

$$\mathbf{S}^m \times \mathbf{S}^n = e^0 \cup e^m \cup e^n \cup e^{m+n}.$$

Here both e^m and e^n are attached by constant maps, so that $e^0 \cup e^m \cup e^n$ is homeomorphic to $\mathbf{S}^m \vee \mathbf{S}^n$. Thus $\mathbf{S}^m \times \mathbf{S}^n$ is, up to homeomorphism, $(\mathbf{S}^m \vee \mathbf{S}^n) \cup e^{m+n}$. The attaching map of the $(m+n)$-cell is a map

$w : \mathbf{S}^{m+n-1} \to \mathbf{S}^m \vee \mathbf{S}^n$ called the *Whitehead product map*. [cf. Exercise 2 of 5.7 for a generalization of this map].

EXERCISES

1. Prove that the composite, and product, of cellular maps is cellular. Is the diagonal map $Q \to Q \times Q$ cellular?
2. Prove that if Q_1, Q_2 are cell-complexes, then so also is $Q_1 \sqcup Q_2$.
3. Let Q, R be cell-complexes, let $i : P \to Q$ be the inclusion of the subcomplex P of Q, and let $f : P \to R$ be cellular. Prove that if K is a cell complex and $i' : R \to K, f' : Q \to K$ are cellular maps such that $i'f = f'i$, then the unique map $g : R\ {}_f\!\sqcup Q \to K$ such that $g\bar{f} = f', g\bar{i} = i'$ is cellular.
4. Read an account of the classification of surfaces [for example in Cairns [11]] and give cell structures for the normal forms of surfaces.

The following exercises outline a part of the theory of infinite cell-complexes.

5. Let Q be a non-empty, not necessarily Hausdorff, space, and suppose given on Q a cell structure $\{h_\lambda\}_{\lambda \in \Lambda}$ in the sense of the definition on p. 124 except that Λ is not supposed finite. We say Q is a *CW-complex* if the following axioms hold: **CW1** A set $C \subset Q^n$ is closed in Q^n if and only if $C \cap Q^{n-1}$ is closed in Q^{n-1} and $h_\lambda^{-1}[C]$ is closed in $\mathbf{E}^n$ for each λ such that $n_\lambda = n$. **CW2** A set $C \subset Q$ is closed in Q if and only if $C \cap Q^n$ is closed in Q^n for each n.

Let $\Lambda_n = \{\lambda \in \Lambda : n_\lambda = n\}$ and let Λ_n have the discrete topology. Define $q : Q^{n-1} \sqcup (\Lambda_n \times \mathbf{E}^n) \to Q^n$ to be $x \rightsquigarrow x$ on Q^{n-1} and $(\lambda, e) \rightsquigarrow h_\lambda e$ on $\Lambda_n \times E^n$. Prove that CW1 is equivalent to: Q^n has the identification topology with respect to q.

In the following exercises, Q denotes a CW-complex with cell structure $\{h_\lambda\}_{\lambda \in \Lambda}$.

6. Prove that a function $f : Q^n \to Y$ is continuous if and only if $f \mid Q^{n-1}$ is continuous and $fh_\lambda : \mathbf{E}^n \to Y$ is continuous for each λ in Λ with $n_\lambda = n$. Prove that a function $f : Q \to Y$ is continuous if and only if $f \mid Q^n$ is continuous for each n. Deduce that $f : Q \to Y$ is continuous if and only if fh_λ is continuous for each λ in Λ.
7. Let C be a closed subset of Q and $g : C \to \mathbf{I}$ any map. Prove that g extends to a map $f : Q \to \mathbf{I}$ and thus prove that Q is normal. Prove that each point of Q is closed, and hence show that Q is Hausdorff. [You may assume the Tietze extension theorem.]
8. Prove that any compact subset of Q is contained in a finite union of cells e_λ. [If $C \subset Q$ meets an infinite number of cells, choose points c_i of C, $i = 1, 2, \ldots,$ which lie in distinct cells e_{λ_i}. Define $g\, c_i = 1/i$ and extend g over Q by Exercise 7.]
9. A CW-complex Q is said to be *locally finite* if each point x of Q has a neighbourhood N such that N is contained in a finite subcomplex of Q. Prove that Q is locally finite if and only if it is, *qua* topological space, locally compact.
10. Prove that a CW-complex is a k-space. Prove that if P, Q are CW-complexes then $P \times Q$, with the cell structure given in the present section, is a CW-complex if and only if $P \times Q$ is a k-space. Hence show that $P \times Q$ is a CW-complex if P or Q is locally finite. [Prove first that the space $P \times_W Q$ of Exercise 12 of 4.3 is always a CW-complex.]

NOTES

We have already mentioned the account of adjunction spaces in Hu [2], who deals also with the case of infinite cell complexes which were introduced in Whitehead [1]. For important applications of these complexes in algebraic topology, see Spanier [2], Dold [3]. The fact that 4.6.6 is false in general has led to a general discussion of product topologies in Brown [1, 2] (cf. also Steenrod [2]), and a theory of quasi-topological spaces in Spanier [2].

The description of a space by attaching cells is only one of many useful methods. Manifolds, for example, are often better described by means of attaching handles (see the survey articles, Wall [1], Smale [1]).

5. Projective and other spaces

5.1 Quaternions

In this section, we construct the algebra **H** of quaternions, and we show that **H** is a field. The word field is here used in a slightly more general sense than is usual, since the multiplication of **H** is non-commutative.

As a set, and in fact as a vector space over **R**, **H** is just $\mathbf{R} \times \mathbf{R}^3$. Thus, if $q \in \mathbf{H}$ then $q = (\lambda, x)$ where λ is real and x is in $\mathbf{R}^3$; we call λ the *real part* of q, x the *vector part* of q and write

$$\text{Re}\,(q) = \lambda, \qquad \text{Ve}\,(q) = x.$$

It is convenient to identify $(\lambda, 0)$ with λ and $(0, x)$ with x, and so to write

$$q = \lambda + x.$$

We shall, as far as possible, keep this convention of using Greek letters for the real part, and Roman letters for the vector part, of a quaternion. Thus the addition and scalar multiplication of the vector space structure of **H** is given by

$$(\lambda + x) + (\lambda' + x') = \lambda + \lambda' + x + x'.$$
$$\lambda'(\lambda + x) = \lambda'\lambda + \lambda' x.$$

We now define a distributive multiplication on **H**. To do this it suffices to define the product xy of vectors, since the product of quaternions $q = \lambda + x$, $r = \mu + y$ must then, by distributivity, be defined by

$$qr = \lambda\mu + \lambda y + \mu x + xy \tag{5.1.1}$$

However, xy will not be a vector—in terms of the usual scalar and vector product in $\mathbf{R}^3$ we set

$$xy = -x \cdot y + x \times y \tag{5.1.2}$$

The scalar and vector product of vectors are bilinear. It follows from this and (5.1.1), (5.1.2) that multiplication of quaternions is bilinear—that is, for any quaternions q, r, s and real number λ

$$q(r + \lambda s) = qr + \lambda(qs),$$
$$(r + \lambda s)q = rq + \lambda(sq). \tag{5.1.3}$$

One of the points of definition (5.1.2) is that it replaces the non-associative vector product by an associative product. By well-known rules

$$\begin{aligned} x(yz) &= x(-y\cdot z + y \times z) \\ &= -(y\cdot z)x - x\cdot y \times z + x \times (y \times z) \\ &= -x\cdot y \times z - (y\cdot z)x + (x\cdot z)y - (x\cdot y)z \end{aligned}$$

and a similar computation gives the same value for $(xy)z$. Hence $x(yz) = (xy)z$. It follows from this and (5.1.1) that for any quaternions q, r, s

$$q(rs) = (qr)s;$$

that is, multiplication of quaternions is associative.

A vector a of $\mathbf{R}^3$ is a unit vector if $a\cdot a = 1$; this is clearly equivalent to $a^2 = -1$. Two vectors a, b are orthogonal, that is $a\cdot b = 0$, if and only if ab is a vector. In such case

$$ab = a \times b = -b \times a = -ba.$$

Conversely, if $ab = -ba$, then by (5.1.2) $a\cdot b = 0$ and so a and b are orthogonal.

The ordered set a, b, c of vectors is said to form a *right-handed orthonormal system* if a, b, c are of unit length, are mutually orthogonal and $a \times b = c$.

5.1.4 *The set of vectors a, b, c is a right-handed orthonormal system if and only if*

(*a*) $a^2 = b^2 = c^2 = -1$,
(*b*) $abc = -1$.

Proof Condition (*a*) holds if and only if a, b, c are of unit length. If further a, b are orthogonal and $a \times b = c$, then

$$-1 = c^2 = (a \times b)c = abc.$$

Conversely, given (*a*) and (*b*), then

$$ab = -abc^2 = c.$$

Thus ab is a vector and so $a\cdot b = 0$, $a \times b = c$. Clearly c is orthogonal to a and b. □

Suppose that a, b, c form a right-handed orthonormal system. Then clearly

$$ab = c = -ba, \qquad bc = a = -cb, \qquad ca = b = -ac. \quad (5.1.5)$$

Now any vector x of $\mathbf{R}^3$ can be written uniquely as

$$x = x_1 a + x_2 b + x_3 c, \quad x_1, x_2, x_3 \in \mathbf{R}.$$

Therefore, the product xy of vectors is determined uniquely by 5.1.4(a), (5.1.5) and the condition of bilinearity – in fact, these rules imply the expression (5.1.2) for xy. Now any quaternion q can be written uniquely as

$$q = q_0 + q_1 a + q_2 b + q_3 c, \qquad q_i \in \mathbf{R}.$$

The rules given allow us to work out the product of q and $q' = q_0' + q_1' a + q_2' b + q_3' c$—we write out the complete formula for the one and only time:

$$\begin{aligned} qq' = {} & q_0 q_0' - q_1 q_1' - q_2 q_2' - q_3 q_3' \\ & + (q_0 q_1' + q_1 q_0' + q_2 q_3' - q_3 q_2')a \\ & + (q_0 q_2' + q_2 q_0' + q_3 q_1' - q_1 q_3')b \\ & + (q_0 q_3' + q_3 q_0' + q_1 q_2' - q_2 q_1')c. \end{aligned} \tag{5.1.6}$$

The real number 1 acts on the quaternions as identity – $1q = q1 = q$ for any quaternion q. We prove that for any non-zero quaternion q there is a quaternion q^{-1} such that $q\,q^{-1} = q^{-1}q = 1$.

Let $q = \lambda + x$. The *conjugate* of q is defined by

$$\bar{q} = \lambda - x.$$

If x, y are vectors then

$$\begin{aligned} \overline{xy} &= x \cdot y - x \times y \\ &= y \cdot x + y \times x \\ &= yx \\ &= \bar{y}\bar{x}. \end{aligned}$$

It follows from this and (5.1.1) that for any quaternions q, r,

$$\overline{qr} = \bar{r}\bar{q} \tag{5.1.7}$$

Notice also that $\overline{q + r} = \bar{q} + \bar{r}$ and that if λ is real then $\overline{\lambda q} = \lambda\bar{q}$.

Let $|x|$ denote, as usual, the square root of $x \cdot x$. Then, if $q = \lambda + x$, we have

$$q\bar{q} = \lambda^2 + x \cdot x = \bar{q}q. \tag{5.1.8}$$

So we may define the *modulus* of q to be

$$|q| = (q\bar{q})^{1/2}.$$

Thus $|q|$ is the Euclidean norm of q when q is considered as an element of $\mathbf{R}^4$ (under the identification $\mathbf{R} \times \mathbf{R}^3 = \mathbf{R}^4$).

Let $q, r \in \mathbf{H}$. Then

$$\begin{aligned} |qr|^2 &= (qr)(\overline{qr}) = qr\bar{r}\bar{q} \\ &= q\,|r|^2\bar{q} = q\bar{q}\,|r|^2 \\ &= |q|^2|r|^2. \end{aligned}$$

This proves the important rule

$$|qr| = |q|\,|r|.$$

Clearly, $q = 0$ if and only if $|q| = 0$. Therefore, if $q \neq 0$

$$q(\bar{q}/|q|^2) = (\bar{q}/|q|^2)q = 1;$$

so we write $q^{-1} = \bar{q}/|q|^2$, and call this quaternion the inverse of q.

The quaternions satisfy all the axioms for a field, the multiplication being in this case non-commutative. We also regard **H** as carrying the structures of vector space over **R**, the conjugation function $q \rightsquigarrow \bar{q}$, the modulus $q \rightsquigarrow |q|$ and also the topology induced by this modulus (this topology is, of course, the usual topology for $\mathbf{R} \times \mathbf{R}^3$). Thus **H**, like **R** and **C**, has a rich structure and this is its advantage and interest.

For any unit vector x the set of quaternions $\lambda + \mu x$ for $\lambda, \mu \in \mathbf{R}$ is a subfield of **H** isomorphic to the complex numbers under the function $\lambda + \mu x \rightsquigarrow \lambda + \mu i$. In particular, if we let i be the vector $(1, 0, 0)$ of **H** and identify i with the complex number i, then we can regard **C** as a subfield of **H**. We emphasize, however, that the elements of **C** do not commute with the elements of **H** since, if j is the vector $(0, 1, 0)$ of **H**, then $ij = -ji$.

There are two generalizations of the quaternions. The *Cayley numbers*, or *octonions*, **O**, are the elements of $\mathbf{R}^8$ with a distributive multiplication with identity which also satisfies $|xy| = |x|\,|y|$, $x, y \in \mathbf{R}^8$. Also, for any $x \neq 0$ in $\mathbf{R}^8$, there is an element x^{-1} such that $xx^{-1} = x^{-1}x = 1$. However, this multiplication is non-associative. An account of the Cayley numbers, and also a proof that the *only multiplications on* $\mathbf{R}^n$ *which are bilinear and satisfy* $|xy| = |x|\,|y|$, *are for* $n = 1, 2, 4, 8$, *and in these cases the resulting algebras are isomorphic to* **R**, **C**, **H**, *and* **O**, is given in A. Albert [1] (cf. also Kurosch [1]).

A multiplication on $\mathbf{R}^n$ is said to have *divisors of zero* if there are non-zero elements x, y of $\mathbf{R}^n$ such that $xy = 0$. It is true that $\mathbf{R}^n$ *has a bilinear multiplication with no divisors of zero only for* $n = 1, 2, 4,$ *or* 8. [cf. J. W. Milnor [1].]

A different type of generalization of the quaternions is the sequence of Clifford algebras C_n, $n \geqslant 1$. These are associative, but have divisors of zero for $n > 2$. They are closely related to orthogonal transformations of $\mathbf{R}^m$—the best account of them is in the paper of M. F. Atiyah, R. Bott, and A. Schapiro [1].

EXERCISES

1. Let q be a quaternion and let y be a vector orthogonal to Ve (q). Prove that $qy = y\bar{q}$.

2. Let q, r be quaternions such that $|q| = |r|$, $\operatorname{Re}(q) = \operatorname{Re}(r)$. Prove that there is a unit quaternion s such that $qs = sr$.
3. Let y be a unit vector. Prove that the mapping $\mathbf{R}^3 \to \mathbf{R}^3$ given by $x \rightsquigarrow yxy$ is reflection in the plane through the origin and perpendicular to y.
4. For any quaternion q, let $L_q : \mathbf{H} \to \mathbf{H}$ be the function $x \rightsquigarrow qx$. Prove that L_q is $\mathbf{R}$-linear and that

$$L_q L_{\bar{q}} = |q|^2 L_1.$$

Let L_q denote also the 4×4 real matrix of L_q with respect to a basis $1, a, b, c$ of $\mathbf{H}$ where a, b, c is a right-handed orthonormal set. Prove that $L_{\bar{q}}$ is the transpose of L_q and deduce that $\det(L_q) = |q|^4$.
5. Since $\mathbf{C}$ is a subfield of $\mathbf{H}$, we can regard $\mathbf{H}$ as a right vector space over $\mathbf{C}$. Prove that $\mathbf{H}$ is of dimension 2 over $\mathbf{C}$ with basis $1, j$ (where $j = (0, 1, 0)$). Prove that the function $L_q : \mathbf{H} \to \mathbf{H}$ of the previous exercise is $\mathbf{C}$-linear and that if $q = z + jw$ ($z, w \in \mathbf{C}$) then L_q has matrix

$$M_q = \begin{bmatrix} z & -\bar{w} \\ w & \bar{z} \end{bmatrix}$$

Prove that $M_{q'} M_q = M_{q'q}$. Hence show that $\mathbf{H}$ is isomorphic to the vector space $\mathbf{C}^2$ over $\mathbf{C}$ with multiplication given by

$$(z', w')(z, w) \doteq (z'z - \bar{w}'w, w'z + \bar{z}'w).$$

6. An integer n is said to be 4-square if there are integers n_1, n_2, n_3, n_4 such that $n = n_1^2 + n_2^2 + n_3^2 + n_4^2$. Prove that the product of 4-square integers is 4-square. [It may be proved also that any prime number is 4-square.]
7. Let a, b, c be quaternions. Prove that the equation

$$aq + qb = c$$

has a unique solution for q if $2a \operatorname{Re}(b) + a^2 + |b|^2 \neq 0$.

5.2 Normed vector spaces again

Let $\mathbf{K}$ denote one of the fields $\mathbf{R}$, $\mathbf{C}$, or $\mathbf{H}$. We write $d = d(\mathbf{K})$ for the dimension of $\mathbf{K}$ as a vector space over $\mathbf{R}$, so that

$$d(\mathbf{R}) = 1, \qquad d(\mathbf{C}) = 2, \qquad d(\mathbf{H}) = 4. \tag{5.2.1}$$

Let V be a vector space over $\mathbf{K}$. The theory of linear dependence, bases and dimension of V is usually given for the case $\mathbf{K}$ is commutative and V is a left vector space over $\mathbf{K}$. However, the change from left to right vector spaces is purely formal, and the usual proofs of the basis theorems do not use the commutativity of the field. We therefore assume this theory as known. The reason for using right vector spaces will be clear later when discussing the matrix of a linear transformation.

An example of a finite dimensional vector space over $\mathbf{K}$ is of course $\mathbf{K}^n$; the *standard ordered basis* of $\mathbf{K}^n$ consists of the elements $e^1, \ldots, e^n$ where the jth coordinate of e^i is δ_{ij}. Any n-dimensional vector space over $\mathbf{K}$ is isomorphic to $\mathbf{K}^n$.

Now $\mathbf{R}$ is a subfield of $\mathbf{K}$ and so $\mathbf{K}^n$ can be considered as a vector space over $\mathbf{R}$. The dimension of this vector space is nd: as vector spaces over $\mathbf{R}$, $\mathbf{K}$ is isomorphic to $\mathbf{R}^d$, whence, by associativity, $\mathbf{K}^n$ is isomorphic to $\mathbf{R}^{nd}$.

Let V be a vector subspace (over $\mathbf{K}$) of $\mathbf{K}^n$. If we consider $\mathbf{K}^n$ as a vector space over $\mathbf{R}$, then V is also a vector subspace over $\mathbf{R}$. We distinguish the two notions of a subspace as $\mathbf{K}$-*subspace* and $\mathbf{R}$-*subspace* respectively. This distinction is necessary since when $\mathbf{K} \neq \mathbf{R}$ not all $\mathbf{R}$-subspaces are $\mathbf{K}$-subspaces.

Thus $\mathbf{C}^2$ is a vector space of $\mathbf{R}$-dimension 4. Any $\mathbf{C}$-subspace of $\mathbf{C}^2$ is of $\mathbf{R}$-dimension 0, 2, or 4, but there are $\mathbf{R}$-subspaces of $\mathbf{C}^2$ of $\mathbf{R}$-dimension 1 or 3. It is not even true that every 2-dimensional $\mathbf{R}$-subspace is a $\mathbf{C}$-subspace. For example, the $\mathbf{R}$-subspace U spanned by $(1, 0)$ and $(0, i)$ contains $(1, 0)$ but not $(i, 0)$. Therefore, U is not a $\mathbf{C}$-subspace of $\mathbf{C}^2$.

For the rest of this section let V be a finite dimensional normed vector space over $\mathbf{K}$. Then V is also a finite dimensional normed vector space over $\mathbf{R}$; therefore, any two norms on V are equivalent [Exercise 11 of 3.5], and any linear function from V to a normed vector space W is continuous. (We are here assuming a theorem not proved in the text—a reader who does not wish to use this theorem may instead assume that V has, for example, the Euclidean norm with respect to some basis.)

If V is n-dimensional over $\mathbf{K}$, then V is nd-dimensional over $\mathbf{R}$, and so [cf. Exercise 11 of 2.8] there are homeomorphisms

$$E(V) \approx \mathbf{E}^{nd}, \quad B(V) \approx \mathbf{B}^{nd}, \quad S(V) \approx \mathbf{S}^{nd-1}.$$

The intersection of an m-dimensional $\mathbf{K}$-subspace U with $S(V)$ is the sphere $S(U)$, which we call a *great* $\mathbf{K}$-*sphere* in $S(V)$; here $S(U)$ is homeomorphic to $\mathbf{S}^{md-1}$.

Let U be a $\mathbf{K}$-subspace of V. We can, by choosing a basis for U and extending it over V, find a linear function $p : V \to V$ such that $pp = p$ and $\operatorname{Im} p = U$. If $q = 1 - p$, then q is linear and $U = q^{-1}[0]$. Therefore U is closed in V. Also, if $U^c = p^{-1}[0]$, then V is the direct sum $U \oplus U^c$—we call U^c a *complementary subspace* of U.

We end this section with a remark on the special case $V = \mathbf{K}^n$ with the Euclidean norm. An element $\varphi \in \mathbf{K}$ can be written as d-tuple $(\varphi_1, \ldots, \varphi_d)$ of elements of $\mathbf{R}$, and an element of $\mathbf{K}^n$ is an n-tuple $(\varphi^1, \ldots, \varphi^n)$ of elements of $\mathbf{K}$. The forementioned isomorphism $\Phi : \mathbf{K}^n \to \mathbf{R}^{nd}$ of vector spaces over R is simply

$$(\varphi^1, \ldots, \varphi^n) \rightsquigarrow (\varphi_1^1, \varphi_2^1, \ldots, \varphi_d^n).$$

Since $|\varphi|^2 = |\varphi_1|^2 + \cdots + |\varphi_d|^2$, this isomorphism is norm preserving, that is, $|\Phi(x)| = |x|$. Therefore, in this case, it is reasonable to *identify* $E(\mathbf{K}^n)$ with $\mathbf{E}^{nd}$, $B(\mathbf{K}^n)$ with $\mathbf{B}^{nd}$, and $S(\mathbf{K}^n)$ with $\mathbf{S}^{nd-1}$.

5.3 Projective spaces

Let V be an $(n + 1)$-dimensional normed vector space over $\mathbf{K}$. We write V_* for $V \setminus \{0\}$.

By a *line* in V_* is meant the set $U_* = U \setminus \{0\}$ for any $\mathbf{K}$-subspace U of V of dimension 1. More precisely, this is a line in V through the origin 0 but with 0 excluded; however we shall have no need of other lines. The exclusion of 0 is a technical convenience.

If $x \in V_*$, we write px for the line in V_* spanned by x, that is,

$$px = \{x\varphi : \varphi \in \mathbf{K}_*\}.$$

If l is any line in V_* and $x \in l$ then $l = px$.

The *projective space* of V is $P(V)$, the set of all lines in V_*. In particular, if $V = \mathbf{K}^{n+1}$ with the Euclidean norm, then $P(V)$ is written $P^n(\mathbf{K})$ and is called *n-dimensional projective space over* $\mathbf{K}$.

We relate this definition of $P^n(\mathbf{K})$ with the common definition in terms of homogeneous coordinates. It is often said that a point in $P^n(\mathbf{K})$ is given by homogeneous coordinates $[x_1, \ldots, x_{n+1}]$ where $x_i \in \mathbf{K}$, not all of the x_i are 0, and the convention is made that for any $\varphi \in \mathbf{K}_*$

$$[x_1\varphi, \ldots, x_{n+1}\varphi] = [x_1, \ldots, x_{n+1}].$$

This is the same as saying that the homogeneous coordinates of a point is the *set* of $(n + 1)$-tuples $(x_1\varphi, \ldots, x_{n+1}\varphi)$ for all $\varphi \in \mathbf{K}_*$. This set is simply px where $x = (x_1, \ldots, x_{n+1})$. Thus the two definitions coincide.

The *fundamental map* is the function

$$\begin{aligned} p : V_* &\to P(V) \\ x &\rightsquigarrow px \end{aligned}$$

Clearly, p is a surjection. We give $P(V)$ the identification topology with respect to V.

5.3.1 *The fundamental map is an open map.*

Proof Let A be open in V_*. We prove that $p^{-1}p[A]$ is open in V_*. Now $p^{-1}p[A]$ is the union of the sets $A\varphi = \{a\varphi : a \in A\}$ for all $\varphi \in \mathbf{K}_*$. So we have only to show that each such $A\varphi$ is open.

Let $x\varphi \in A\varphi$ so that $x \in A$. Then $B(x, r) \subset A$ for some $r > 0$. This implies that $B(x\varphi, r\,|\varphi|) \subset A\varphi$—hence $A\varphi$ is open. □

5.3.2 *If U is any* **K**-*subspace of V, then $P(U)$ is a closed subspace of $P(V)$.*

Proof Any line in U_* is also a line V_*—therefore $P(U)$ is a subset of $P(V)$. Further, the fundamental map $p_V : V_* \to P(V)$ sends U_* onto $P(U)$. Now U_* is a closed, saturated subset of V_*. Therefore by 4.3.1 (Corollary 1) the restriction $p_V \mid U_*, P(U)$ is an identification map. □

5.3.3 *Let U, W be* **K**-*subspaces of V such that $V = U \oplus W$ and W is of* **K**-*dimension* 1. *Then $U^\dagger = P(V) \setminus P(U)$ is homeomorphic to U.*

Proof Let w be a non-zero element of W. Each element of V can be written uniquely in the form $x + w\varphi$ where $x \in U$ and $\varphi \in \mathbf{K}$; and $x + w\varphi$ belongs to U if and only if $\varphi = 0$.

The inverse image of $U^\dagger$ under the fundamental map p is the open set $V_* \setminus U_*$. Let $r : V_* \setminus U_* \to U$ be defined by $r(x + w\varphi) = x\varphi^{-1}$, $x \in U$, $\varphi \in \mathbf{K}$. Then r is continuous and $r(v) = r(v\psi)$ for any $v \in V_* \setminus U_*$, $\psi \in \mathbf{K}_*$. Therefore r defines a map $r' : U^\dagger \to U$ such that $r'p = r$.

Let $q : U \to U^\dagger$ be defined by $x \rightsquigarrow p(x + w)$. Then q is continuous and $r'q = 1, qr' = 1$ [cf. Fig. 5.1.] □

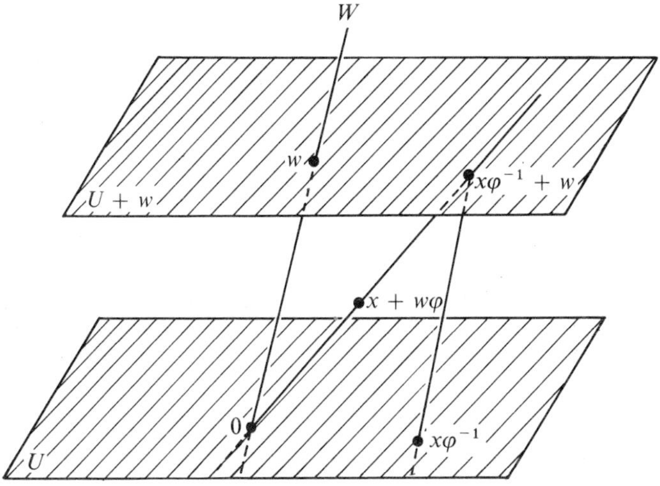

Fig. 5.1

Remark As a special case of the last result, let $V = \mathbf{K}^{n+1}$, $U = \mathbf{K}^n \times \{0\}$, and let us identify $P(U)$ with $P^n(\mathbf{K})$. We can then state 5.3.3 as: $P^n(\mathbf{K})$ is obtained from the n-dimensional space U by adding 'at infinity' a projective $(n-1)$-space $P^{n-1}(\mathbf{K})$. The reason for the words 'at infinity' here are that if φ tends to 0, then $p(x + w\varphi)$ tends to the point px of $P(U)$, while the corresponding point $x\varphi^{-1}$ of U 'tends to infinity'.

5.3.4 *$P(V)$ is a Hausdorff space.*

Proof By 5.3.3, it is sufficient to prove that any two (distinct) points of $P(V)$ belong to a set $U^\dagger$ for U a **K**-subspace of V of dimension n.

Let px, py be distinct points of $P(V)$. Let $x^1, \ldots, x^{n+1}$ be a basis for V such that $x^1 = x$, $x^2 = x + y$, and let U be the **K**-subspace spanned by $x^2, x^3, \ldots, x^n$. Clearly U is of dimension n, and $x, y \in V_* \setminus U_*$, so that $px, py \in U^\dagger$. □

Remark The preceding results extend on the whole to the case that V is of infinite dimension. The main difficulty that occurs is that a subspace U of V need not be closed in V, and that even if U is closed there may not be a continuous projection map $p : V \to V$ such that $pp = p$ and $\operatorname{Im} p = U$. However, it is a consequence of the Hahn–Banach theorem [cf. Simmons [1]] that, if U is a finite dimensional subspace of V, then such a projection map does exist, and this is the essential fact for the previous proofs.

5.3.5 *$P(V)$ is a path-connected, compact space.*

Proof If $\dim V = 1$, then $P(V)$ has only one point and so is path-connected. If $\dim V > 1$, then V_* is path-connected, and hence so also is V.

The sphere $S(V)$ is homeomorphic to $\mathbf{S}^{d(n+1)-1}$ and so is compact. If px is a point of $P(V)$ then $px = p(x\,|x|^{-1})$, and $x\,|x|^{-1} \in S(V)$. Therefore p maps $S(V)$ onto $P(V)$, whence $P(V)$ is compact. □

The map $h_V = p \mid S(V) : S(V) \to P(V)$ is called the *Hopf map of V*. Its role in describing the structure of projective spaces is shown by the next result.

5.3.6 *Let U be a* **K**-*subspace of V of dimension n. There is a homeomorphism*

$$P(V) \to P(U) {}_h\sqcup E(U).$$

Proof We have already seen that $U^\dagger = P(V) \setminus P(U)$ is homeomorphic to U; but it is difficult from this to see how $U^\dagger$ is attached to $P(U)$ since there is no more of U to describe the attaching. However there is a homeomorphism

$$g : B(U) \to U,$$
$$x \rightsquigarrow x(1 - \|x\|)^{-1}$$

Let $q : U \to U^\dagger$ be the map $x \rightsquigarrow p(x + w)$ of 5.3.3, where w is a point of $V \setminus U$. Then if $x \in B(U)$

$$\begin{aligned} qg(x) &= p(x(1 - \|x\|)^{-1} + w) \\ &= p(x + w(1 - \|x\|)), \end{aligned}$$

and this suggests considering the map

$$\begin{aligned} f : E(U) &\to P(V) \\ x &\rightsquigarrow p(x + w(1 - \|x\|)). \end{aligned}$$

Clearly, $f \mid S(U), P(U)$ is the Hopf map $x \rightsquigarrow p(x)$, while

$$f \mid B(U), P(V) \setminus P(U) = qg$$

which is a homeomorphism. Since $E(U)$ is compact, and $P(V)$ is Hausdorff, the theorem follows from 4.5.8.□

As an important corollary we have:

5.3.7 *Projective n-space $P^n(\mathbf{K})$ has a cell structure in which*

$$P^n(\mathbf{K}) = e^0 \cup e^d \cup e^{2d} \cup \cdots \cup e^{nd}.$$

Proof The proof is by induction. The result is clearly true for $n = 0$, since $P^0(\mathbf{K})$ consists of a single point. Suppose the result is true for $n - 1$. Let $U = \mathbf{K}^n \times \{0\}$, let $k : \mathbf{E}^{dn} \to E(U)$ be a homeomorphism, and let $f : E(U) \to P^n(\mathbf{K})$ be the map constructed in the proof of 5.3.6. Then we can take $fk : \mathbf{E}^{dn} \to P^n(\mathbf{K})$ as characteristic map for the dn-cell of $P^n(\mathbf{K})$. □

The case $n = 1$ of 5.3.7 is particularly simple. A cell complex $e^0 \cup e^d$ is homeomorphic to $\mathbf{S}^d$; hence there is a homeomorphism

$$P^1(\mathbf{K}) \to \mathbf{S}^d.$$

The attaching maps of the n-cell in 5.3.7 is the Hopf map $h : \mathbf{S}^{nd-1} \to P^{n-1}(\mathbf{K})$ (provided we identify $P^{n-1}(\mathbf{K})$ and $P(\mathbf{K}^n \times \{0\})$). The inverse images by h of points of $P^{n-1}(\mathbf{K})$ are exactly the intersection of $S(\mathbf{K}^n) = \mathbf{S}^{nd-1}$ with lines in $\mathbf{K}^n$; each such sphere is a great $\mathbf{K}$-sphere, and *a fortiori* a great $\mathbf{R}$-sphere, homeomorphic to $\mathbf{S}^{d-1}$; these spheres are disjoint and cover $\mathbf{S}^{nd-1}$. This is represented symbolically by the diagram

$$\mathbf{S}^{d-1} \to \mathbf{S}^{nd-1} \xrightarrow{h} P^{n-1}(\mathbf{K}).$$

In particular, let $n = 2$. We have a covering of $\mathbf{S}^{2d-1}$ by disjoint great $\mathbf{K}$-spheres homeomorphic to $\mathbf{S}^{d-1}$. The identification space determined by this covering is homeomorphic to $P^1(\mathbf{K})$ and so to $\mathbf{S}^d$. Therefore we have a diagram

$$\mathbf{S}^{d-1} \to \mathbf{S}^{2d-1} \to \mathbf{S}^d$$

in which the map $\mathbf{S}^{2d-1} \to \mathbf{S}^d$ is also called a Hopf map, and is written h.

The previous constructions can be followed through in detail to give a formula for $h : \mathbf{S}^{2d-1} \to \mathbf{S}^d$, but it is a rather complicated one. It is better to define maps a, b

$$\mathbf{S}^d \overset{b}{\leftarrow} \mathbf{E}^d \overset{a}{\to} P^1(\mathbf{K})$$
$$(s\varphi, 1 - 2t^2) \leftsquigarrow t\varphi \rightsquigarrow p(\varphi, \sqrt{(1 - t^2)})$$

where $s = 2t\sqrt{(1 - t^2)}$, $0 \leqslant t \leqslant 1$, $\varphi \in \mathbf{K}$ and $|\varphi| = 1$. Then $ax = ay$ if and only if $bx = by$. Therefore a and b define a homeomorphism $P^1(\mathbf{K}) \to \mathbf{S}^d$. We leave the reader to check that the composite $\mathbf{S}^{2d-1} \to P^1(\mathbf{K}) \to \mathbf{S}^d$ is

$$\begin{aligned} &\mathbf{S}^{2d-1} \to \mathbf{S}^d \\ &(\varphi, \psi) \rightsquigarrow (2\varphi\bar{\psi}, |\psi|^2 - |\varphi|^2) \end{aligned} \tag{5.3.8}$$

where φ, ψ are elements of $\mathbf{K}$ such that $|\varphi|^2 + |\psi|^2 = 1$.

In particular, let $\mathbf{K} = \mathbf{R}$, and identify $\mathbf{R}^2$ with $\mathbf{C}$. Then the map (5.3.8) is identical with $\mathbf{S}^1 \to \mathbf{S}^1$, $z \rightsquigarrow -iz^2$.

EXERCISES

1. Let V be a normed vector space over $\mathbf{K}$ and let x, y in V be of modulus 1. Prove that if $r = \text{dist}\,(x, py)$, and $s = r/(r + 2)$, then the two sets $B(x, s)$, $B(y, s)$ are mapped into disjoint open neighbourhoods of px, py by $p : V_* \to P(V)$.
2. If U is a $\mathbf{K}$-subspace of V of dimension $m + 1$, then $P(U)$ is called an m-dimensional projective subspace of $P(V)$. Prove that if P_1, P_2 are projective subspaces of $P(V)$, then, under the ordering of subspaces by inclusion, P_1, P_2 have a least upper bound $P_1 \vee P_2$ and a greatest lower bound $P_1 \wedge P_2$, and that

$$\dim P_1 + \dim P_2 = \dim P_1 \wedge P_2 + \dim P_1 \vee P_2.$$

3. Prove that if $V = \mathbf{R}^n$, then V_* is homeomorphic to $\mathbf{R} \times \mathbf{S}^{n-1}$.
4. Let $H_{n,p,q}$ denote the 'quadric' in $\mathbf{R}^n$ defined by the equation

$$x_1^2 + x_2^2 + \cdots + x_p^2 - x_{p+1}^2 - \cdots - x_{p+q}^2 = 1 \quad (p + q \leqslant n).$$

Prove that $H_{n,p,q}$ is homeomorphic to $\mathbf{S}^{p-1} \times \mathbf{R}^{n-p}$.
5. Let $H'_{n,p}$ denote the 'quadric' in $P^n(\mathbf{R})$ defined by the equation in homogeneous coordinates

$$x_0^2 + x_1^2 + \cdots + x_{p-1}^2 - x_p^2 - \cdots - x_n^2 = 0 \quad (1 \leqslant p \leqslant n).$$

Prove that $H'_{n,1}$ and $H'_{1,n}$ are homeomorphic to $\mathbf{S}^{n-1}$; for $2 \leqslant p \leqslant n - 1$, $H'_{n,p}$ is homeomorphic to the subspace obtained by identifying each point (y, z) of $\mathbf{S}^{p-1} \times \mathbf{S}^{n-p}$ with its opposite $(-y, -z)$. Prove that every point of $H'_{n,p}$ has a neighbourhood homeomorphic to $\mathbf{R}^{n-1}$. Prove also that $H'_{3,2}$ is homeomorphic to $\mathbf{S}^1 \times \mathbf{S}^1$.
6. A *topological group* consists of a topological space G and a group structure on

G such that the function $G \times G \to G$, $(x, y) \rightsquigarrow xy^{-1}$, is continuous. Prove that $\mathbf{K}_*$ and $S(\mathbf{K})$ are topological groups.

7. An *action* (or *operation*) of the topological group G on the right of a space X is a function $X \times G \to X$, written $(x, g) \rightsquigarrow xg$, such that if e is the identity of G then (i) $xe = x$ for all x in X, (ii) if $x \in X$, $g, h \in G$, then $x(gh) = (xg)h$. Given such an action, the *orbit space* of X is the space X/G whose elements are the equivalence classes under the relation $x \sim y \Leftrightarrow$ there is a g in G such that $xg = y$; the topology of X/G is the identification topology with respect to the projection $p : X \to X/G$. Prove that p is an open map. Show that 5.3.1 is a consequence (when $G = \mathbf{K}_*$, $X = \mathbf{V}_*$).

8. Let U, $U^\dagger$ be as in 5.3.3. Prove that there is a commutative diagram

$$\begin{array}{ccc} U^\dagger \times \mathbf{K}_* & \xrightarrow{s} & V_* \\ \Big\downarrow p_1 & & \Big\downarrow p \\ U^\dagger & \xrightarrow[i]{} & P(V) \end{array}$$

in which s is a homeomorphism, i is the inclusion and p_1 is the projection of the product.

5.4 Isometries of inner product spaces

The object of this section is to give a brief description of some spaces of isometries—these spaces are among the central objects of mathematics.

Since there are very good treatments of inner product spaces available (e.g., Halmos [1], Hoffman–Kunze [1], Simmons [1]) we shall state without proofs the results that we need. But first we want to record a remark about matrices over a non-commutative field.

Let V be a (right) vector space over $\mathbf{K}$, let $f : V \to V$ be a linear function and let $v^1, \ldots, v^n$ be a basis for V. Then for $i = 1, \ldots, n$ we can write

$$fv^j = \sum_{i=1}^{n} v^i f_{ij}$$

where the elements f_{ij} belong to $\mathbf{K}$. The function $(i, j) \rightsquigarrow f_{ij}$ is an $n \times n$ matrix $\tilde{f}$ over $\mathbf{K}$. Suppose further $g : V \to V$ is linear with matrix $\tilde{g}$. Then it is easy to check that

$$gf\,v^j = \sum_{h,i} v^h g_{hi} f_{ij}$$

and it follows that $\widetilde{gf}$, the matrix of gf, is the product $\tilde{g}\tilde{f}$ of the matrices of g and f. This result is false for left vector spaces over a non-commutative field.

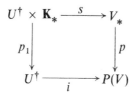

Let V be a (right) vector space over $\mathbf{K}$. An *inner product* on V is a function

$$V \times V \to \mathbf{K}$$
$$(x, y) \rightsquigarrow (x \mid y)$$

satisfying the following axioms: (all $x, y, z \in V$, $\varphi \in \mathbf{K}$)

IPS 1 $(x \mid y + z) = (x \mid y) + (x \mid z)$,
IPS 2 $(x \mid y\,\varphi) = (x \mid y)\varphi$,

IPS 3 $\overline{(x \mid y)} = (y \mid x)$,
IPS 4 $x \neq 0 \Rightarrow (x \mid x) > 0$.

These are the usual axioms for an inner product with due allowance made for the fact that we have right instead of left vector spaces. It is easy to prove from the axioms that $|\;|$ defined by $|x| = \sqrt{(x \mid x)}$ is a norm on V. The *standard inner product* on $\mathbf{K}^n$ is defined by

$$(x \mid y) = \sum_{i=1}^{n} \overline{\varphi}_i \psi_i$$

for $x = (\varphi_1, \ldots, \varphi_n)$, $y = (\psi_1, \ldots, \psi_n)$.

Let $x, y \in V$. We say x, y are *orthogonal* if $(x \mid y) = 0$. A subset X of V is an *orthogonal set* if any two distinct elements of X are orthogonal; and X is an *orthonormal set* if it is orthogonal and each x in X satisfies $|x| = 1$. In particular, an *orthonormal basis* for a subspace U of V is an orthonormal set which is also a basis for U. If X is a subset of V, then $X^{\perp}$ is the subspace of V of all y such that $(y \mid x) = 0$ for all x in X.

We now state without proof the basic results on inner product spaces.

5.4.1 If U is a vector subspace of V with a finite orthonormal basis, then V is the direct sum of U and $U^{\perp}$.

5.4.2 If V is finite dimensional then V has an orthonormal basis.

Let W be another inner product space. A function $f : V \to W$ is called an *isometry* if (*a*) f is a (linear) isomorphism of vector spaces, (*b*) for all x, y in V, $(fx \mid fy) = (x \mid y)$.

5.4.3 Let $f : V \to W$ be a linear isomorphism. Then f is an isometry $\Leftrightarrow$ $|fx| = |x|$ for all $x \in V$.

5.4.4 Let $v^1, \ldots, v^n$ be an orthonormal basis for V. (*a*) If $f : V \to W$ is linear then $fv^1, \ldots, fv^n$ is an orthonormal basis for W if and only if f is an isometry. (*b*) If $w^1, \ldots, w^n$ is an orthonormal basis for W, then there is a unique isometry $f : V \to W$ such that $fv^i = w^i$, $i = 1, \ldots, n$.

5.4.5 If V is n-dimensional, then there is an isometry $f : V \to \mathbf{K}^n$.

5.4.6 Let $f : V \to V$ be linear, let $v^1, \ldots, v^n$ be an orthonormal basis for V and let A be the matrix of f with respect to this basis. Then f is an isometry if and only if $\bar{A}^t A = I$ (where I is the unit matrix).

The set of all isometries $V \to V$ is written $G(V)$—it is clear that $G(V)$ is a group under composition.

Let $M_n(\mathbf{K})$ denote the (right) vector space of all $n \times n$ matrices over $\mathbf{K}$. Then $M_n(\mathbf{K})$ is isomorphic to $\mathbf{K}^{n^2}$ and this isomorphism determines an inner product structure on $M_n(\mathbf{K})$—thus $M_n(\mathbf{K})$ becomes a topological space. With this topology, the product function $M_n(\mathbf{K}) \times M_n(\mathbf{K}) \to M_n(\mathbf{K})$ which sends $(A, B) \rightsquigarrow AB$ is continuous since it is $\mathbf{R}$-bilinear [Example 7 of 2.8].

The group $G(\mathbf{K}^n)$ will be identified with the topological subspace of $M_n(\mathbf{K})$ of matrices A such that $\bar{A}^t A = I$. The group $G(\mathbf{K})$ is particularly simple. If $[\varphi]$ is a 1×1 matrix, then

$$[\varphi] \in G(\mathbf{K}) \Leftrightarrow \bar{\varphi}\varphi = 1 \Leftrightarrow |\varphi| = 1$$

Therefore $G(\mathbf{K})$ is homeomorphic to the sphere $S(\mathbf{K}) = \mathbf{S}^{d-1}$

If $\mathbf{K}$ is commutative, that is, if $\mathbf{K} = \mathbf{R}$ or $\mathbf{C}$, the determinant function $\det : M_n(\mathbf{K}) \to \mathbf{K}$ is defined. If $A \in G(\mathbf{K})$, then $\bar{A}^t A = I$ and it follows easily that $\varphi = \det A$ satisfies $\bar{\varphi}\varphi = 1$. Hence det defines a morphism of groups

$$\det : G(\mathbf{K}^n) \to S(\mathbf{K}).$$

The kernel of this morphism, that is the set of A in $G(\mathbf{K}^n)$ with determinant $+1$, is called the *special* group of isometries and is written $SG(\mathbf{K}^n)$.

The groups $G(\mathbf{K}^n)$ are given particular names in the three cases $\mathbf{K} = \mathbf{R}, \mathbf{C}$, or $\mathbf{H}$.

(*a*) $G(\mathbf{R}^n)$ is called the *orthogonal group* and is written $\mathrm{O}(n)$. In this case det is a morphism $\mathrm{O}(n) \to \{-1, +1\}$, and the special orthogonal group $\mathrm{SO}(n)$ is also called the group of rotations of $\mathbf{R}^n$.

(*b*) $G(\mathbf{C}^n)$ is called the *unitary group* and is written $\mathrm{U}(n)$. In this case, det is a morphism $\mathrm{U}(n) \to \mathbf{S}^1$ whose kernel is $\mathrm{SU}(n)$.

(*c*) $G(\mathbf{H}^n)$ is called the *symplectic group* and is written $\mathrm{Sp}(n)$. (There is another family of groups, the group of *spinors* and *special spinors*—these groups are written $\mathrm{Pin}(n)$ and $\mathrm{Spin}(n)$ and are closely related to $\mathrm{O}(n)$ and $\mathrm{SO}(n)$ respectively; cf. Atiyah–Bott–Schapiro [1].)

These three families of groups are related. In fact $\mathbf{C}^n$ is isomorphic as *normed vector space over* $\mathbf{R}$ to $\mathbf{R}^{2n}$; hence, by 5.4.3, any isometry $f : \mathbf{C}^n \to \mathbf{C}^n$ defines an isometry $\lambda_n f : \mathbf{R}^{2n} \to \mathbf{R}^{2n}$. So λ_n defines an injective morphism of groups

$$\lambda_n : \mathrm{U}(n) \to \mathrm{O}(2n).$$

5.4.7 *There is an isomorphism of groups*

$$\mu : \mathbf{S}^1 \to \mathrm{SO}(2).$$

Proof Let μ be the composite of λ_1 with the isomorphism $\mathbf{S}^1 \to \mathrm{U}(1)$ defined previously. Let $\mathbf{R}^2$ have $\mathbf{R}$-basis the complex numbers 1, i. If $z \in \mathbf{S}^1$, then $\mu(z)$ is the function $\mathbf{R}^2 \to \mathbf{R}^2$, $x \rightsquigarrow zx$. So if $z = \cos\alpha + i\sin\alpha$, we find that $\mu(z)$ has matrix

$$\begin{bmatrix} \cos\alpha & -\sin\alpha \\ \sin\alpha & \cos\alpha \end{bmatrix}$$

On the other hand, it is easily checked (using the equations $A^{-1} = A^t$, $\det A = 1$) that any element A of SO(2) must be of this form. □

A similar argument works for **H** and **C**: since $\mathbf{H}^n$ is isomorphic as normed vector space over **C** to $\mathbf{C}^{2n}$, there is an injective morphism of groups $\mathrm{Sp}(n) \to \mathrm{U}(2n)$.

5.4.8 *There is an isomorphism of groups*

$$\mathbf{S}^3 \to \mathrm{SU}(2).$$

The proof is similar to that of 5.4.7 and is left to the reader.

Next we show how to determine SO(3) in terms of $\mathbf{S}^3$. We first need to show that every element of SO(3) is a rotation about an axis.

5.4.9 *Let $f \in \mathrm{SO}(3)$. Then there is an ortho-normal basis a, b, c for $\mathbf{R}^3$ such that with respect to this basis f has matrix*

$$\begin{bmatrix} 1 & 0 & 0 \\ 0 & \cos\alpha & -\sin\alpha \\ 0 & \sin\alpha & \cos\alpha \end{bmatrix}.$$

Proof Let A be the matrix of f with respect to some orthonormal basis. Then $A^tA = I$ and $\det A = 1$. Hence $A^t(A - I) = I - A$ and so

$$\begin{aligned} \det(A - I) &= \det A^t(A - I) \\ &= \det(I - A) \\ &= -\det(A - I). \end{aligned}$$

Therefore $\det(A - I) = 0$. Hence there is a non-zero element a of $\mathbf{R}^3$, such that $(A - I)a = 0$, that is, $fa = a$. Replacing a by $a/|a|$ if necessary, we may suppose $|a| = 1$.

Let b, c be an orthonormal basis of the plane U through the origin and orthogonal to a. Since f is an isometry and $fa = a$, we have $f[U] \subset U$. Therefore $g = f \mid U, U$ is defined and is an isometry. The equation $fa = a$

implies that $\det g = \det f = 1$. Therefore g has matrix with respect to b, c

$$\begin{bmatrix} \cos\alpha & -\sin\alpha \\ \sin\alpha & \cos\alpha \end{bmatrix}$$

and the matrix for f follows. □

The form of the matrix of 5.4.9 shows that f is a rotation through angle α about the axis a.

5.4.10 *There is a surjective morphism*

$$v : \mathbf{S}^3 \to \mathrm{SO}(3)$$

with kernel the quaternions $+1, -1$.

Proof Let q be a unit quaternion, and let $x \in \mathbf{R}^3$. Define

$$r = q\,x\,\bar{q}.$$

Then

$$\begin{aligned} 2\,\mathrm{Re}\,(r) &= r + \bar{r} \\ &= q\,x\,\bar{q} + q\,\bar{x}\,\bar{q} \\ &= q\,(x + \bar{x})\,\bar{q} \\ &= 0 \quad \text{since } \mathrm{Re}\,(x) = 0. \end{aligned}$$

Therefore r is a vector. It follows that $x \rightsquigarrow q\,x\,\bar{q}$ defines a function $v_q : \mathbf{R}^3 \to \mathbf{R}^3$—clearly v_q is linear. It is also an isomorphism since it has inverse $x \rightsquigarrow \bar{q}\,x\,q$.

Let $x \in \mathbf{R}^3$. Then

$$|v_q x| = |q\,x\,\bar{q}| = |q|\,|x|\,|\bar{q}| = |x|.$$

Therefore v_q is an isometry.

In fact, we can find a formula for v_q.

Suppose $q = \cos\alpha + a\sin\alpha$ where a is a unit vector. Let a, b, c be a right-handed orthonormal system in $\mathbf{R}^3$. Then an easy check shows that

$$\begin{aligned} v_q a &= a \\ v_q b &= b\cos 2\alpha + c\sin 2\alpha \\ v_q c &= -b\sin 2\alpha + c\cos 2\alpha. \end{aligned} \qquad (*)$$

Therefore v_q is rotation about the axis a through an angle 2α. In particular, v_q belongs to SO(3) and any element of SO(3) is of the form v_q for some q in $\mathbf{S}^3$.

The function v is a homomorphism, since if $q, r \in \mathbf{S}^3$, then

$$\begin{aligned} v_{qr}\,(x) &= qr\,x\,\overline{qr} \\ &= qr\,x\,\bar{r}\bar{q} \\ &= v_q v_r(x). \end{aligned}$$

It follows that $v_q = v_r \Leftrightarrow v_{qr^{-1}} = 1$. If $q = \cos\alpha + a\sin\alpha$, then it follows from (*) that $v_q = 1$ if and only if $\alpha = n\pi, n \in \mathbf{Z}$. Therefore, $v_q = 1 \Leftrightarrow q = \pm 1$, and so $v_q = v_r \Leftrightarrow q = \pm r$. □

5.4.10 (*Corollary 1*) *There is a homeomorphism*

$$\mathrm{SO}(3) \to P^3(\mathbf{R}).$$

Proof The space SO(3) is Hausdorff since it is a subspace of the normed vector space $M_3(\mathbf{R})$. Since $\mathbf{S}^3$ is compact, and v is continuous, the space SO(3) is obtained, like $P^3(\mathbf{R})$, by identifying antipodal points of $\mathbf{S}^3$. □

This result gives, of course, a cell structure for SO(3). In fact, cell structures have been given for all the coset spaces $G(\mathbf{K}^n)/G(\mathbf{K}^m)$ $0 \leqslant m < n$; for an excellent account of this, see Steenrod–Epstein [1].

EXERCISES

1. Prove the assertions 5.4.1–5.4.6.
2. Prove that $G(\mathbf{K}^n)$ is a closed, bounded subset of $M_n(\mathbf{K})$. Deduce that $G(\mathbf{K}^n)$ is compact. [Use the equation $A^{-t}A = I$.]
3. Let a_n be the point $(0, \ldots, 0, 1)$ of $\mathbf{K}^n$. Define

$$\rho_n : G(\mathbf{K}^n) \to S(\mathbf{K}^n), \quad f \rightsquigarrow fa_n.$$

Prove that ρ_n is continuous and surjective, and that $\rho_n^{-1}[a_n]$ is a subgroup of $G(\mathbf{K}^n)$ isomorphic to $G(\mathbf{K}^{n-1})$. Prove also that if $\mathbf{K}$ is commutative then ρ_n restricts to a continuous surjection $q_n : SG(\mathbf{K}^n) \to S(\mathbf{K}^n)$ such that $q_n^{-1}[a_n]$ is a subgroup of $SG(\mathbf{K}^n)$ isomorphic to $SG(\mathbf{K}^{n-1})$.
4. Prove that the groups SO(n), SU(n), U(n), Sp(n) are connected. [Use the previous result and induction.]
5. Let U, V be two vector subspaces of $\mathbf{K}^n$ of the same dimension. Prove that there is an isometry $\sigma : \mathbf{K}^n \to \mathbf{K}^n$ such that $\sigma[U] = V$. Prove also that if $x, y \in \mathbf{S}^{n-1}$, then there is a rotation σ of $\mathbf{R}^n$ such that $\sigma x = y$.
6. Let U be a subspace of $\mathbf{R}^n$ of dimension $(n - 1)$, so that $\mathbf{R}^n = U \oplus U^\perp$. The function $x + y \rightsquigarrow x - y$ $(x \in U, y \in U^\perp)$ is called *reflection* in the hyperplane U. Prove that such a reflection is an isometry of $\mathbf{R}^n$, and also that any isometry of $\mathbf{R}^n$ with U as its set of fixed points is reflection in U.
7. Prove that all elements of O(n) are products of reflections. [Use induction.]

5.5 Simplicial complexes

Although cell complexes form a highly useful class of spaces, they are not restrictive enough for many purposes. In this section we discuss the simplicial complexes—these are cell complexes but with a particular kind of attaching map of the cells. For applications of simplicial complexes we refer the reader to Hilton–Wylie [1], Lefschetz [1], Cairns [1], or Spanier

[3]. The purpose of this section is mainly to link our results with the theories described in these books.

A subset A of $\mathbf{R}^k$ is called an *n-simplex* if there are points $a_0, \ldots, a_n$ in A such that any point a of A can be written *uniquely* in the form

$$a = t_0 a_0 + \cdots + t_n a_n$$

where $$0 \leqslant t_i,\ t_0 + \cdots + t_n = 1.$$

In such case A is said to be *spanned* by $a_0, \ldots, a_n$ and the numbers $t_0, \ldots, t_n$ are called the *barycentric coordinates* of the point $a = t_0 a_0 + \cdots + t_n a_n$. (The reason for the latter name is that a is the centre of gravity, or barycentre, of the particle system with weight t_i at a_i, $i = 1, \ldots, n$.) The points $a_0, \ldots, a_n$ are called the *vertices* of A—we will see below that the vertices are determined by the set A.

5.5.1 *Let A be an n-simplex with vertices $a_0, \ldots, a_n$. Let $b_0, \ldots, b_m$ be points of A and let $s_0, \ldots, s_m$ be positive real numbers whose sum is* 1. *Then the point $b = s_0 b_0 + \cdots + s_m b_m$ belongs to A, and $b = a_{i_0}$ if and only if $b_0 = \cdots = b_m = a_{i_0}$.*

Proof Since $b_j \in A$, we can write

$$b_j = \sum_{i=0}^{n} t_{ji} a_i, \qquad t_{ji} \geqslant 0, \qquad \sum_i t_{ji} = 1.$$

It follows that $$b = \sum_{i,j} s_j t_{ji} a_i.$$

But $s_j t_{ji} \geqslant 0$ and

$$\sum_{i,j} s_j t_{ji} = \sum_j s_j \sum_i t_{ji} = \sum_j s_j = 1.$$

Therefore $b \in A$.

Suppose that $b = a_{i_0}$. Then for $i \neq i_0$

$$\sum_j s_j t_{ji} = 0.$$

Since the s_j are positive, it follows that $t_{ji} = 0$, $i \neq i_0$, and so

$$b_0 = \cdots = b_m = a_{i_0}.$$

Conversely, this last condition clearly implies $b = a_{i_0}$. □

5.5.1 (*Corollary 1*) *If A is a simplex spanned by both $a_0, \ldots, a_n$ and $b_0, \ldots, b_m$, then $m = n$ and the b_j are a rearrangement of the a_i.*

Proof Since $b_0, \ldots, b_m$ are vertices of A, we can write uniquely

$$a_i = \sum_j s_j b_j, \qquad s_i \geqslant 0, \quad s_0 + \cdots + s_m = 1.$$

By 5.5.1, if $s_{j_i} \neq 0$ then $b_j = a_i$. Hence there is a unique j_i such that $b_{j_i} = a_i$. □

If A is an n-simplex then n is determined by A—we call n the *dimension* of A.

Another obvious consequence of 5.5.1 is that an n-simplex A is convex, and is in fact the smallest convex subset of $\mathbf{R}^k$ containing the vertices of A.

Let A be the simplex with vertices $a_0, \ldots, a_n$. A *face* of A is a simplex spanned by any subset of $\{a_0, \ldots, a_n\}$. For example, the vertices $a_0, \ldots, a_{i-1}, a_{i+1}, \ldots, a_n$ span an $(n-1)$-simplex which is often written $\partial_i A$—it is the face opposite the i'th vertex.

Suppose that $a = t_0 a_0 + \cdots + t_n a_n$, where the t_i are barycentric coordinates, and suppose $t_0 \neq 1$. Let

$$a' = (1 - t_0)^{-1}(t_1 a_1 + \cdots + t_n a_n).$$

Then a' belongs to the face $\partial_0 A$ of A opposite a_0 and

$$a = t_0 a_0 + (1 - t_0)a'.$$

Thus the points of A are the points of the line segments joining a_0 to points of the $(n-1)$-simplex $\partial_0 A$. This gives an inductive description of an n-simplex and justifies the following pictures of n-simplexes for $n \leqslant 3$.

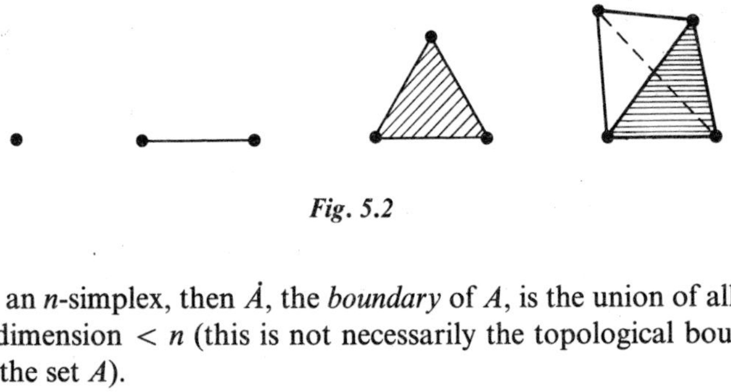

Fig. 5.2

If A is an n-simplex, then $\dot{A}$, the *boundary* of A, is the union of all faces of A of dimension $< n$ (this is not necessarily the topological boundary Bd A of the set A).

5.5.2 *If A is an n-simplex, then there is a homeomorphism $A \to \mathbf{E}^n$ which maps $\dot{A}$ homeomorphically onto $\mathbf{S}^{n-1}$.*

Proof We first construct a homeomorphism from A to an n-simplex B in $\mathbf{R}^n$. Let B have vertices $b_0, \ldots, b_n$ where $b_0 = n^{-1/2}(-1, \ldots, -1)$ and $b_1, \ldots, b_n$ is the standard basis of $\mathbf{R}^n$ [cf. Fig. 5.3]. If A has vertices $a_0. \ldots, a_n$ define

$$f : A \to B, \qquad t_0 a_0 + \cdots + t_n a_n \rightsquigarrow t_0 b_0 + \cdots + t_n b_n;$$

clearly f is a homeomorphism which maps $\dot{A}$ onto $\dot{B}$.

A homeomorphism $g : B \to \mathbf{E}^n$ is now constructed by radial projection from the origin. □

It follows from 5.5.2 that instead of using $(\mathbf{E}^n, \mathbf{S}^{n-1})$ as the standard model from which to construct cell complexes, we could use instead $(A, \dot{A})$ for any n-simplex A—this is the method we shall use when showing that a simplicial complex can be given a cell structure.

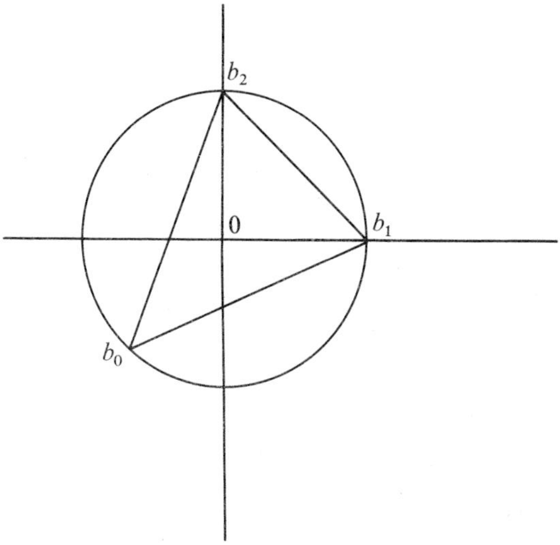

Fig. 5.3

One further point will explain the utility of simplicial methods. A map $f : A \to B$ from an n-simplex A to an m-simplex B is called *linear* if the barycentric coordinates of fa are linear functions of the barycentric coordinates of a in A. It is obvious that an m-simplex is *linearly* homeomorphic to an n-simplex if and only if $m = n$. The corresponding statement without the word linearly is true but is much more difficult to prove—it constitutes in fact the Invariance of Dimension.

A *simplicial complex* K is a finite set of simplices of $\mathbf{R}^k$ (for some k) with the following properties:

SIM 1 If $A \in K$ then any face of A belongs to K.

SIM 2 Any two simplices of K meet in a common face (possibly empty).

The space of K is $|K|$, the union of all the simplices of K, with the topology as a subspace of $\mathbf{R}^k$.

In Fig. 5.4, (a) and (b) are pictures of simplicial complexes but (c) is not; however (c) can be made into a simplicial complex by adding extra simplices.

If K is a simplicial complex, then K^n is the subset of K of all simplices of dimensions $\leqslant n$.

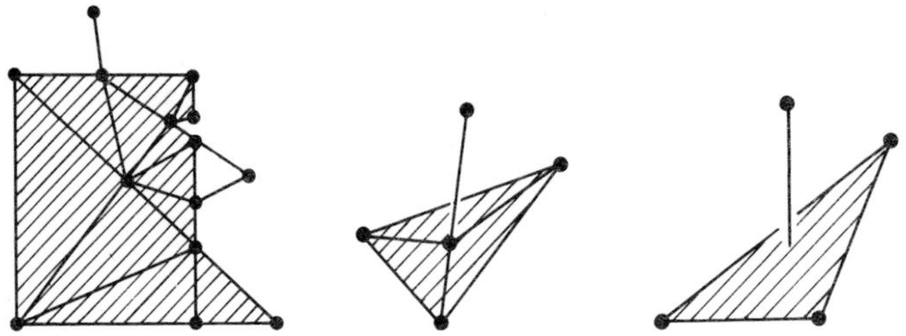

(*a*) (*b*) (*c*)

Fig. 5.4

5.5.3 *If K is a simplicial complex, then $|K|$ has a cell structure whose open cells are $A \setminus \dot{A}$ for each simplex A of K.*

Proof It is clear from 5.5.2 that $A \setminus \dot{A}$ is an open cell, for each simplex A of K. Also if A is of dimension n, then we can take A itself as a model of the n-cell; the inclusion $A \to |K|$ is then a characteristic map for A which maps $\dot{A}$ into the $(n-1)$-skeleton of $|K|$. □

5.6 Bases and sub-bases for open sets; initial topologies

The aim of this section is the construction of the initial topology on X with respect to a family of functions from X; this important construction is a kind of 'dual' to that of final topologies in chapter 4. Before proceeding with the definitions we need some of the theory of bases and sub-bases.

If $\mathscr{B}$ is a base for the neighbourhoods of a topological space X then $\mathscr{B}$ defines the topology on X completely (since a subset N of X is a neighbourhood of a point x of X if and only if N contains a set B of $\mathscr{B}(x)$). We now consider the converse problem: let X be a set and $\mathscr{B} : x \rightsquigarrow \mathscr{B}(x)$ a function assigning to each x in X a non-empty set of subsets of X; under what conditions is $\mathscr{B}$ a base for the neighbourhoods of a topology on X?

5.6.1 *For a function $\mathscr{B}$ as above, $\mathscr{B}$ is a base for the neighbourhoods of a topology on X if and only if the following conditions hold:*

(*a*) *If $x \in X$ and $B \in \mathscr{B}(x)$ then $x \in B$.*

(*b*) *If $B, B' \in \mathscr{B}(x)$ then $B \cap B'$ contains a set of $\mathscr{B}(x)$.*

(*c*) *If $B \in \mathscr{B}(x)$ then B contains a set M such that $x \in M$ and also if $y \in M$ then M contains a set B' of $\mathscr{B}(y)$.*

Proof The topology defined by $\mathscr{B}$, if it exists, is such that N is a neighbourhood of x if and only if N contains a set of $\mathscr{B}(x)$. Then (*a*), (*b*), and (*c*) are simply restatements of the Axioms N1, N3, and N4 respectively for neighbourhoods. (Notice, by the way, that (*a*) is a consequence of (*c*).) □

In the case when each element of $\mathscr{B}(x)$ is an open neighbourhood of x in the topology defined by $\mathscr{B}$, then we call $\mathscr{B}$ an *open base* for the neighbourhoods of X: the conditions for $\mathscr{B}$ to be an open base for the neighbourhoods of a topology on X are clearly 5.6.1 (*a*), (*b*), and also (*c'*): *if* $B \in \mathscr{B}(x)$ *then* $B \in \mathscr{B}(y)$ *for each* y *in* B.

We now consider similar questions for open sets.

5.6.2 *Let X be a topological space and* $\mathscr{U}$ *an open cover of X. The following conditions are equivalent:*

(*a*) *each open set of X is a union of elements of* $\mathscr{U}$,
(*b*) *for each x in X, the set* $\mathscr{U}(x)$ *of elements of* $\mathscr{U}$ *which contain x is a base for the neighbourhoods of x.*

Proof (*a*) ⇒ (*b*) If N is a neighbourhood of x, then N contains an open set U such that $x \in U$. Since U is a union of elements of $\mathscr{U}$ there is set U_x in $\mathscr{U}$ such that $x \in U_x$. Hence $x \in U_x \subset N$.
(*b*) ⇒ (*a*) If U is an open set of X, then for each x in U there is an element U_x of $\mathscr{U}(x)$ such that $x \in U_x \subset U$. Hence $U = \bigcup_{x \in U} U_x$. □

If $\mathscr{U}$ is an open cover of X satisfying either of the conditions (*a*), (*b*) of 5.6.2, then we say $\mathscr{U}$ is a *base for the open sets* of the topology of X, or, simply, a base for the topology of X. Conversely, given a set $\mathscr{U}$ of subsets of X, we wish to know under what circumstances $\mathscr{U}$ is a base for a topology on X—if this topology exists it will clearly be unique.

5.6.3 *A cover* $\mathscr{U}$ *of X by subsets of X is a base for a topology on X if and only if for each U, V in* $\mathscr{U}$ *and x in* $U \cap V$ *there is a W in* $\mathscr{U}$ *such that*

$$x \in W \subset U \cap V.$$

Proof The forward implication is easy, by 5.6.1 (*b*).

For the converse implication, let $\mathscr{U}(x)$ be for each x in X the set of elements of $\mathscr{U}$ which contain x. Then it is immediate that $x \rightsquigarrow \mathscr{U}(x)$ is an open base for the neighbourhoods of a topology on X. □

EXAMPLES 1. Let X, Y be topological spaces. The sets $U \times V$ for U open in X, V, and Y form a base for the product topology on $X \times Y$.
2. Let X be a metric space. The open balls $B(x, r)$ for all x in X and $r > 0$ form a base for the metric topology on X.
3. The intervals $]a, b[$ for a, b in $\mathbf{Q}$ form a base for the usual topology of $\mathbf{R}$.

A generalization of the notion of base for a topology is that of *sub-base*:

this is a set $\mathscr{V}$ of subsets of a topological space X such that the set of finite intersections of elements of $\mathscr{V}$ is a base for the topology of X.

5.6.4 *If X is a set and $\mathscr{V}$ any set of subsets of X, then $\mathscr{V}$ is a sub-base for a unique topology on X.*

Proof Let $\mathscr{U}$ be the set of finite intersections of elements of $\mathscr{V}$. Then $X \in \mathscr{U}$ (since X is the intersection of the empty set of elements of $\mathscr{V}$!) and so $\mathscr{U}$ covers X. Also, the intersection of two elements of $\mathscr{U}$ again belongs to $\mathscr{U}$—so $\mathscr{U}$ is a base for a topology $\mathscr{T}$ on X, by 5.6.3.

Any topology on X which has $\mathscr{V}$ as a sub-base has $\mathscr{U}$ as a base, and therefore coincides with $\mathscr{T}$. This proves uniqueness of the topology. □

We shall next characterize continuity of functions in terms of bases and sub-bases. However, for the applications of our results that we have in mind, it is helpful to have a more general kind of function than that considered before.

Let X, Y be sets. By a *function f out of X into Y*, written

$$f : X \rightarrowtail Y,$$

we shall mean a triple consisting of X, Y and a subset F of $X \times Y$ with the property that if $(x, y), (x, y') \in F$, then $y = y'$. If $(x, y) \in F$, we write $y = fx$. The *domain* $\mathscr{D}f$ of f is the set of x in X such that fx is defined—thus we have extended the definition of function given in the Appendix by allowing the domain of f to be any subset of X.

If $f : X \rightarrowtail Y, g : Y \rightarrowtail Z$ are functions as above, then the composite $gf : X \rightarrowtail Z$ has domain the set of x in X such that $fx \in \mathscr{D}g$, and gf sends $x \rightsquigarrow gfx$. The definitions of $f[A]$ and $f^{-1}[A]$ for a set A and function $f : X \rightarrowtail Y$ apply without change.

Let X, Y be topological spaces and $f : X \rightarrowtail Y$ a function. For our purposes it is convenient to say that f is *continuous* if $f^{-1}[U]$ is open in X for each open set U of Y. Since $\mathscr{D}f = f^{-1}[Y]$, this implies that $\mathscr{D}f$ is open in X. It is easy to prove that the composite of such continuous functions is again continuous.

5.6.5 *Let $f : X \rightarrowtail Y$ be a function where X, Y are topological spaces. The following conditions are equivalent:*
(a) f is continuous,
(b) if $\mathscr{U}$ is a base for the topology of Y, then $f^{-1}[U]$ is open in X for each U in $\mathscr{U}$,
(c) if $\mathscr{V}$ is a sub-base for the open sets of Y, then $f^{-1}[V]$ is open in X for each V in $\mathscr{V}$.

Proof The implications $(a) \Rightarrow (b) \Rightarrow (c)$ follow from the definition of continuity.

The implication $(b) \Rightarrow (a)$ follows from the (easily verified) fact that the inverse image of a union of sets is the union of the inverse images. The implication $(c) \Rightarrow (b)$ follows from the fact that the inverse image of an intersection of sets is the intersection of their inverse images. □

Suppose now that we are given a set X, a family $(X_\lambda)_{\lambda \in \Lambda}$ of topological spaces and for each λ in Λ a function

$$f_\lambda : X \rightarrowtail X_\lambda$$

A topology $\mathscr{I}$ on X is *initial with respect to* (f_λ) if it has the following property: for any topological space Y a function $k : Y \to X_\mathscr{I}$ is continuous if and only if the composite $f_\lambda k : Y \to X_\lambda$ is continuous for each λ in Λ.

5.6.6 *If $\mathscr{I}$ is an initial topology on X with respect to (f_λ) then $\mathscr{I}$ is the coarsest of the topologies $\mathscr{J}$ on X such that each $f_\lambda : X_\mathscr{J} \to X_\lambda$ is continuous.*

Proof Since the identity function $1 : X_\mathscr{I} \to X_\mathscr{I}$ is continuous, it follows that $f_\lambda = f_\lambda 1$ is continuous. Suppose $\mathscr{J}$ is any topology on X such that each $f_\lambda : X_\mathscr{J} \to X_\lambda$ is continuous. Let $k : X_\mathscr{J} \to X_\mathscr{I}$ be the identity function. Then $f_\lambda k = f_\lambda : X_\mathscr{J} \rightarrowtail X_\lambda$, and so k is continuous. □

It follows from 5.6.6 that there is at most one initial topology on X with respect to (f_λ).

5.6.7 *The initial topology on X with respect to (f_λ) exists and is the topology which has as a sub-base the sets*

$$f_\lambda^{-1}[U], \textit{ all } U \textit{ open in } X_\lambda, \textit{ all } \lambda \textit{ in } \Lambda.$$

Proof Let $\mathscr{I}$ be the topology on X with the above sub-base. We prove that $\mathscr{I}$ is initial.

Let $k : Y \rightarrowtail X_\mathscr{I}$ be a function where Y is a topological space, and suppose first that k is continuous. By 5.6.5 each $f_\lambda : X_\mathscr{I} \rightarrowtail X_\lambda$ is continuous and hence $f_\lambda k : Y \rightarrowtail X_\lambda$ is continuous.

Suppose, conversely, that each $f_\lambda k : Y \rightarrowtail X_\lambda$ is continuous. Then again, because the sets $f_\lambda^{-1}[U]$ for U open in X_λ form a sub-base for $\mathscr{I}$, the function k is continuous. □

In the first two of the following examples the initial topologies are taken with respect to functions whose domains are X itself.

Examples 4. Let X be a subset of the topological space X_1 and let $i : X \to X_1$ be the inclusion. The initial topology on X with respect to i has a sub-base the sets $i^{-1}[U]$ for U open in X_1. But $i^{-1}[U] = U \cap X$; so the initial topology with respect to i is simply the relative topology on X.

More generally, if X_1 is a topological space and $i : X \to X_1$ is an injection, then the initial topology with respect to i is that which makes i a homeomorphism into.

5. Let $(X_\lambda)_{\lambda \in \Lambda}$ be a family of topological spaces, and let X be the product of their underlying sets. The *product topology* on X is the initial topology with respect to the family of projections $p_\lambda : X \to X_\lambda$.

Suppose that $U \subset X_\lambda$. Then $p_\lambda^{-1}[U]$ consists of all points x in X such that $x_\lambda \in U$ (the other coordinates of x being unrestricted). That is, $p_\lambda^{-1}[U]$ is the product

$$\prod_{\mu \in \Lambda} U'_\mu \quad \text{where} \quad U'_\mu = \begin{cases} X_\mu, & \mu \neq \lambda \\ U, & \mu = \lambda_1, \ldots, \lambda_n. \end{cases}$$

A finite intersection of such sets, say $p_{\lambda_1}^{-1}[U_1] \cap \cdots \cap p_{\lambda_n}^{-1}[U_n]$, is the product

$$\prod_{\mu \in \Lambda} U''_\mu \quad \text{where} \quad U''_\mu = \begin{cases} X_\mu, & \mu \neq \lambda_1, \ldots, \lambda_n \\ U_i, & \mu = \lambda_1, \ldots, \lambda_n \end{cases}$$

Thus if Λ is finite, say $\Lambda = \{1, \ldots, n\}$, then a base for the open sets of X consists of all products $U_1 \times \cdots \times U_n$ for U_i open in X_i, and the product topology is that defined in 2.3. However, if $\Lambda = \{1, 2, \ldots\}$, then a base for the open sets of X consists of all products $U_1 \times U_2 \times \cdots$ in which U_i is open in X_i and $U_i = X_i$ for all but a finite number of i.

6. Initial topologies of the general type given above are used widely in the theory of manifolds. Suppose given a set X, a set $\mathscr{U}$ of subsets of X whose union is X, and for each U in $\mathscr{U}$ a function $f_U : X \rightarrowtail V$ with domain U and which maps U bijectively to an open subset of a given normed vector space V (a typical case is when $V = \mathbf{R}^n$). It is not hard to see that if X is given the initial topology with respect to $(f_U)_{U \in \mathscr{U}}$, then each U in $\mathscr{U}$ is open in X and mapped by f_U homeomorphically into V. So each x in X has a neighbourhood homeomorphic to an open subset of the normed vector space V. Such a space is called a *manifold modelled on* V—these spaces are important in many branches of topology, geometry, and analysis.

There is a useful 'transitive law' for initial topologies.

5.6.8 *Suppose there is given a set X; a family*

$$(f_\lambda : X \rightarrowtail X_\lambda)_{\lambda \in \Lambda}$$

of functions out of X; and for each λ in Λ a family

$$(g_{\lambda\mu} : X_\lambda \rightarrowtail X_{\lambda\mu})_{\mu \in M_\lambda}$$

of functions out of X_λ. If X_λ has the initial topology with respect to $(g_{\lambda\mu})_{\mu \in M_\lambda}$, then the initial topologies on X with respect to (f_λ) and $(g_{\lambda\mu} f_\lambda)$ coincide.

Proof Let $k : Y \rightarrowtail X$ be a function where Y is a topological space. Since X_λ has the initial topology the functions $g_{\lambda\mu} f_\lambda k$, $\mu \in M_\lambda$, are continuous if and only if $f_\lambda k$ is continuous; the result follows easily. $\square$

5.6.8 (*Corollary 1*) *Let $i : X \to X_1$ be an inclusion function and let $f_\lambda : X_1 \rightarrowtail X_\lambda$, $\lambda \in \Lambda$ be a family of functions where X_λ is a topological space. If X_1 has the initial topology with respect to (f_λ), then the initial topology on X with respect to $(f_\lambda i)$ is the subspace topology.*

EXERCISES

1. A space X is *second countable*, or *satisfies the second axiom of countability*, if there is a countable base for the open sets of X. Prove that if a space is second countable then it is separable and first countable. Prove also that a separable metric space is second countable.
2. Prove that any open cover of a second countable space has a countable sub-cover.
3. Let $\mathscr{U}$ be an open cover of X. For each U in $\mathscr{U}$ let $i_U : X \rightarrowtail X$ be the inclusion $x \rightsquigarrow x$, $x \in U$. Prove that X has the initial topology with respect to the family $(i_U)_{U \in \mathscr{U}}$.
4. Let X be a set, $(X_\lambda)_{\lambda \in \Lambda}$ a family of topological spaces and $(f_\lambda : X \rightarrowtail X_\lambda)_{\lambda \in \Lambda}$ a family of functions. Let Z be the topological product of the family $(X_\lambda)_{\lambda \in \Lambda}$ and let $f : X \rightarrowtail Z$ be the unique function such that $p_\lambda f = f_\lambda$, $\lambda \in \Lambda$. Prove that the initial topology with respect to f is the same as the initial topology with respect to (f_λ).

[The purpose of the following exercises is partly to supplement existing accounts. The reader who has difficulty with any of them will obtain help from proofs of similar results in Hu [2] or R. Brown [2].]

5. Let $\mathscr{C}$ be a set of functions $Y \to X$ where Y, X are spaces. If $C \subset Y$, $U \subset X$ let $M_{\mathscr{C}}(C, U)$ denote the set of functions f in $\mathscr{C}$ such that $f[C] \subset U$ (when $\mathscr{C}$ can be understood from the context we abbreviate this to $M(C, U)$). The *compact-open* topology on $\mathscr{C}$ is that which has as a sub-basis for the open sets the sets $M_{\mathscr{C}}(C, U)$ for all compact subsets C of Y and open subsets U of X. In the following, $\mathscr{C}$ will have the compact-open topology. Prove that (i) if $\mathscr{D}$ is a subset of $\mathscr{C}$ and $\mathscr{D}$ has the compact-open topology, then $\mathscr{D}$ is a subspace of $\mathscr{C}$; (ii) if X is Hausdorff, then $\mathscr{C}$ is Hausdorff; (iii) if $\mathscr{C}$ contains all constant functions $Y \to X$, and $\mathscr{C}$ is Hausdorff, then X is Hausdorff; (iv) if Y is discrete, and $\mathscr{C}$ consists of all functions $Y \to X$, then $\mathscr{C}$ is homeomorphic to the topological product $\prod_{y \in Y} X$ of the family $y \rightsquigarrow X$, $y \in Y$.

[Throughout the following exercises, X^Y will denote the space of k-continuous functions $Y \to X$ [Exercise 14 of 4.3] with the compact-open topology.]

6. Let $f : W \to X$ be a continuous function and $g : Y \to Z$ be a k-continuous function. Prove that f and g induce by composition continuous functions $f^Y : W^Y \to X^Y$ and $X^g : X^Z \to X^Y$. Prove that if f is a homeomorphism into, then so also is f^Y, and that X^g is a homeomorphism into if g is surjective and *compact covering*, that is, if for each compact set B of X there is a compact set C of Y such that $g[C] = B$.
7. A space Y is called *Hausdorff like*, which we abbreviate to *hl*, if each compact subspace of Y is locally compact. Prove that the following kinds of spaces are *hl* (i) Hausdorff spaces, (ii) spaces in which each point has a base of closed neighbour-

hoods [cf. the remark on terminology on p. 118], (iii) finite spaces, (iv) spaces in which each compact subset is finite.

*8. Let $\mathscr{U}$ be a sub-base for the open sets of X. Prove that if Y is *hl* then the sets $M(C, U)$ for C a compact subset of Y and U an element of $\mathscr{U}$ form a sub-base for the open sets of X^Y.

9. Prove that there is a continuous bijection $\sigma : (X \times Y)^Z \to X^Z \times Y^Z$ given by $f \rightsquigarrow (p_1 f, p_2 f)$ where p_1, p_2 are the projections of the product $X \times Y$. Prove also that σ is a homeomorphism if Z is an *hl*-space.

10. The *evaluation map* $\varepsilon : X^Y \times Y \to X$ is defined by $(f, y) \rightsquigarrow fy$. A topology $\mathscr{T}$ on the set X^Y is called *compactly-admissible* if, when the set X^Y has this topology, then $\varepsilon \mid X^Y \times C$ is continuous for all compact subsets C of Y. Prove that the compact-open topology on X^Y is coarser than any compactly-admissible topology, and is itself compactly-admissible if Y is an *hl*-space. [Cf. Kelley [1] Theorem 7.5.] Hence, if Y is a *hl*-space then $\varepsilon : X^Y \times Y \to X$ is k-continuous (where X^Y has of course the compact-open topology).

11. Let $f : Z \times Y \to X$ be k-continuous. Prove that if $z \in Z$ then the function $f_z : Y \to X, y \rightsquigarrow f(z, y)$ is k-continuous. Prove also that the function $e(f) : Z \to X^Y$, $z \rightsquigarrow f_z$, is k-continuous, and that the *exponential map* $e : X^{Z \times Y} \to (X^Y)^Z, f \rightsquigarrow e(f)$, is injective.

12. Prove that the exponential map $e : X^{Z \times Y} \to (X^Y)^Z$ has the following properties. (i) If Z is an *hl*-space then e is continuous. (ii) If Y is an *hl*-space then e is surjective. (iii) If Z and Y are *hl*-spaces then e is a homeomorphism.

13. Let $p : P \to X$ be a k-identification map [Exercise 15 of 4.3] and let Y be an *hl*-space. Prove that $p \times 1 : P \times Y \to X \times Y$ is a k-identification map. [Assume $g : X \times Y \to Z$ is such that $h = g(p \times 1)$ is k-continuous, and show that $e(h) : P \to Z^Y$ factors through $p : P \to X$.]

14. Prove that if $f : W \to X$ is k-continuous and Y is an *hl*-space, then $f^Y : W^Y \to X^Y$ is k-continuous. Prove also that if $g : Y \to Z$ is a k-identification map and Y, Z are *hl*-spaces, then $h = X^g : X^Z \to X^Y$ is a *k-homeomorphism into*, that is h is injective, k-continuous, and $h \mid X^Z$, Im h has a k-continuous inverse.

15. Let $F(Y, X)$ denote the set of continuous functions $Y \to X$ with the compact-open topology. Prove that $F(kY, X) = X^Y$ and that the inclusion $F(Y, X) \to X^Y$ embeds $F(Z, F(Y, X))$ as a subspace of $(X^Y)^Z$, which subspace we also write as $F(Z, F(Y, X))$. Prove that if Y is an *hl*-space and $f : Z \times Y \to X$ is k-continuous, then $e(f) \in F(Z, F(Y, X))$ if and only if $f \mid \{z\} \times Y, f \mid Z \times C$ are continuous for each z in Z and compact subset C of Y.

16. Prove that the exponential map restricts to a function $e : F(Z \times Y, X) \to F(Z, F(Y, X))$ which is bijective if Y is locally compact, or if Z and Y are first countable and Y is also an *hl*-space. Hence prove that if $p : P \to X$ is an identification map, then $p \times 1 : P \times Y \to X \times Y$ is an identification map if Y is locally compact, or if Y is an *hl*-space and P, X, Y are all first countable.

17. Let $Z \times_s Y$ denote the space $(Z \times Y)_\Sigma$ [cf. Exercise 11 of 4.2] where Σ is the set of subspaces $\{z\} \times Y, Z \times C$ of $Z \times Y$ for all $z \in Z$ and compact subsets C of Y. Prove that if Z, Y, X are *hl*-spaces then the natural map $(Z \times_s Y) \times_s X \to Z \times_s (Y \times_s X)$ is a homeomorphism. [Use facts such as that the exponential map

from $F((Z \times_s Y) \times_s X, W)$ to $F(Z \times_s Y, F(X, W))$ is a homeomorphism for all W.]

5.7 Joins

The join $X * Y$ of two topological spaces X, Y arises in a natural way when X, Y are disjoint subspaces of some normed vector space V. Consider the line segments $[x, y]$ for x in X, y in Y and suppose X, Y are so placed in V that two such line segments $[x, y]$, $[x', y']$ meet, if they meet

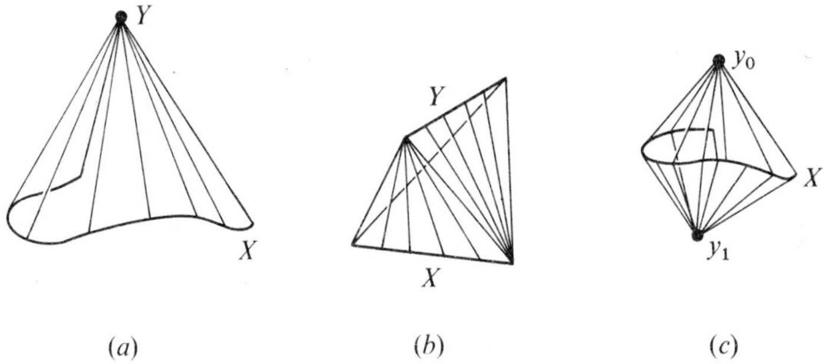

Fig. 5.5

at all, only in end points. The union of the line segments $[x, y]$, $x \in X$, $y \in Y$ with its topology as a subspace of V, is then called the *join* of X and Y and is written $X * Y$—the points of $X * Y$ are thus the sums

$$rx + sy, \qquad x \in X, y \in Y, \qquad r, s \geqslant 0, \qquad r + s = 1.$$

This construction has several awkward features. First of all it is only defined for subsets of a normed space. Second, it is not even defined for all such subsets—for example, to define $\mathbf{S}^1 * \mathbf{S}^1$ it is necessary to embed two copies of $\mathbf{S}^1$ in some space in such a way that the above curious condition on line segments holds. Thirdly, it is not obvious that the resulting space is independent of the embeddings.

We shall generalize the above construction in two ways. First, we give a canonical definition for topological spaces. Second, we give the definition for the join of n spaces $X_1, \ldots, X_n$.

Let $X_1, \ldots, X_n$ be topological spaces. The *join*

$$X = X_1 * \cdots * X_n$$

shall as a set consist of all $2m$-tuples, $1 \leqslant m \leqslant n$,

$$x = (r_{i_1}, x_{i_1}, \ldots, r_{i_m}, x_{i_m})$$

where $1 \leqslant i_1 < i_2 < \cdots < i_m \leqslant n$, $r_{i_j} > 0$, $r_{i_1} + \cdots + r_{i_m} = 1$, and $x_{i_j} \in X_{i_j}$. This looks rather formidable—we therefore write x in the form

$$x = r_1 x_1 + \cdots + r_n x_n$$

where $r_i \geqslant 0$, $r_1 + \cdots + r_n = 1$, $x_i \in X_i$ and it is agreed that if $r_i = 0$ then the term $r_i x_i$ is to be ignored. This convention agrees well with our earlier description of $X * Y$ for subspaces of a normed vector space. Notice also that the conditions on the r_i say that the point $(r_1, \ldots, r_n)$ belongs to the $(n-1)$-simplex Δ^{n-1} in $\mathbf{R}^n$ spanned by the standard basis vectors of $\mathbf{R}^n$.

We have also to define a topology on X—we do this by means of initial topologies as defined in the previous section. Consider the functions

$$\begin{array}{ll} \xi_i : X \to \mathbf{I} & \eta_i : X \rightarrowtail X_i \\ r_1 x_1 + \cdots + r_n x_n \rightsquigarrow r_i & r_1 x_1 + \cdots + r_n x_n \rightsquigarrow x_i \end{array}$$

Here η_i has domain $\xi_i^{-1}]0, 1]$. If $x \in X$, we call the points $\xi_i x$ and $\eta_i x$ (when defined) the *coordinates* of x; the functions ξ_i, η_i are called the *coordinate functions*. If $f : Z \to X$ is a function, then the functions $\xi_i f$, $\eta_i f$ are called the *coordinates* of f.

The *join topology* on $X = X_1 * \cdots * X_n$ is the initial topology with respect to the coordinate functions. Thus a function $f : Z \to X$ is continuous if and only if its coordinates are continuous; so with this topology we are well placed for deciding continuity of function *into* the join. However the only functions out of the join of whose continuity we can be assured are the coordinate functions ξ_i, η_i. The difficulties this leads to will be mentioned later.

As an application of this topology we prove:

5.7.1 *Let $1 \leqslant i \leqslant n$. There is a natural homeomorphism*

$$a : (X_1 * \cdots * X_i) * (X_{i+1} * \cdots * X_n) \to X_1 * \cdots * X_n.$$

Proof Let a be the function which sends

$$\begin{aligned} x &= r(r_1 x_1 + \cdots + r_i x_i) + s(r_{i+1} x_{i+1} + \cdots + r_n x_n) \\ &\rightsquigarrow r r_1 x_1 + \cdots + r r_i x_i + s r_{i+1} x_{i+1} + \cdots + s r_n x_n \end{aligned}$$

where of course $x_i \in X_i$, $(r, s) \in \Delta^1$, $(r_1, \ldots, r_i) \in \Delta^{i-1}$, $(r_{i+1}, \ldots, r_n) \in \Delta^{n-i-1}$. The coordinates of a (considered as a function into $X_1 * \cdots * X_n$) are given by

$$\xi_j : x \rightsquigarrow \begin{cases} r r_j, & j \leqslant i \\ s r_j, & j > i \end{cases} \qquad \eta_j : x \rightsquigarrow x_j, \quad j = 1, \ldots, n$$

where x is as above; notice that if $r = 0$ then the term $r(r_1 x_1 + \cdots + r_i x_i)$ does not occur, but we interpret $r\, r_j$ as 0.

Suppose that $j \leqslant i$. Then η_j is the composite of the coordinate functions

$$x \rightsquigarrow r_1x_1 + \cdots + r_ix_i \rightsquigarrow x_j$$

and so η_j is continuous. The points x for which $r \neq 0$ form an open set and on this set ξ_j is the product of the coordinate function $x \rightsquigarrow r$ and the composite $x \rightsquigarrow r_1x_1 + \cdots + r_ix_i \rightsquigarrow r_j$; hence ξ_j is continuous on this set. If $r = 0$, then ξ_j is given by $x \rightsquigarrow 0$, and the continuity of ξ_j at such points is proved in a similar way to that of the sandwich rule [Exercise 1 of 1.2]; we leave details to the reader.

The inverse of a sends the point $y = r_1y_1 + \cdots + r_ny_n$ $(y_i \in X_i, (r_1, \cdots, r_n) \in \Delta^{n-1})$ to the point

$$r(r_1y_1 + \cdots + r_iy_i) + s(r_{i+1}y_{i+1} + \cdots + r_ny_n)$$

where r is $(r_1 + \cdots + r_i)^{-1}$ when this is defined, and is otherwise 0. The coordinates of a^{-1}, as a function into the 2-fold join, are thus

$$y \rightsquigarrow (r_1 + \cdots + r_i)^{-1}, \qquad y \rightsquigarrow (r_{i+1} + \cdots + r_n)^{-1}$$
$$y \rightsquigarrow r_1y_1 + \cdots + r_iy_i, \qquad y \rightsquigarrow r_{i+1}y_{i+1} + \cdots + r_ny_n.$$

The first two functions are each defined on an open set and are, from their form, continuous. The third function is a function into an i-fold join and it is continuous because its coordinates are continuous; similarly, the fourth function is continuous.

It follows that a^{-1} is continuous, and so a is a homeomorphism. □

In future we shall leave detailed proofs of continuity which are similar to the above as exercises to the reader.

5.7.1 (*Corollary 1*) *There is a homeomorphism*

$$X_1 * (X_2 * X_3) \to (X_1 * X_2) * X_3.$$

Proof This is immediate from 5.7.1 since both spaces are homeomorphic to $X_1 * X_2 * X_3$. □
(We leave the reader to work out a formula for this homeomorphism.)

5.7.2 *If $X_1, \ldots, X_n$ are Hausdorff, then so also is $X_1 * \cdots * X_n$.*

Proof By 5.7.1 and induction it is sufficient to prove the result for the case $n = 2$. We say sets A, A' *separate* x, x' if A, A' are disjoint neighbourhoods of x, x' respectively. Let

$$x = r\,x_1 + s\,x_2, \qquad x' = r'x_1' + s'x_2'$$

be points of $X_1 * X_2$. If $r \neq r'$, then the sets $\xi_1^{-1}[U]$, $\xi_1^{-1}[U']$ for any sets U, U' which separate r, r' in **I**, separate x, x'. Suppose $r = r' \neq 0$, $x_1 \neq x_1'$. Let V_1, V_1' separate x_1, x_1'. Then $\eta_1^{-1}[V_1]$, $\eta_1^{-1}[V_1']$ separate

x, x'. A similar argument applies for the remaining case

$$s = s' \neq 0, x_2 \neq x_2'. \quad \square$$

We define the function

$$\begin{aligned} J : X_1 \times \cdots \times X_n \times \Delta^{n-1} &\to X_1 * \cdots * X_n \\ (x_1, \ldots, x_n, r_1, \ldots, r_n) &\rightsquigarrow r_1 x_1 + \cdots + r_n x_n. \end{aligned}$$

5.7.2 (*Corollary 1*) *If $X_1, \ldots, X_n$ are compact and Hausdorff, then J is an identification map.*

Proof Clearly J is always surjective; it is also continuous because its components are continuous. The $(n-1)$-simplex Δ^{n-1} is a closed, bounded subset of $\mathbf{R}^n$ and so is compact. Therefore J is a continuous surjection from a compact space to a Hausdorff space, and so J is an identification map. $\square$

Without the condition of compactness 5.7.2 (Corollary 1) is false [cf. Exercise 3], and so we have two topologies for the join, one convenient for maps *into* the join, the other convenient for maps *from* the join and in general these topologies are distinct. There are three ways out of this difficulty, namely to replace topological spaces and continuous maps either by (*a*) Hausdorff spaces and k-continuous maps [cf. Brown [2]], or (*b*) Hausdorff k-spaces and continuous maps [cf. Steenrod [2]], or (*c*) quasi-topological spaces and quasi-continuous maps [cf. Spanier [2] and a forthcoming book by Dyer and Eilenberg]. However a discussion of any of these topics would take us too far afield, and so we shall often assume when dealing with joins that the spaces concerned are compact and Hausdorff.

We now give some particular examples. Let e denote the unique point of $\mathbf{E}^0$. The following function

$$\begin{aligned} X \times \mathbf{I} &\to X * \mathbf{E}^0 \\ (x, r) &\rightsquigarrow rx + (1 - r)e \end{aligned}$$

is continuous because its coordinates are continuous; also it shrinks $X \times \{0\}$ to the single point e of $X * E^0$. So this map induces a continuous bijection

$$CX \to X * \mathbf{E}^0$$

Hence when X is compact and Hausdorff, CX is homeomorphic to $X * \mathbf{E}^0$ [cf. Fig. 5.5(*a*)].

Suppose now $Y = \mathbf{S}^0$ [cf. 5.5(*c*)]. For convenience, let us denote the points 1, -1 of $\mathbf{S}^0$ by e_+, e_- respectively. The function

$$\begin{aligned} X \times \mathbf{I} &\to X * \mathbf{S}^0 \\ (x, r) &\rightsquigarrow \begin{cases} (2 - 2r)x + (2r - 1)e_+, & r \geqslant \frac{1}{2} \\ 2rx + (1 - 2r)e_-, & r \leqslant \frac{1}{2} \end{cases} \end{aligned}$$

is continuous because its coordinates are continuous; also it shrinks $X \times \{1\}$ to the point e_+, and $X \times \{0\}$ to the point e_- of $X * \mathbf{S}^0$. Hence this map induces a continuous bijection

$$SX \to X * \mathbf{S}^0$$

which is a homeomorphism if X is compact and Hausdorff.

An alternative way of dealing with CX and SX is to use the bijections $CX \to X * \mathbf{E}^0$, $SX \to X * \mathbf{S}^0$, together with the join topologies, to define topologies on CX and SX. These topologies we call the *coarse* topologies. As pointed out before, the coarse topologies are convenient for maps into, rather than from, these spaces.

The join is also convenient when dealing with spheres. From the previous paragraphs and 5.7.1 we see that the n-fold join of $\mathbf{S}^0$ with itself is homeomorphic to the $(n - 1)$'th suspension of $\mathbf{S}^0$; by 4.4 this n-fold join is homeomorphic to $\mathbf{S}^{n-1}$.

There is another relationship between joins and spheres. The sphere $\mathbf{S}^{p+q+1}$ may be taken as consisting of all pairs (x, y) such that

$$x \in \mathbf{R}^{p+1}, \quad y \in \mathbf{R}^{q+1} \quad \text{and} \quad |x|^2 + |y|^2 = 1.$$

With this coordinatization, we define

$$\begin{aligned} k : \mathbf{S}^{p+q+1} &\to \mathbf{S}^p * \mathbf{S}^q \\ (x, y) &\rightsquigarrow rx' + sy' \end{aligned}$$

where $x' = x/|x|$, $y = y/|y|$, $r\pi/2 = \sin^{-1}|x|$, $s\pi/2 = \sin^{-1}|y|$. Notice that these definitions imply that $r + s = 1$, that x' is defined only for $x \neq 0$ (and so only for $r \neq 0$), and that y' is defined for $y \neq 0$ (and so only for $s \neq 0$). It follows that k is well-defined; k is continuous because its coordinates are continuous. Also k is a bijection because it has inverse

$$rx' + sy' \rightsquigarrow (x' \sin r\pi/2, y' \sin s\pi/2)$$

where, as usual, $x' \sin r\pi/2$, $y' \sin s\pi/2$ are taken as 0 when, respectively, r, s are 0. Since all the spaces concerned are compact and Hausdorff, it follows that k is a homeomorphism. We shall see that this result is related to the existence of the homeomorphism of 4.4.6.

5.7.3 *If A, B are subspaces of X, Y respectively, then $A * B$ is a subspace of $X * Y$.*

Proof This is an easy consequence of the transitive law for initial topologies. □

5.7.3 (*Corollary 1*) *The maps*

$$\begin{array}{ll} X \to X * Y, & Y \to X * Y \\ x \rightsquigarrow 1.x & y \rightsquigarrow 1.y \end{array}$$

are homeomorphisms into.

Proof This follows from 5.7.3 since, for example, the image of X under the given map is the subspace $X * \varnothing$ of $X * Y$. □

We shall use the maps of 5.7.3 (*Corollary 1*) to identify X, Y with the corresponding subspaces of $X * Y$.

Let us consider again the picture of $X * Y$ [Fig. 5.6]. By the *top half* of $X * Y$ we mean the subspace of points $rx + sy$ such that $s \geqslant \frac{1}{2}$. By the *bottom-half* of $X * Y$ we mean the subspace of points $rx + sy$ such that $r \geqslant \frac{1}{2}$. It is intuitively clear from the picture that the top-half of $X * Y$ is bijective with $CX \times Y$, and the bottom-half of $X * Y$ is bijective with $X \times CY$. Actually we prove the following result.

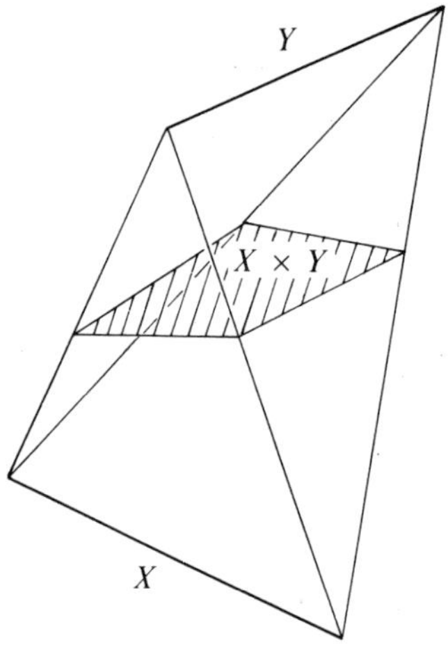

Fig. 5.6

5.7.4 *There is a homeomorphism*

$$v : X * Y * \mathbf{E}^0 \to (X * \mathbf{E}^0) \times (Y * \mathbf{E}^0)$$

which restricts to a homeomorphism

$$X * Y \to (X * \mathbf{E}^0) \times Y \cup X \times (Y * \mathbf{E}^0).$$

Proof The unique point of $\mathbf{E}^0$ is written e.

Consider first the case $X = \{x\}$, $Y = \{y\}$. We have then to produce, in essence, a homeomorphism.

$$\Delta^1 * E^0 \to \Delta^1 \times \Delta^1$$

that is, a homeomorphism from a triangle to a square. This is

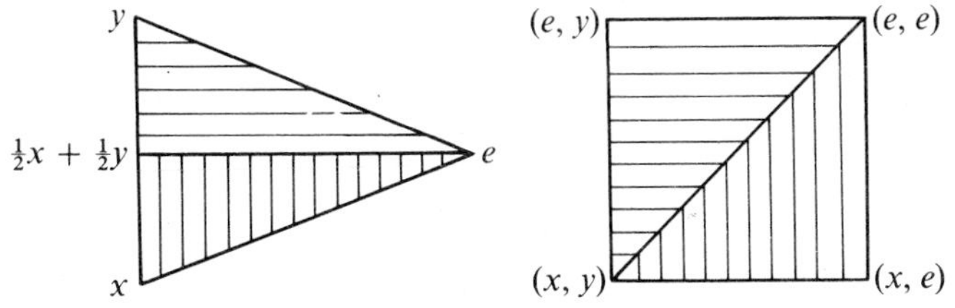

Fig. 5.7

done by splitting the triangle into two and mapping each half linearly onto one of the triangles into which the square is divided by a diagonal [Fig. 5.7].

We now consider the general case. Points of $X * Y * \mathbf{E}^0$ are $rx + sy + te$ where $(r, s, t) \in \Delta^2$, $x \in X$, $y \in Y$; points of $(X * \mathbf{E}^0) \times (Y * \mathbf{E}^0)$ are $(ux + ve, u'y + v'e)$ where $(u, v), (u', v') \in \Delta^1$, $x \in X, y \in Y$. We require a function which sends

$$x \rightsquigarrow (x, e), \qquad y \rightsquigarrow (e, y)$$
$$e \rightsquigarrow (e, e), \qquad \tfrac{1}{2}x + \tfrac{1}{2}y \rightsquigarrow (x, y).$$

Hence for $r \leqslant s$ (and so for $r \leqslant \frac{1}{2}$) we have

$$\begin{aligned} rx + sy + te &= 2r(\tfrac{1}{2}x + \tfrac{1}{2}y) + (s - r)y + te \\ &\rightsquigarrow 2r(x, y) + (s - r)(e, y) + t(e, e) \\ &= (2rx + (s - r + t)e, (r + s)y + te) \end{aligned}$$

while for $s \leqslant r$ (and so for $s \leqslant \frac{1}{2}$) we have

$$rx + sy + te \rightsquigarrow ((r + s)x + te, \quad 2sy + (r - s + t)e).$$

These formulae agree for $r = s$ and so define v. The detailed proof of continuity of v (as a function into the product) is left to the reader.

The inverse of v is defined as follows. If $u' \leqslant u$ then

$$\begin{aligned} (ux + ve, u'y + v'e) &= u'(x, y) + v(e, e) + (u - u')(x, e) \\ &\rightsquigarrow u'(\tfrac{1}{2}x + \tfrac{1}{2}y) + ve + (u - u')x \\ &= (u - \tfrac{1}{2}u')x + \tfrac{1}{2}u'y + ve \end{aligned}$$

while, if $u \leqslant u'$, then

$$(ux + ve, u'y + v'e) \rightsquigarrow \tfrac{1}{2}ux + (u' - \tfrac{1}{2}u)y + v'e.$$

Again the proof of continuity is left to the reader.

Notice also that $v(rx + sy + te) \in (X * \mathbf{E}^0) \times Y \cup X \times (Y * \mathbf{E}^0)$ if and only if t is either 0, or $r - s$ for $r \leqslant s$, or $s - r$ for $s \leqslant r$, and one of the last two conditions can hold only if $r = s$ (since $t \geqslant 0$), that is, only if $t = 0$; but $t = 0$ if and only if $rx + sy + te \in X * Y$. This proves the last part of the result. □

5.7.4 (*Corollary 1*) *If X, Y are compact Hausdorff, there is a homeomorphism*

$$\mu : C(X * Y) \to CX \times CY$$

which restricts to a homeomorphism

$$X * Y \to CX \times Y \cup X \times CY.$$

Proof This is immediate from 5.7.4, the associativity of the join and the fact that $Z * \mathbf{E}^0$ is homeomorphic to CZ if Z is compact and Hausdorff. □

Since $\mathbf{S}^{p+q+1}$ is homeomorphic to $\mathbf{S}^p * \mathbf{S}^q$, and $C\mathbf{S}^n$ is homeomorphic to $\mathbf{E}^{n+1}$, there is also a homeomorphism

$$\mathbf{S}^{p+q+1} \to \mathbf{E}^{p+1} \times \mathbf{S}^q \cup \mathbf{S}^p \times \mathbf{E}^{q+1},$$

a fact we have proved by a different method in 4.4.

EXERCISES

1. Prove that if K, L are cell complexes, then the topological space $K * L$ can be given the structure of a cell complex.
2. Let X, Y be compact, Hausdorff spaces. Prove that if SX, SY have base point the 'top' vertex, then there is a homeomorphism

$$SX \times SY \to (SX \vee SY) \,_w\sqcup\, C(X * Y)$$

for some map $w : X * Y \to SX \vee SY$. (This map w is called the *Whitehead product* map.) [Let $p : CX \to SX = CX/X$ be the identification map. Use 4.5.8. on the composite $C(X * Y) \xrightarrow{\mu} CX \times CY \xrightarrow{p \times p} SX \times SY$.]
3. Show that the map $J : X_1 \times X_2 \times \Delta^1 \to X_1 * X_2$ of 5.7.2 (Corollary 1) is not in general an identification map. [Use 5.7.3.]
4. Let X, Y be compact and Hausdorff. Let $p : X \times \{1\} \times Y \to X$ be the projection. Prove that $X * Y$ is homeomorphic to

$$X \,_p\sqcup\, (CX \times Y).$$

5. Let Z_n denote the space obtained from

$$\mathbf{E}^n \times \mathbf{S}^{n-1} \cup \mathbf{S}^{n-1} \times \mathbf{E}^n$$

by the identifications $(x, y) = (y, x)$ $(x \in \mathbf{E}^n, y \in \mathbf{S}^{n-1})$. Prove that Z_n is homeomorphic to $\mathbf{S}^{n-1} * P^{n-1}(\mathbf{R})$. [Define $f : \mathbf{E}^n \times P^{n-1}(\mathbf{R}) \to Z_n$ as follows. Let $x \in \mathbf{E}^n$, $y \in P^{n-1}(\mathbf{R})$. Draw a line through x parallel to y. This will meet S^{n-1} in z, z' say where z, z' are named so that z is at least as near to x as z' is. Define $f(x, y)$ to be the equivalence class of $(2x - z, z)$. Now use Exercise 4 and 4.5.8.]

6. The *symmetric square* of a space X is obtained from $X \times X$ by identifying (x, y) with (y, x) for all x, y in X. Prove that if A_n denotes the symmetric square of $\mathbf{E}^n$, then A_n is homeomorphic to CZ_n, where Z_n is as in the previous exercise. Deduce that the symmetric square of $\mathbf{S}^n$ is homeomorphic to a mapping cone $C(f)$ for a map $f : \mathbf{S}^{n-1} * P^{n-1}(\mathbf{R}) \to \mathbf{S}^n$.

7. Let $p : P \to X$ be a function from a topological space P. The *k-identification topology* on X is that which makes $p : kP \to X$ an identification map. Let CX, SX have this topology with respect to the usual maps from $X \times \mathbf{I}$ and let the join $X_1 * \cdots * X_n$ have the *weak topology*, that is the k-identification topology with respect to the map J of 5.7.2 (corollary 1). Prove that if all spaces concerned are Hausdorff, then the maps a, v of 5.7.1, 5.7.4 are k-homeomorphisms. Use these topologies, and the notion of k-homeomorphism, to generalize Exercises 2, 4 above to the non-compact (but Hausdorff) case.

5.8 The smashed product

We recall that a pointed space is a topological space X and a point x_0 of X, called the base point, such that $\{x_0\}$ is closed in X. We shall find it convenient to denote both the base point x_0 of a pointed space X and the set $\{x_0\}$ by $\cdot$. Note also that we use the same symbol for a pointed space and the underlying topological space.

The *wedge* of pointed spaces X, Y is the subspace of the product $X \times Y$ defined by

$$X \vee Y = X \times \cdot \cup \cdot \times Y.$$

(It is easy to prove that $X \vee Y$ is homeomorphic to the space obtained from $X \sqcup Y$ by identifying the two base points to a single point.) Thus the wedge consists of the 'axes' of the product $X \times Y$.

The *smashed product* of pointed spaces X, Y is the identification space

$$X \divideontimes Y = X \times Y / X \vee Y$$

with the set $X \vee Y$ taken as base point of $X \divideontimes Y$. (This space has also been called the *reduced join*, *smash product*, *collapsed product*; other notations used are $X \wedge Y$, $X \bowtie Y$.)

5.8.1 *If X, Y are Hausdorff spaces then so also is $X \divideontimes Y$.*

Proof Clearly $X \vee Y$ is closed in $X \divideontimes Y$ and $(X \times Y) \setminus (X \vee Y)$ is an open subset of $X \divideontimes Y$. Hence any two points of this subset are separated in $X \divideontimes Y$. To complete the proof we have only to show that the base point of $X \divideontimes Y$ is separated from any other point (x, y).

Let $U_{\bullet}$, U be disjoint open neighbourhoods of $\cdot$, x respectively, and let $V_{\bullet}$, V be disjoint open neighbourhoods of $\cdot$, y respectively. Then $U_{\bullet} \times Y \cup X \times V_{\bullet}$, $U \times V$ are disjoint open neighbourhoods of $X \vee Y$, (x, y) respectively. Also these neighbourhoods are saturated with respect to the identification map $X \times Y \to X \divideontimes Y$. Therefore their images separate $\cdot$ and (x, y). □

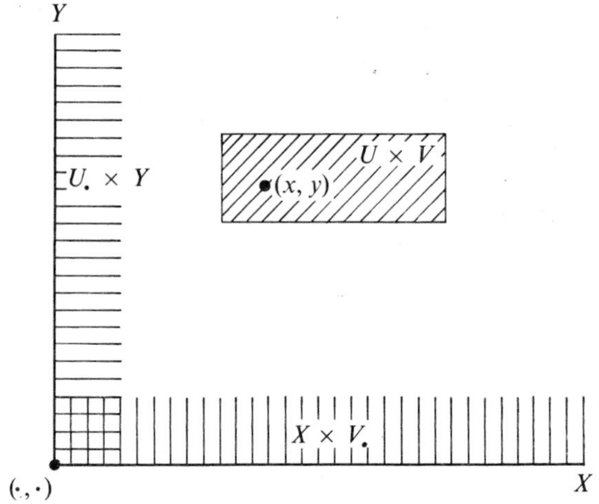

Fig. 5.8

5.8.2 *For any pointed spaces X, Y, Z there is a bijection*

$$a: X \divideontimes (Y \divideontimes Z) \to (X \divideontimes Y) \divideontimes Z$$

which is continuous if X is locally compact, or Y and Z are compact and Hausdorff. Further a^{-1} is continuous if Z is locally compact or X and Y are compact and Hausdorff. Hence a is a homeomorphism if X and Z are locally compact, or if two of X, Y, Z are compact and Hausdorff.

Proof Each of the composites

$$X \times Y \times Z \xrightarrow{1 \times p} X \times (Y \divideontimes Z) \xrightarrow{p} X \divideontimes (Y \divideontimes Z)$$

$$X \times Y \times Z \xrightarrow{p \times 1} (X \divideontimes Y) \times Z \xrightarrow{p} (X \divideontimes Y) \divideontimes Z$$

shrinks to a point the subspace

$$\cdot \times Y \times Z \cup X \times \cdot \times Z \cup X \times Y \times \cdot$$

of $X \times Y \times Z$. The existence of the bijection a is immediate. By 4.6.6, $1 \times p$ is an identification map if X is locally compact, and $p \times 1$ is an identification map if Z is locally compact. By 3.6.3 (corollary 1), $1 \times p$ is closed if Y and Z are compact and Hausdorff, and hence, in this case, $1 \times p$ is an identification map. A similar result holds if X and Y are compact and Hausdorff. The continuity of a and a^{-1} in the various cases follows (since also a compact Hausdorff space is locally compact). □

For any pointed space X, the *reduced suspension* of X is

$$\Sigma X = X \divideontimes \mathbf{S}^1.$$

The reason for this name is that if $p' : \mathbf{I} \to \mathbf{S}^1$ is the map $t \rightsquigarrow e^{2\pi i t}$, then the composite

$$X \times \mathbf{I} \xrightarrow{1 \times p'} X \times \mathbf{S}^1 \xrightarrow{p} X \divideontimes \mathbf{S}^1$$

is an identification map ($1 \times p'$ is an identification map by 3.6.3 (corollary 1)) which shrinks to a point the subspace $X \times \dot{\mathbf{I}} \cup \cdot \times \mathbf{I}$ of $X \times \mathbf{I}$. But the suspension SX is obtained from $X \times \mathbf{I}$ by shrinking each of $X \times \{0\}$, $X \times \{1\}$ to a point. It follows that ΣX is homeomorphic to

$$SX/\cdot \times \mathbf{I}.$$

We shall prove below that $\Sigma\mathbf{S}^{n-1}$ is homeomorphic to $\mathbf{S}^n$ where $\mathbf{S}^{n-1}$ has base point $e_{n-1} = (1, 0, \ldots, 0)$. In the case $n = 1$, the following picture shows how $\mathbf{S}^2$ is represented as $\mathbf{S}^1 \divideontimes \mathbf{S}^1$—each point of $\mathbf{S}^2$ is described by coordinates (x, y) for $x, y \in \mathbf{S}^1$, with $(x, y) = e_2$ the base point of $\mathbf{S}^2$ if x or y is e_1.

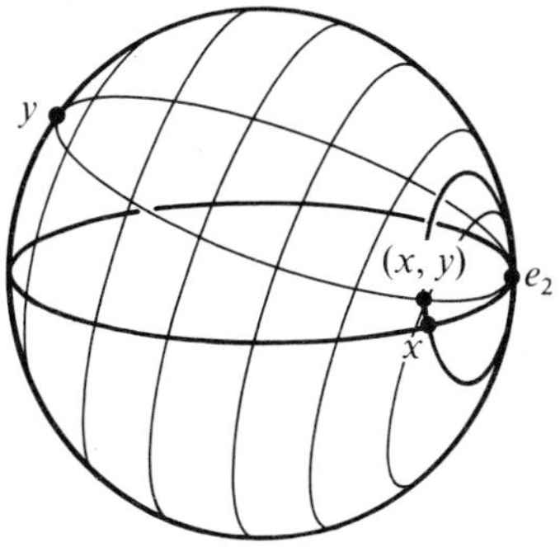

Fig. 5.9

More generally we prove

5.8.3 *There is a homeomorphism*

$$\mathbf{S}^m \divideontimes \mathbf{S}^n \to \mathbf{S}^{m+n}.$$

Proof By the last part of 4.7, the product $\mathbf{S}^m \times \mathbf{S}^n$ has a cell structure

$$e^0 \cup e^m \cup e^n \cup e^{m+n}$$

in which the subcomplex $e^0 \cup e^m \cup e^n$ is $\mathbf{S}^m \vee \mathbf{S}^n$. It follows that $\mathbf{S}^m \divideontimes \mathbf{S}^n$ has a cell structure $e^0 \cup e^{m+n}$. The attaching map of the $(m + n)$-cell is constant, and so $e^0 \cup e^{m+n}$ is homeomorphic to $\mathbf{S}^{m+n}$. □

In the case when X, Y are compact and Hausdorff there is a simple relation between the join $X * Y$ and $\Sigma(X \divideontimes Y)$. Let R be the subspace of $X * Y$ of points $rx + sy$ where x, or y, or both x and y, are at the base point.

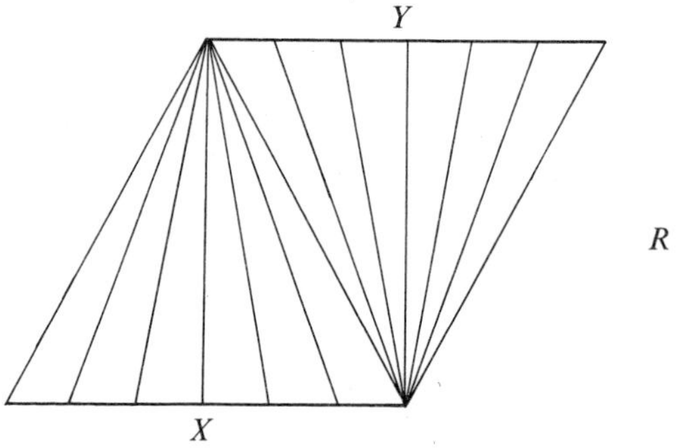

Fig. 5.10

5.8.4 *There is a bijection*

$$k:\Sigma(X \divideontimes Y) \to (X * Y)/R$$

which is continuous and which is a homeomorphism if X, Y are compact and Hausdorff.

Proof The composite

$$X \times Y \times \Delta^1 \to X * Y \to (X * Y)/R$$

is continuous and shrinks to a point the subspace

$$\cdot \times Y \times \Delta^1 \cup X \times \cdot \times \Delta^1 \cup X \times Y \times \dot{\Delta}^1$$

of $X \times Y \times \Delta^1$. The identification map

$$X \times Y \times \Delta^1 \to X \times Y \times \mathbf{S}^1 \to (X \divideontimes Y) \times \mathbf{S}^1 \to \Sigma(X \divideontimes Y)$$

has the same effect. □

The proof of the last part of **5.8.4** is incomplete since it is necessary to prove $(X * Y)/R$ is Hausdorff. This is not difficult, given Exercise 5 of **4.3**.

EXERCISES

1. Prove that any space with base point X is homeomorphic to $X \circledast \mathbf{S}^0$.
2. Prove that for any spaces with base point X, Y the spaces $X \circledast Y$ and $Y \circledast X$ are homeomorphic.
3. Prove that if K, L are cell complexes with base points vertices of K, L respectively, then $K \circledast L$ can be given the structure of a cell complex.
4. Let the smashed product be given the k-identification topology with respect to the natural map from the product. Prove that if X, Y, Z are Hausdorff, then the map a of 5.8.2 is a k-homeomorphism. Generalize 5.8.4 in a similar way.

NOTES

Some results on projective spaces, and their generalization, the Grassmannians, are given in Bourbaki [1] Chapter 6. The cell-structure of Stiefel manifolds (which include the various groups given here) is given in Steenrod–Epstein [1]. The study of orthogonal groups leads naturally to the study of Lie groups (Chevalley [1]), topological groups (Pontrjagin [1], Bourbaki [1] Chapter 4, Montgomery–Zippin [1]) and an extensive theory of fibre bundles (Steenrod [1], Husemoller [1]). Both the join and the smashed product are important constructions in algebraic topology (cf. James [1] and Husemoller [1]; in the latter book the smashed product is called the reduced product).

6. The fundamental groupoid

The topological invariants discussed in chapter 3 were not very subtle. One reason for this lies in the nature of the invariants; a set of numbers is not a subtle mathematical structure, and in order to probe further we need better structures to model the geometric properties of spaces.

The invariants we want come from a study of paths in a space. The paths on X with their addition form a *category*—this is an algebraic object, defined by Eilenberg–Maclane [1], which is basic to the study of much recent mathematics. From the category PX of paths on X we obtain another category πX which, because of its extra properties, is called a *groupoid*. This *fundamental groupoid* πX and its various subgroupoids give a useful algebraic model of X.

We have now used the word 'model' twice—a precise expression of the idea is given by the notion of functor which is a sort of homomorphism between categories. Any functor of topological spaces, such as the fundamental groupoid, gives rise to a topological invariant of any space X. Algebraic topology can be defined as the study of functors from the category of topological spaces to a category of algebraic objects (e.g., groups, rings, groupoids, etc.).

The study of categories and functors has been called 'abstract nonsense' partly because of the low ratio of theorems to definitions, and partly because the techniques are each of them not difficult. But such a description is not intended to decry either the charm or the utility of the subject.

6.1 Categories

A category $\mathscr{C}$ consists of
(a) a set $\mathrm{Ob}(\mathscr{C})$, called the set of *objects* of $\mathscr{C}$,
(b) for each x, y in $\mathrm{Ob}(\mathscr{C})$ a set $\mathscr{C}(x, y)$ called the set of *morphisms in* $\mathscr{C}$ *from* x *to* y,
(c) a function, called *composition*, which to each g in $\mathscr{C}(y, z)$ and to each f in $\mathscr{C}(x, y)$ assigns an element gf in $\mathscr{C}(x, z)$; that is, composition is a function

$$\mathscr{C}(y, z) \times \mathscr{C}(x, y) \to \mathscr{C}(x, z).$$

These terms must satisfy the axioms:

CAT 1 (Associativity) If $h \in \mathscr{C}(z, w)$, $g \in \mathscr{C}(y, z)$, $f \in \mathscr{C}(x, y)$ then

$$h(gf) = (hg)f.$$

CAT 2 (Existence of identities) For each x in $\mathrm{Ob}(\mathscr{C})$ there is an element 1_x in $\mathscr{C}(x, x)$ such that if $g \in \mathscr{C}(w, x)$, $f \in \mathscr{C}(x, y)$ then

$$1_x g = g, \qquad f 1_x = f.$$

We shall always assume that the various sets $\mathscr{C}(x, y)$ are disjoint and shall write $\mathscr{C}$ also for the union of these sets: thus 'f is an element of $\mathscr{C}$' and '$f \in \mathscr{C}$' both mean '$f \in \mathscr{C}(x, y)$ for some objects x, y of $\mathscr{C}$'. Then the structure of a category can be roughly stated as: a category is a set $\mathscr{C}$ with a multiplication which is associative and has two-sided identities, but such that the multiplication is partial, i.e., is not everywhere defined.

If $f \in \mathscr{C}(x, y)$ then we also write $f : x \to y$ or $x \xrightarrow{f} y$; the notation $x \to y$ simply denotes some element of $\mathscr{C}(x, y)$. For each x in $\mathrm{Ob}(\mathscr{C})$ the identity in $\mathscr{C}(x, x)$ is unique, since if 1_x, $1'_x$ are both identities in $\mathscr{C}(x, x)$ then

$$1_x = 1_x 1'_x = 1'_x.$$

It is usually convenient to abbreviate 1_x to 1 (this means that from equations such as $fg = 1$, $gf = 1$ we cannot deduce $fg = gf$ since 1 may denote different identities in each equation).

The definition of a category, and the notation, is suggested by Examples 2–5 below. In the first example, we show that the paths in a topological space form a category in which composition of paths is written as addition.

Examples 1. Let X be a topological space. The category PX of paths on X has the set X as its set of objects, and for any x, y in X the set $PX(x, y)$ is the set of paths in X from x to y. Composition of paths b, a is written, as before, $b + a$. The identity in $PX(x, x)$ is the zero path 0_x. Finally, addition of paths is associative since if c, b, a are paths of lengths r, q, p respectively, then $c + (b + a)$ is defined if and only if $(c + b) + a$ is defined, and both paths are given by

$$t \rightsquigarrow \begin{cases} at, & 0 \leqslant t \leqslant p \\ b(t - p), & p \leqslant t \leqslant p + q \\ c(t - p - q), & p + q \leqslant t \leqslant p + q + r. \end{cases}$$

2. Let $\mathscr{S}et$ be the category whose objects are all sets and whose morphisms $X \to Y$ are simply the functions from X to Y, and whose composition is the usual composition of functions. The axioms for a category are clearly satisfied, the identity in $\mathscr{S}et\,(X, X)$ being the identity function 1_X.

3. Let $\mathscr{T}op$ be the category whose objects are all topological spaces and

whose morphisms $X \to Y$ are the maps, that is, the continuous functions, $X \to Y$. Again the axioms are obviously satisfied.

Here we already see the double use of the idea of category. (*a*) General statements about topological spaces and continuous functions can in many cases be regarded as statements of an algebraic character about the category $\mathcal{T}$op, and this is often convenient, particularly when it brings out analogies between constructions for topological spaces and constructions for other mathematical objects. (*b*) In the case of PX, the paths on X, this category is regarded as an algebraic object in its own right, as much worthy of study as rings, groups, or fields.

4. The category of all groups and all morphisms (i.e., homomorphisms) of groups is written $\mathcal{G}$.

5. On the other hand, if G is a group, then G is also a category with one object—namely, the identity 1 of G—with morphisms $1 \to 1$ the elements of G, and with composition the multiplication of G. Actually, for this construction one needs only that G is a *monoid*, that is, the multiplication is associative and has a two-sided identity.

Let $\mathscr{C}$, $\mathscr{D}$ be categories. We say $\mathscr{D}$ is a *subcategory* of $\mathscr{C}$ if

(*a*) each object of $\mathscr{D}$ is an object of $\mathscr{C}$, i.e., $\text{Ob}(\mathscr{D}) \subset \text{Ob}(\mathscr{C})$

(*b*) for each x, y in $\text{Ob}(\mathscr{D})$, we have $\mathscr{D}(x, y) \subset \mathscr{C}(x, y)$

(*c*) composition of morphisms in $\mathscr{D}$ is the same as that for $\mathscr{C}$, and

(*d*) for each x in $\text{Ob}(\mathscr{D})$ the identity in $\mathscr{D}(x, x)$ is the identity in $\mathscr{C}(x, x)$.

The subcategory $\mathscr{D}$ of $\mathscr{C}$ is called *full* if

$$\mathscr{D}(x, y) = \mathscr{C}(x, y)$$

for all objects x, y of $\mathscr{D}$; and $\mathscr{D}$ is a *wide* subcategory if $\text{Ob}(\mathscr{D}) = \text{Ob}(\mathscr{C})$. For example, we can obtain full subcategories of any category $\mathscr{C}$ by taking $\text{Ob}(\mathscr{D})$ to be any class of objects of $\mathscr{C}$, and then defining $\mathscr{D}(x, y) = \mathscr{C}(x, y)$ for all x, y in $\text{Ob}(\mathscr{D})$. In this way, we obtain the full subcategories of $\mathcal{T}$op whose objects are all Hausdorff spaces, all metrizable spaces, or all compact spaces. On the other hand, we obtain a wide subcategory of $\mathcal{T}$op by suitably restricting the maps, for example to be open, or closed, or identification maps. In fact, any property $\mathscr{P}$ of continuous functions defines a wide subcategory of $\mathcal{T}$op (in which the morphisms $X \to Y$ are the continuous functions $X \to Y$ which have property $\mathscr{P}$) provided only that any identity map has property $\mathscr{P}$, and that the composite of two maps with property $\mathscr{P}$ also has property $\mathscr{P}$.

Let $\mathscr{C}$ be a category, and suppose f, g are morphisms in $\mathscr{C}$ such that $gf = 1$, an identity morphism. Then we call g a *left-inverse* of f and f a *right-inverse* of g; we also say that g is a *retraction*, and f a *co-retraction*.

6.1.1 *Let $f : x \to y$, $g_1, g_2 : y \to x$ be morphisms in $\mathscr{C}$ such that*

$$g_1 f = 1_x, \qquad f g_2 = 1_y.$$

Then $g_1 = g_2$. If, further, $gf = 1_x$, then $g = g_1$.

Proof $g_1 = g_1 1_y = g_1(fg_2) = (g_1 f)g_2 = 1_x g_2 = g_2$. Similarly, $g = g_2$ and so $g = g_1$. □

This result can be stated: if f has a left and a right inverse, then f has a unique two-sided inverse. Such a morphism f is called *invertible*, or an *isomorphism*, and the unique inverse of f is written f^{-1} or, when using additive notation, $-f$. If there is an isomorphism $x \to y$, then we say x and y are isomorphic. It is easy to prove that the relation 'x is isomorphic to y' is an equivalence relation on the objects of $\mathscr{C}$.

A category in which every morphism is an isomorphism is called a *groupoid*. For example, a group, regarded as a category with one object, is also a groupoid.

The category PX of paths on X is not a groupoid since if a is a path in X of positive length then there is no path b such that $b + a$ is a zero path. This is an awkward feature of PX. Another awkward feature is that even for quite simple spaces (e.g., $X = \mathbf{I}$) $PX(x, y)$ can be uncountable. In the next section we shall show how to avoid both of these difficulties by constructing from PX the fundamental groupoid πX.

EXERCISES

1. Prove that (i) the composite of retractions is a retraction, (ii) the composite of co-retractions is a co-retraction, (iii) the composite of isomorphisms is an isomorphism.
2. Let a, b, c be morphisms such that ba, cb are defined and are isomorphisms. Prove a, b, c are isomorphisms.
3. A *graph* Γ is a set $\text{Ob}(\Gamma)$ and for each x, y in $\text{Ob}(\Gamma)$ a set $\Gamma(x, y)$ called the set of *edges* from x to y—the sets $\Gamma(x, y)$ are supposed disjoint. A *path* in Γ from x to y consists of either the empty sequence $\varnothing \to \Gamma(x, x)$ if $x = y$ or a sequence $(a_n, \ldots, a_1)$ such that (i) $a_i \in \Gamma(x_i, x_{i+1})$, (ii) $x_1 = x$, $x_{n+1} = y$; the set of paths from x to y is written $P\Gamma(x, y)$. These paths are multiplied by the rule that if

$$a = (a_n, \ldots, a_1), \quad b = (b_m, \ldots, b_1)\,(a \in P\Gamma(x, y), \quad b \in P\Gamma(y, z))$$

then $ba = (b_m, \ldots, b_1, a_n, \ldots, a_1)$; also the empty sequence in $P\Gamma(x, x)$ is to act as identity. Prove that $P\Gamma$ is a category. (This is the category *freely generated* by Γ.)
4. Let $\mathscr{C}$ be a category. Prove that a category $\mathscr{C}^{\text{op}}$, the opposite or *dual* of $\mathscr{C}$, is defined as follows. (i) $\text{Ob}(\mathscr{C}^{\text{op}}) = \text{Ob}(\mathscr{C})$, (ii) if $x, y \in \text{Ob}(\mathscr{C}^{\text{op}})$ then $\mathscr{C}^{\text{op}}(x, y) = \mathscr{C}(y, x)$ (however, the elements of $\mathscr{C}^{\text{op}}(x, y)$ are written f^* for each f in $\mathscr{C}(y, x)$), (iii) the composition in $\mathscr{C}^{\text{op}}$ is defined by $g^* f^* = (fg)^*$.

5. Let $\mathscr{C}$ be a category. A morphism $f : C \to D$ in $\mathscr{C}$ is called *monic* (and a *mono*) if for all A in $\mathrm{Ob}(\mathscr{C})$ and $g, h : A \to C$ in $\mathscr{C}$, the relation $fg = fh$ implies $g = h$; f is called *epic* (and an *epi*) if for all B in $\mathrm{Ob}(\mathscr{C})$ and all $g, h : D \to B$, the relation $gf = hf$ implies $g = h$ [cf. Exercises A.1.2, A.1.3, of the Appendix]. Prove that (i) a co-retraction is monic and a retraction is epic, (ii) an isomorphism is both epic and monic, (iii) the composition of monos is monic, the composition of epis is epic. Give an example of a category in which some morphism is epic and monic but not an isomorphism.
6. Prove that f in $\mathscr{C}$ is monic $\Leftrightarrow f^*$ in $\mathscr{C}^{\mathrm{op}}$ is epic.
7. An object P of a category $\mathscr{C}$ is called a *point* in $\mathscr{C}$ if $\mathscr{C}(X, P)$ has exactly one element for all objects X of $\mathscr{C}$; and P is a *copoint* if $\mathscr{C}(P, X)$ has exactly one element for all objects X of $\mathscr{C}$. Prove that all points in $\mathscr{C}$ are isomorphic, as are all copoints. If P is both a point and a copoint, then P is called a *zero object*. Prove that (i) the categories $\mathscr{S}$et and $\mathscr{T}$op have points and copoints, but no zero, (ii) the category of groups, and the category of vector spaces over a given field, both have a zero object.
8. Prove that in the category $\mathscr{G}$ of groups, a morphism $f : G \to H$ is monic if and only if it is injective; less trivially, f is epic if and only if f is surjective. [Suppose f is not surjective and let $K = \mathrm{Im}\, f$. If the set of cosets H/K has two elements, then K is normal in H and it is easy to prove f is not epic. Otherwise there is a permutation γ of H/K whose only fixed point is K. Let $\pi : H \to H/K$ be the projection and choose a function $\theta : H/K \to H$ such that $\pi\theta = 1$. Let $\tau : H \to K$ be such that $x = (\tau x)(\theta\pi x)$ for all x in H and define $\lambda : H \to H$ by $x \rightsquigarrow (\tau x)(\theta\gamma\pi x)$. The morphisms α, β of H into the group P of all permutations of H, defined by $\alpha(h)(x) = hx$, $\beta(h) = \lambda^{-1}\alpha(h)\lambda$ satisfy $\alpha h = \beta h$ if and only if $h \in K$. Hence $\alpha f = \beta f$].
9. Prove that in the category of Hausdorff spaces and continuous functions, a map $f : X \to Y$ is epic if and only if $\mathrm{Im}\, f$ is a dense subset of Y.
10. For sets X, Y define a *relation from X to Y* to be a triple (X, Y, R) where R is a subset of $X \times Y$. If R is a subset of $X \times Y$, S is a subset of $Y \times Z$, let SR be the subset of $X \times Z$ of pairs (x, z) such that for some y in Y, $(x, y) \in R$ and $(y, z) \in S$. Using this product, define the composite of a relation from X to Y and a relation from Y to Z, and prove that sets and relations between sets form a category containing $\mathscr{S}$et as a wide subcategory. [An obvious question seems to be: what do the conditions epic, monic, iso imply about a relation in this category?]
11. Let $\mathscr{C}$ be a category and $f : C \to D$ a morphism in $\mathscr{C}$. Prove that f induces functions for each X in $\mathrm{Ob}(\mathscr{C})$

$$f_X : \mathscr{C}(X, C) \to \mathscr{C}(X, D) \qquad f^X : \mathscr{C}(D, X) \to \mathscr{C}(C, X)$$
$$g \rightsquigarrow fg \qquad\qquad\qquad g \rightsquigarrow gf.$$

Prove that the following conditions are equivalent: (i) f is an isomorphism, (ii) f_X is a bijection for each X in $\mathrm{Ob}(\mathscr{C})$, (iii) f^X is a bijection for each X in $\mathrm{Ob}(\mathscr{C})$. Prove also that f is monic if f_X is injective for all X, and f is epic if and only if f^X is injective for all X. Under what conditions is f^X surjective for all X, f_X surjective for all X?

6.2 Construction of the fundamental groupoid

The fundamental groupoid πX will be a groupoid such that $\pi X(x, y)$ is a set of equivalence classes of $PX(x, y)$. In order to define the equivalence relation, we consider first two paths a, b in $PX(x, y)$ of the same length r. A *homotopy rel end points of length q from a to b* is a map

$$F : [0, r] \times [0, q] \to X$$

such that

$$\begin{aligned} F(s, 0) &= a(s), \quad & F(s, q) &= b(s), \quad & s &\in [0, r] \\ F(0, t) &= x, \quad & F(r, t) &= y, \quad & t &\in [0, q] \end{aligned} \tag{6.2.1}$$

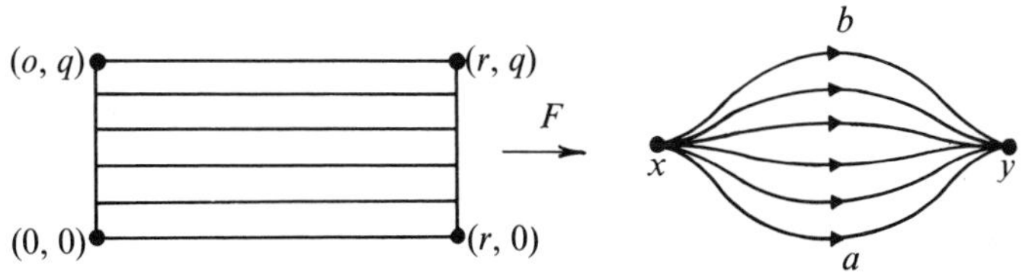

Fig. 6.1

Notice that for each t in $[0, q]$ the path $F_t : s \rightsquigarrow F(s, t)$ is a path in $PX(x, y)$; the family (F_t) can be thought of as a 'continuous family of paths' between $F_0 = a$ and $F_1 = b$. Alternatively, we can think of F as a 'deformation' of a into b.

We use the notation $F : a \sim b$ to mean that F is a homotopy rel end points from a to b (of some length). There is a unique homotopy of length 0 from a to a. If $F : a \sim b$ is a homotopy of length q, then $-F$ defined by $(s, t) \rightsquigarrow F(s, q - t)$ is a homotopy $b \sim a$. If $F : a \sim b$, $G : b \sim c$ are of length q, q' respectively where a, b, c are of length r, then the *sum* of F and G

$$\begin{aligned} G + F : [0, r] \times [0, q + q'] &\to X \\ (s, t) &\rightsquigarrow \begin{cases} F(s, t), & 0 \leqslant t \leqslant q \\ G(s, t - q), & q \leqslant t \leqslant q + q' \end{cases} \end{aligned}$$

is continuous by the glueing rule [2.5.12] and is a homotopy $a \sim c$.

Two paths a, b of the same length are called *homotopic* rel end points, written $a \sim b$, if there is a homotopy $F : a \sim b$. We abbreviate homotopic rel end points to homotopic since in the case of paths we have no need of other homotopies. Then it is clear from the previous paragraph that the relation $a \sim b$ is an equivalence relation.

Let $F : [0, r] \times [0, q] \to X$ be a homotopy $a \sim b$. Then there is a homotopy $F' : a \sim b$ of length 1, namely

$$\begin{aligned} F' : [0, r] \times \mathbf{I} &\to X \\ (s, t) &\rightsquigarrow F(s, qt). \end{aligned}$$

So for the rest of this chapter we restrict attention to homotopies of length 1.

It is convenient to have a flexible notation for homotopies. We think of a homotopy F (of length 1) as a function $t \rightsquigarrow F_t$ where F_t is a path, and then abbreviate $t \rightsquigarrow F_t$ to F_t. This enables us to say, for example, that if F_t is a homotopy $a \sim b$, then F_{1-t} is a homotopy $b \sim a$.

For any real number $r \geqslant 0$ and x in X, let r_x denote the constant path at x of length r. When no confusion can be caused, we abbreviate r_x to r. In particular, for any path a and $r \geqslant 0$, the paths $a + r$, $r + a$ are well defined.

We can now state the basic lemmas on homotopies of paths.

6.2.2 *Let $a, b \in PX(x, y)$, $c, d \in PX(y, z)$ where $|a| = |b|$, $|c| = |d|$.*
(a) If $a \sim b$, then $-a \sim -b$.
(b) If $a \sim b$ and $c \sim d$, then $c + a \sim d + b$.
(c) For any $r \geqslant 0$, $a + r \sim r + a$.

Proof (*a*) Let F be a homotopy $a \sim b$. Then

$$(s, t) \rightsquigarrow F(|a| - s, t)$$

is a homotopy $-a \sim -b$.
(*b*) Let $F : a \sim b$, $G : c \sim d$. Then

$$H : [0, |c| + |a|] \times \mathbf{I} \to X$$

$$(s, t) \rightsquigarrow \begin{cases} F(s, t), & s \leqslant |a| \\ G(s - |a|, t), & |a| \leqslant s \end{cases}$$

is a homotopy $c + a \sim d + b$ [cf. Fig. 6.2].

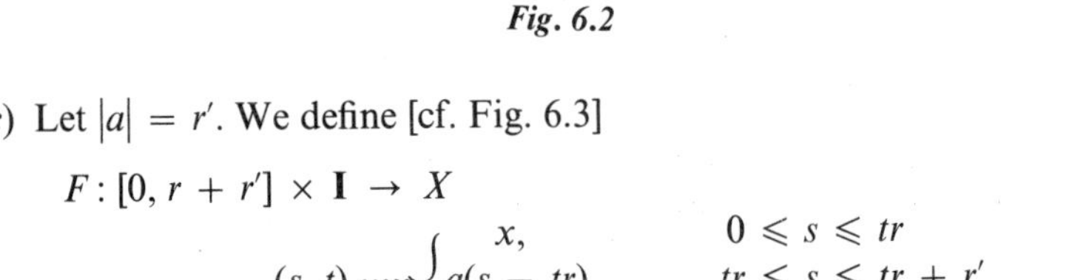

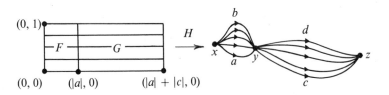

Fig. 6.2

(*c*) Let $|a| = r'$. We define [cf. Fig. 6.3]

$$F : [0, r + r'] \times \mathbf{I} \to X$$

$$(s, t) \rightsquigarrow \begin{cases} x, & 0 \leqslant s \leqslant tr \\ a(s - tr), & tr \leqslant s \leqslant tr + r' \\ y, & tr + r' \leqslant s \leqslant r + r'. \end{cases} \qquad \square$$

It should be noticed that in the homotopy H of 6.2.2 (*b*) the point y is fixed (that is, $H(|a|, t) = y$ for all t in $\mathbf{I}$). However, there are homotopies $c + a \sim d + b$ which do not have this property. This fact is exploited in the next result.

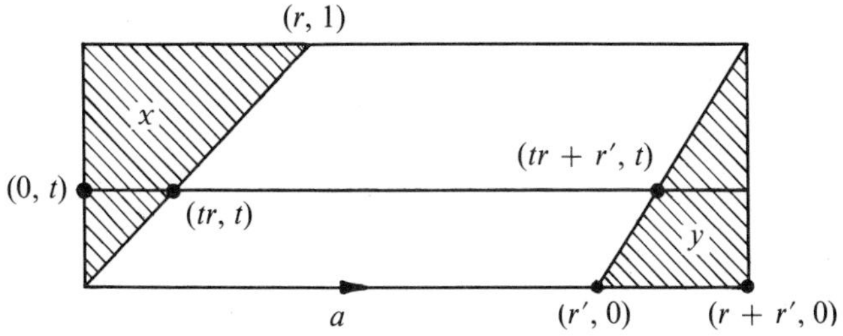

Fig. 6.3

6.2.3 *If $a \in PX(x, y)$ and $|a| = r$, then*

$$-a + a \sim 2r_x, \qquad a - a \sim 2r_y$$

Proof It is a little simpler to define a homotopy $2r_x \sim -a + a$. We define [cf. Fig. 6.4]

$$F : [0, 2r] \times \mathbf{I} \to X$$
$$(s, t) \rightsquigarrow \begin{cases} a(s), & 0 \leqslant s \leqslant rt \\ a(2rt - s), & rt \leqslant s \leqslant 2rt \\ x, & 2rt \leqslant s \leqslant 2r. \end{cases}$$

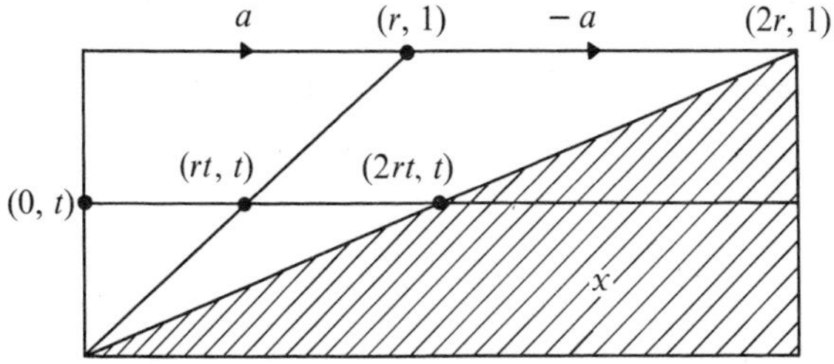

Fig. 6.4

Clearly F is well-defined, continuous, and a homotopy $2r_x \sim -a + a$. It follows that $-a + a \sim 2r_x$. On replacing a by $-a$ we find $a - a \sim 2r_y$. □

The path F_t of the proof of 6.2.3 is depicted for various t in Fig. 6.5. Of course, a and $-a$ should really be superimposed.

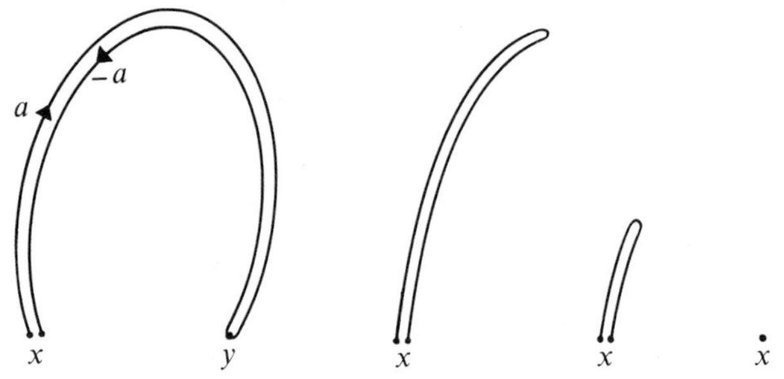

Fig. 6.5

We now define an equivalence relation between paths of various lengths. Let $a, b \in PX(x, y)$. We say a, b are *equivalent* if there are real numbers, $r, s \geqslant 0$ such that $r + a$, $s + b$ are homotopic (in which case, r, s must satisfy $|a| + r = |b| + s$). This relation is obviously reflexive and symmetric (since homotopy is reflexive and symmetric). It is also transitive; for given homotopies $r + a \sim s + b$, $s' + b \sim t + c$ (where a, b, c are paths and $r, s, s', t \geqslant 0$) then there are homotopies

$$s' + r + a \sim s' + s + b = s + s' + b \sim s + t + c.$$

The definition of equivalence of paths is non-canonical (that is, it involves choices) and this is perhaps unaesthetic. An alternative, but equivalent, definition is suggested in Exercise 5.

EXAMPLES 1. Let a in $P(x, y)$ be of length $r \geqslant 0$. Let $r' > 0$. Then a is equivalent to a path of length r', namely, the path $b : s \rightsquigarrow a(sr/r')$. A specific homotopy $r' + a \sim r + b$ is given by

$$F : (s, t) \rightsquigarrow \begin{cases} a(rs/\lambda_t), & 0 \leqslant s \leqslant \lambda_t \\ y, & \lambda_t \leqslant s \leqslant r + r' \end{cases}$$

where $\lambda_t = r(1 - t) + r't$ [cf. Fig. 6.6].

This argument does not show that any a in $PX(x, x)$ is equivalent to a path of length 0 since, if we take $r' = 0$ in the formula for $F(s, t)$, then F is continuous if and only if a is constant. However, any constant path r is equivalent to a zero path since $r = r + 0$.

2. Equivalent paths of the same length are in fact homotopic. This can be proved by constructing for any $r, s \geqslant 0$ a homeomorphism $G : [0, r] \times \mathbf{I} \to [0, r + s] \times \mathbf{I}$ which is the identity on $\{0\} \times \mathbf{I} \cup [0, r] \times \dot{\mathbf{I}}$ and which

maps $\{r\} \times \mathbf{I}$ homeomorphically onto $[r, r + s] \times \dot{\mathbf{I}} \cup \{r + s\} \times \mathbf{I}$. So if $F: [0, r + s] \times \mathbf{I} \to X$ is a homotopy $s + a \sim s + b$ where a, b have length r, then the composite FG is a homotopy $a \sim b$. However this result, although it is interesting and will be used in chapter 7, is not essential to

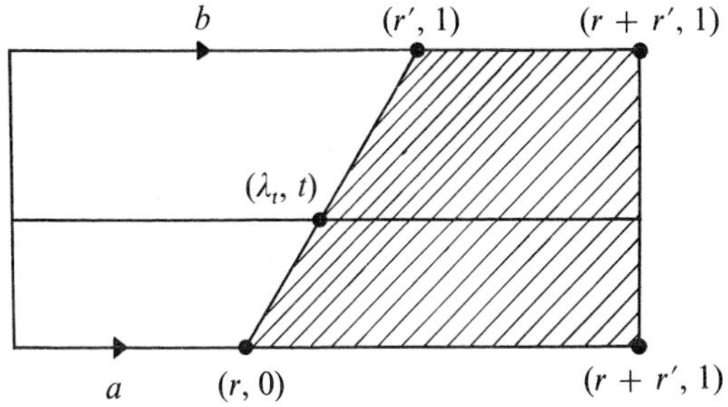

Fig. 6.6

this chapter. So we only state that in Fig. 6.7 G will be the identity on the shaded areas, and on the remaining part will map each line segment $[z, w]$ linearly onto $[z, w']$. Further details are left to the reader.

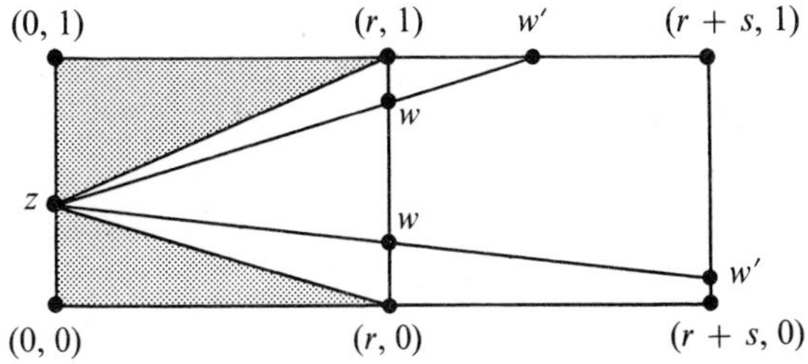

Fig. 6.7

If two paths a, b are equivalent we write $a \sim b$—the last two examples show that this will cause no confusion. The equivalence classes of paths from x to y are called *roads* from x to y, and the set of these roads is written $\pi X(x, y)$. The class of the zero path at x is written 0, or to avoid ambiguity, 0_x. By Example 1, the zero road 0_x includes all constant paths at x.

6.2.4 *A negative and sum of roads is defined by*

$$-\text{cls } a = \text{cls }(-a), \qquad a \in PX(x, y), \quad b \in PX(y, z)$$
$$\text{cls } b + \text{cls } a = \text{cls }(b + a).$$

Proof Suppose a, a' are equivalent paths in $PX(x, y)$; then there are constant paths r, r' such that $r + a, r' + a'$ are homotopic. Hence

$$r - a \sim -a + r = -(r + a) \sim -(r' + a') \sim r' - a'.$$

Therefore $-\text{cls } a$ is well-defined.

Suppose further that b, b' are equivalent paths in $PX(y, z)$ and that $s + b, s' + b'$ are homotopic, where $s, s' \geqslant 0$. Then

$$\begin{aligned} r + s + b + a &\sim r + s' + b' + a \\ &\sim s' + b' + r + a \\ &\sim s' + b' + r' + a' \\ &\sim s' + r' + b' + a'. \quad \square \end{aligned}$$

6.2.5 *Addition of roads is associative. Further if* $\alpha \in \pi X(x, y)$ *then*

$$\alpha + 0_x = 0_y + \alpha = \alpha$$

$$-\alpha + \alpha = 0_x, \alpha - \alpha = 0_y.$$

Proof The first statement is obvious, since addition of paths is associative. The equations $\alpha + 0_x = 0_y + \alpha = \alpha$ are immediate from the relations $a + 0_x = 0_y + a = a$ for paths a in $P(x, y)$.

The last equations are immediate from 6.2.3 and the fact that the zero road at x contains all constant paths at x. $\square$

We have now shown that πX is a groupoid whose objects are the points of X and whose morphisms $x \to y$ are the roads from x to y—this groupoid is called the *fundamental groupoid* of X.

EXAMPLES 3. If X consists of a single point, then πX has one object x say and $\pi X(x, x)$ consists only of the zero road. More generally, if the path-components of X consist of single points, then

$$\pi X(x, y) = \begin{cases} \varnothing, & x \neq y \\ \{0_x\}, & x = y. \end{cases}$$

A groupoid with this property is called *discrete*.

4. Let X be a convex subset of a normed vector space, and let a, b be two paths in X from x to y of the same length r. Then a and b are homotopic, since

$$\begin{aligned} F: [0, r] \times \mathbf{I} &\to X \\ (s, t) &\rightsquigarrow (1 - t)a(s) + b(s) \end{aligned}$$

is a homotopy $a \sim b$. It follows easily that any two paths from x to y are equivalent; so $\pi X(x, y)$ has exactly one element for all x, y in X. A groupoid with this property is called *1-connected*, and also a *tree* groupoid; and if πX is a tree groupoid we say X is *1-connected* (this, of course, implies path

connected). For example, the unit interval **I** is 1-connected, since it is a convex subset of **R**.
5. If each path-component of X is 1-connected then, for each x, y in X, $\pi X(x, y)$ contains not more than one element; we then say X, and also πX, is *simply-connected*. Thus X is simply-connected means that any two paths in X with the same end points are equivalent. And a groupoid G is simply-connected if $G(x, y)$ has not more than one element for all objects x, y of G.
6. We have at this stage no techniques for showing that $\pi X(x, y)$ can ever contain more than one element. That is, we cannot yet show that $PX(x, y)$ can contain non-equivalent paths. However, anyone who has tied elastic round sticks will find it reasonable to suppose that the two paths in $\mathbf{R}^2 \setminus \mathbf{E}^2$ shown in Fig. 6.8 are not equivalent. A proof of this fact will appear later when we have techniques for computing the fundamental groupoid.

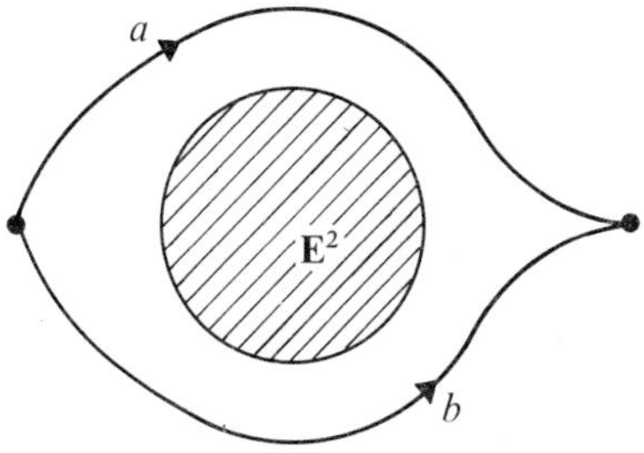

Fig. 6.8

EXERCISES

In these exercises, X is assumed to be a topological space.
1. Let $x, y, z \in X$, and let a in $PX(x, y)$, b in $PX(y, z)$ be paths of length 1. Define the path $b\,.\,a$ of length 1 in $PX(x, z)$ by

$$(b.a)(t) = \begin{cases} a(2t), & 0 \leqslant t \leqslant \frac{1}{2} \\ b(2t-1), & \frac{1}{2} \leqslant t \leqslant 1. \end{cases}$$

Prove that $b.a$ is well-defined, and that $b.a \sim b + a$. Let 1 denote as usual a constant path of length 1. Prove that $a.1 \sim a \sim 1.a$, that $a.(-a) \sim 1, (-a).a \sim 1$, and that if also $c \in P(z, w)$ then

$$c.(b.a) \sim (c.b).a.$$

2. If $\alpha : \mathbf{I} \to \mathbf{R}^{\geqslant 0}$ is continuous, let

$$J(\alpha) = \{(s, t) \in \mathbf{R}^2 : 0 \leqslant t \leqslant 1, \quad 0 \leqslant s \leqslant \alpha(t)\}.$$

Prove that $J(\alpha)$ is compact.

Let a, b be paths from x to y in X. A *wavy homotopy* $F : a \rightsquigarrow b$ is a map

$F : J(\alpha) \to X$ for some map $\alpha : \mathbf{I} \to \mathbf{R}^{\geqslant 0}$ such that $\alpha(0) = |a|$, $\alpha(1) = |b|$ and for $0 \leqslant t \leqslant 1$.

$$\begin{aligned} F(0, t) &= x, & F(s, 0) &= a(s), \quad 0 \leqslant s \leqslant |a| \\ F(\alpha(t), t) &= y, & F(s, 1) &= b(s), \quad 0 \leqslant s \leqslant |b|. \end{aligned}$$

Prove that $\sim\sim$ is an equivalence relation, and that $a \sim\sim b$ if and only if a and b are equivalent.

3. Let a, b be paths from x to y of the same length r. A *weak homotopy* $F : a \sim_W b$ is a function $F : [0, r] \times \mathbf{I} \to X$ satisfying the usual conditions for a homotopy $a \sim b$ except that F is only *separately* continuous, that is, for each s the function $t \rightsquigarrow F(s, t)$ is continuous, and for each t the function $s \rightsquigarrow F(s, t)$ is continuous.

Let $a, b : \mathbf{I} \to \mathbf{S}^1$ be the paths $s \rightsquigarrow e^{2\pi i s}$, $s \rightsquigarrow 1$, respectively. Prove that the map $(s, t) \rightsquigarrow \exp(2\pi i\, s^t)$, $(s \neq 0)$, $(0, t) \rightsquigarrow 1$ is a weak homotopy $b \sim_W a$. [This example shows that weak homotopy gives an uninteresting theory.]

4. Let a, b be paths of length r in X, Y respectively and let c be the path in $X \times Y$, $s \rightsquigarrow (a(s), b(s))$. Prove that c is equivalent to $b' + a'$ where a' is the path $s \rightsquigarrow (a(s), b(0))$ and b' is the path $s \rightsquigarrow (a(r), b(s))$.

5. If $a : [0, r] \to X$ is a path in X, let $\bar{a}$ be the map

$$\begin{aligned} \mathbf{R}^{\geqslant 0} &\to X \\ s &\rightsquigarrow \begin{cases} as, & 0 \leqslant s \leqslant r \\ ar, & r \leqslant s. \end{cases} \end{aligned}$$

Prove that two paths a, b in $PX(x, y)$ are equivalent if and only if there is a map $F : \mathbf{R}^{\geqslant 0} \times \mathbf{I} \to X$ such that there is an $r_0 \geqslant 0$ such that for all $(s, t) \in \mathbf{R}^{\geqslant 0} \times \mathbf{I}$

$$\begin{aligned} F(s, 0) &= \bar{a}s; \qquad F(s, 1) = \bar{b}s; \\ F(0, t) &= x \\ F(s, t) &= y \quad \text{for all } s \geqslant r_0. \end{aligned}$$

6. Let Y be a subspace of X and $i : Y \to X$ the inclusion. Assume that if a, b are paths in Y with the same end points, then the paths ia, ib are equivalent in X. Let $P(X, Y)$ be the set of paths in X whose end points lie in Y. For $a, b \in P(X, Y)$ write $a \approx b$ if there are real numbers $r, s \geqslant 0$ and a map $F : [0, u] \times \mathbf{I} \to X$ such that

$$F_0 = r + a, \qquad F_1 = s + b,$$
$$F_t(0),\ F_t(u) \in Y.$$

Prove that $\approx$ is an equivalence relation on $P(X, Y)$ and that the set of equivalence classes is a groupoid.

7. Define an equivalence relation $\equiv$ among paths in PX as follows. (i) If a is a path and $r \geqslant 0$, then $a + r \equiv r + a \equiv a$. (ii) If $a \in PX(x, y)$ then $a - a \equiv 0_y$, $-a + a \equiv 0_x$. (iii) If $a, b \in PX(x, y)$, then $a \equiv b$ if we can write $a = a_n + \cdots + a_1$, $b = b_n + \cdots + b_1$ and $a_i \equiv b_i$, $i = 1, \ldots, n$ by rules (i) and (ii). Prove that addition of paths induces an addition of equivalence classes by which these equivalence classes form a groupoid. [This should perhaps be called the *semi-fundamental groupoid of* X.]

6.3 Properties of groupoids

In this section, we discuss the basic properties of groupoids and apply the results to the fundamental groupoid. We shall use additive notation and, in particular, $-a$ will denote the inverse (or negative) of an element a of a groupoid G. This enables us to write $a - b$ for $a + (-b)$ (when defined); and to write $a - b - c$ for both $(a - b) - c$ and $a - (c + b)$. Of course we shall also in this chapter write the operation in a group as addition, but this does *not* mean that the groups which arise will all be commutative—we do not suppose $a + b = b + a$.

Let G be a groupoid. A *subgroupoid* of G is a subcategory H of G such that $a \in H \Rightarrow -a \in H$; that is, H is a subcategory which is also a groupoid. We say H is *full* (*wide*) if H is a full (wide) subcategory. We can construct full subgroupoids H of G by taking H to be the full subcategory of G on any subset of $\mathrm{Ob}(G)$. In particular, the full subgroupoid of G on one object of G is written $G\{x\}$—this groupoid has one zero and $a + b$ is defined for all a, b in $G\{x\}$. A groupoid with only one object is called a *group* [cf. Example 5, p. 172] and, in particular, $G\{x\}$ is called the *object group* (or *vertex group*) at x.

If X is a topological space and $x \in X$, then $\pi X\{x\}$ is a group called the *fundamental group of X at x* (this group will often be written $\pi(X, x)$). More generally, if A is any set then the full subgroupoid of πX on the set $A \cap X$ is written πXA. The elements of πXA are all roads, that is all equivalence classes of paths in X, joining points of $A \cap X$ (the use of πXA for sets A not contained in X will prove convenient later).

A groupoid G is *connected* if $G(x, y)$ is non-empty for all objects x, y of G. In particular, πX is connected if and only if X is path-connected.

Let x_0 be an object of G, and let Cx_0 be the full subgroupoid of G on all objects y of G such that $G(x, y)$ is not empty. If x, y are objects of Cx_0, then $G(x, y)$ is non-empty since it contains elements $b + a$ for b in $G(x_0, y)$, a in $G(x, x_0)$. It follows that Cx_0 is connected. Clearly, Cx_0 is the maximal, connected subgroupoid of G with x_0 as an object, and so we call Cx_0 the *component* of G containing x_0.

6.3.1 *Let x, y, x', y' be objects of the connected groupoid G. There is a bijection.*

$$\varphi : G(x, y) \to G(x', y')$$

which if $x = y$, $x' = y'$ can be chosen to be an isomorphism of groups.

Proof Since G is connected we can choose $a : x \to x', b : y \to y'$ in G. We define

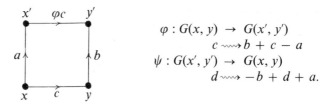

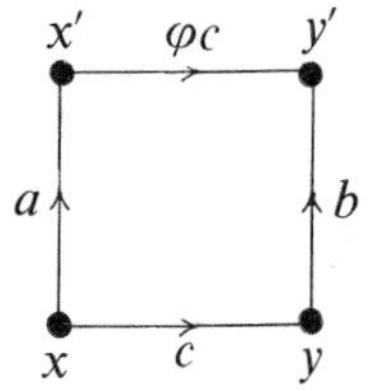

$$\begin{aligned}\varphi : G(x, y) &\to G(x', y')\\ c &\rightsquigarrow b + c - a\\ \psi : G(x', y') &\to G(x, y)\\ d &\rightsquigarrow -b + d + a.\end{aligned}$$

Clearly $\varphi\psi = 1, \psi\varphi = 1$ and so φ is a bijection.

If $x = y$, $x' = y'$ then let $a = b$ so that φ sends $a \rightsquigarrow a + c - a$. If $c, c' \in G(x, x)$ (which is $G\{x\}$) then

$$\begin{aligned}\varphi c + \varphi c' &= a + c - a + a + c' - a\\ &= a + c + c' - a\\ &= \varphi(c + c').\end{aligned}$$

Therefore φ is an isomorphism. □

Thus the object groups of a connected groupoid are all isomorphic. For this reason we shall sometimes speak loosely of *the* object group of a connected groupoid. The isomorphism $G\{x\} \to G\{x'\}$ which sends $c \rightsquigarrow a + c - a$ is written a_*. If $x = x'$, then a_* is simply an inner automorphism of $G\{x\}$.

6.3.2 *Let x, x' belong to the same component of G. Then $a_* = b_* : G\{x\} \to G\{x'\}$ for all $a, b : x \to x'$ if and only if $G\{x\}$ is abelian.*

Proof We first note that

$$(-b + a)_* = (-b)_* a_* = (b_*)^{-1} a_*.$$

Also if $c : x \to x$, then

$$(b_*)^{-1}(b + c)_* = c_*.$$

Thus $b_*^{-1} a_*$ is an inner automorphism of $G\{x\}$ and every inner automorphism of $G\{x\}$ is of this form. But the inner automorphisms of a group are trivial if and only if the group is abelian. □

These definitions and results apply immediately to the fundamental groupoid πX. The components of πX are the groupoids πX_0 for X_0 a path-component of X. If α is a road in πX from x to x' then α determines an isomorphism $\alpha_* : \pi(X, x) \to \pi(X, x')$ of fundamental groups, and α_* is independent of the choice of α if and only if $\pi(X, x)$ is abelian.

We now give some examples of groupoids.

EXAMPLES 1. Let X be any set. The *discrete groupoid on X* is also written X; it has X as its set of objects one zero for each element of X, and no other elements. Notice that if X is given the discrete topology, then this groupoid is πX.

2. Let G be a tree groupoid, so that $G(x, y)$ has exactly one element say a_{yx} for all objects x, y of G. Then $a_{zy} + a_{yx}$ is the unique element of $G(x, z)$ and so we have the addition rule

$$a_{zy} + a_{yx} = a_{zx}. \qquad (*)$$

Conversely, given any set X we can form an essentially unique tree groupoid G such that $\mathrm{Ob}(G) = X$ by choosing distinct elements a_{yx} for each x, y in X and then taking $G(x, y)$ to consist solely of a_{yx}, with the addition rule (*). Notice that a tree groupoid with n objects has n^2 elements of which n are zeros. Tree groupoids with two objects 0, 1 will be important, and will be denoted ambiguously by I, the unique element of $I(0, 1)$ being written ι.

If X is 1-connected and $A \subset X$, then πXA is a tree groupoid. In particular, the fundamental groupoid of $\mathbf{I}$ on the set $\{0, 1\}$ is exactly I—in symbols

$$\pi\mathbf{I}\{0, 1\} = I.$$

3. Let G be a connected groupoid and T a tree groupoid which is a wide subgroupoid of G (we recall that wide means that T, G have the same objects). Let x_0 be an object of G and for each object x of G let τ_x be the unique element of $T(x_0, x)$. If $a \in G(x, y)$, then there is a unique element a' of $G\{x_0\}$ such that

$$a = \tau_y + a' - \tau_x.$$

If, further, $b \in G(y, z)$ then

$$\begin{aligned} b + a &= \tau_z + b' - \tau_y + \tau_y + a' - \tau_x \\ &= \tau_z + b' + a' - \tau_x. \end{aligned}$$

Therefore $\qquad (b + a)' = b' + a'.$

This shows that G can be recovered from T and $G\{x_0\}$.

4. A groupoid G is *totally disconnected* if

$$G(x, y) = \varnothing \quad \text{for } x \neq y.$$

Such a groupoid is determined entirely by the family $(G\{x\})$, $x \in \mathrm{Ob}(G)$, of groups. If X is a space, and A consists of exactly one point in each path-component of X, then πXA is totally disconnected.

EXERCISES

1. Let E be a subset of $X \times X$. Let $\mathscr{C}$ be defined by $\mathrm{Ob}(\mathscr{C}) = X$ and for each x, y in X, let

$$\mathscr{C}(x, y) = \begin{cases} \varnothing, & (x, y) \notin E \\ \{(y, x)\}, & (x, y) \in E. \end{cases}$$

Prove that if E is a transitive relation on X, then an associative, partial multiplication on $\mathscr{C}$ is defined by the rule

$$(z, y)(y, x) = (z, x) \quad \text{whenever } (x, y), (y, z) \in E.$$

Prove that if, further, E is reflexive then $\mathscr{C}$ is a category and that if E is an equivalence relation then $\mathscr{C}$ is a groupoid.
2. Find conditions on a connected groupoid G for G to have a point, copoint.
3. To what extent can 6.3.1, 6.3.2 be generalized to categories?
*4. Let $\mathscr{A}$ be the category of complete, Archimedean, ordered fields in which the morphisms $K \to L$ are the functions $f : K \to L$ such that $f(x + y) = fx + fy$, $f(xy) = fx\,fy$, for all $x, y \in K$. Prove that $\mathscr{A}$ is a 1-connected groupoid.

6.4 Functors and morphisms of groupoids

It is usual when defining an algebraic object to define the mappings or morphisms of that object—that is, it is usual to define a category of the given objects. Now we agree to regard categories themselves as a particular algebraic object. Therefore we must define the mappings or morphisms of categories—these morphisms are known as *functors*.

Groupoids are special cases of categories and functors of groupoids will be called *groupoid morphisms*. Actually, the main line of our applications is to the case of groupoids. However, we will have applications of functors of categories; further, many of our results are of general interest for categories and have applications outside the topics of this book. Some of these applications will be indicated in exercises, and in this way we hope to show how these ideas run into the main stream of mathematics.

Let $\mathscr{C}$, $\mathscr{D}$ be categories. A *functor* $\Gamma : \mathscr{C} \to \mathscr{D}$ assigns to each object x of $\mathscr{C}$ an object Γx of $\mathscr{D}$ and to each morphism $f : x \to y$ in $\mathscr{C}$ a morphism $\Gamma f : \Gamma x \to \Gamma y$ in $\mathscr{D}$; Γf is often called the morphism *induced* by f. These must satisfy the axioms.

FUN 1 If $1 : x \to x$ is the identity in $\mathscr{C}$ then $\Gamma 1 : \Gamma x \to \Gamma x$ is the identity in $\mathscr{D}$; that is, $\Gamma 1_x = 1_{\Gamma x}$.

FUN 2 If $f : x \to y$, $g : y \to z$ are morphisms in $\mathscr{C}$, then

$$\Gamma(gf) = \Gamma g\,\Gamma f.$$

Clearly we have an *identity functor* $1 : \mathscr{C} \to \mathscr{C}$ and if $\Gamma : \mathscr{C} \to \mathscr{D}$, $\Delta : \mathscr{D} \to \mathscr{E}$ are functors, then we can form the composite functor $\Delta\Gamma : \mathscr{C} \to \mathscr{E}$. Thus we can form the category $\mathscr{C}$at of all categories and functors. (There is a logical difficulty here, since it seems that $\mathscr{C}$at being a category must be one of its own objects, and from this one can obtain a contradiction by considering the category of all categories which do not have themselves as objects. For a brief mention of ways round this difficulty, see the Glossary, under *class*.)

Before giving examples of functors we prove one elementary result.

6.4.1 *Let $\Gamma : \mathscr{C} \to \mathscr{D}$ be a functor. Then Γf is a retraction, co-retraction or isomorphism if f is respectively a retraction, co-retraction or isomorphism.*

Proof A relation $gf = 1$ implies $\Gamma g\, \Gamma f = 1$. □

EXAMPLES 1. If X is a space, then PX is a category. Suppose $f : X \to Y$ is a map of spaces. If a is a path in X from x to y then the composite fa is a path in Y from fx to fy. If a is a zero path, then so also is fa. If $b + a$ is defined in X then $fb + fa$ is defined in Y and

$$fb + fa = f(b + a).$$

Therefore f determines a functor $Pf : PX \to PY$.

2. Let us proceed further with the last example. If f is the identity $X \to X$, then so also is $Pf : PX \to PY$. Further, it is easy to check that if $f : X \to Y$, $g : Y \to Z$ are maps then $P(gf) = Pg\, Pf$. Thus P is a functor $\mathscr{T}\text{op} \to \mathscr{C}\text{at}$. From 6.4.1 we deduce that if X is homeomorphic to Y, then PX is isomorphic to PY. Thus PX is a topological invariant of X (though not a very tractable one).

3. For any space X, there is a functor $p : PX \to \pi X$ which is the identity on objects and sends each path in X to its equivalence class.

4. Let G, H be groupoids. A functor $G \to H$ will also be called a *morphism* of groupoids. So we obtain the category $\mathscr{G}\text{d}$ of all groupoids and morphisms of groupoids.

6.4.2 *The fundamental groupoid is a functor*

$$\pi : \mathscr{T}\text{op} \to \mathscr{G}\text{d}.$$

Proof Let $f : X \to Y$ be a map of spaces and $Pf : PX \to PY$ the corresponding functor of path categories. Suppose first of all that a, b are two homotopic paths in X of length r from x to x'. Then there is a map $F : [0, r] \times \mathbf{I} \to Y$ such that $F_0 = a$, $F_1 = b$ and $F(0, t) = x$, $F(r, t) = x'$ for all $t \in \mathbf{I}$. It is easily checked that the composite $fF : [0, r] \times \mathbf{I} \to Y$ is a homotopy $fa \sim fb$.

If a, b are equivalent paths in X from x to x', then there are constant paths r, s such that $r + a$, $s + b$ are homotopic, whence

$$r + fa = f(r + a) \sim f(s + b) = s + fb$$

and so fa is equivalent to fb. Thus we have a well defined function

$$\begin{aligned} \pi f : \pi X &\to \pi Y \\ \text{cls}\, a &\rightsquigarrow \text{cls}\, fa \end{aligned}$$

and it is clear that πf is a morphism of groupoids. The verification of the functorial relations $\pi 1 = 1$, $\pi(gf) = \pi g \, \pi f$, is left to the reader. □

6.4.2 (*Corollary* 1) *If X is homeomorphic to Y, then πX is isomorphic to πY.*

Proof This is immediate from 6.4.1 and 6.4.2. □

Of course, before 6.4.2 (corollary 1) can be used it is necessary to be able to compute πX—techniques for this are given in chapter 8.

EXAMPLE 5. The fundamental group $\pi(X, x)$ is only defined for a space X and point x of X, so that the fundamental group is in no sense a functor $\mathscr{T}\text{op} \to \mathscr{G}$. In order to obtain a functor, one introduces the category $\mathscr{T}\text{op}_{\bullet}$ of *pointed spaces* (or *spaces with base point*). A *pointed space* is a pair (X, x) where $x \in X$ and X is a topological space. A *pointed map* $(X, x) \to (Y, y)$ is determined by the two pointed spaces and a map $f : X \to Y$ such that $fx = y$ (but such a pointed map is usually written f). To any pointed space (X, x) we can assign the fundamental group $\pi(X, x)$ and to any pointed map $f : (X, x) \to (Y, y)$ we can assign $f_* : \pi(X, x) \to \pi(Y, y)$, the restriction of $\pi f : \pi X \to \pi Y$ to the appropriate object groups—this defines the fundamental group functor $\mathscr{T}\text{op}_{\bullet} \to \mathscr{G}$. Since it is easier to specify a group than a groupoid, the fundamental group is often the useful topological invariant (of a pointed space). For example, if X, Y are path-connected spaces, and X, Y have fundamental groups $\mathbf{Z}$, $\mathbf{Z}_2$ respectively, then we know immediately that X is not homeomorphic to Y.

6.4.3 *Let $f : \mathscr{C} \to \mathscr{D}$ be a functor. Then f is an isomorphism $\Leftrightarrow$ the functions*

$$\text{Ob}(\mathscr{C}) \to \text{Ob}(\mathscr{D}), \qquad \mathscr{C}(x, y) \to \mathscr{D}(fx, fy), \qquad x, y \in \text{Ob}(\mathscr{C})$$

induced by f are all bijections.

Proof The implication $\Rightarrow$ is easy since an inverse g of f induces an inverse to each of the functions induced by f.

For the converse, define $g : \mathscr{D} \to \mathscr{C}$ to be the given inverse of f on $\text{Ob}(\mathscr{D})$ and on each $\mathscr{D}(x', y')$. Then it is easy to check that g is a functor, and is an inverse to the functor f. □

Consider now the case of groupoids. If $f : G \to H$ is a morphism, then f induces a morphism of object groups $G\{x\} \to H\{fx\}$ which is written f_x or, simply, f. If f is an isomorphism then so also is each f_x; but it is not true that if each f_x is an isomorphism and f is bijective on objects then f is an isomorphism—for example, G could be totally disconnected and H connected.

The groupoid I gives rise to some simple and useful morphisms. Let G be any groupoid and let $a \in G(x, y)$. Then a morphism $\hat{a} : I \to G$ is defined on objects by $\hat{a}0 = x$, $\hat{a}1 = y$, and on non-zero elements by

$$\hat{a}\iota = a, \qquad \hat{a}(-\iota) = -a$$

where ι is the unique element of $I(0, 1)$. The check that $\hat{a}$ is a morphism is easy since I has only four elements, two of them zeros. Notice that $\hat{a}$ is the only morphism $I \to G$ which sends ι to a.

Products of categories

Let $\mathscr{C}_1, \mathscr{C}_2$ be categories. The *product* $\mathscr{C}_1 \times \mathscr{C}_2$ is defined to have as objects all pairs (x_1, x_2) for x_1 in $\mathrm{Ob}(\mathscr{C}_1)$, x_2 in $\mathrm{Ob}(\mathscr{C}_2)$ and to have as elements the pairs (a_1, a_2) for a_1 in $\mathscr{C}_1$, a_2 in $\mathscr{C}_2$—thus the set $\mathscr{C}_1 \times \mathscr{C}_2$ is just the cartesian product of the two sets. Also, if $a_1 : x_1 \to y_1$ in $\mathscr{C}_1$, $a_2 : x_2 \to y_2$ in $\mathscr{C}_2$, then we take in $\mathscr{C}_1 \times \mathscr{C}_2$

$$(a_1, a_2) : (x_1, x_2) \to (y_1, y_2).$$

The composition is defined as one would expect by

$$(b_1, b_2)(a_1, a_2) = (b_1 a_1, b_2 a_2)$$

whenever both $b_1 a_1, b_2 a_2$ are defined. It is very easy to show that $\mathscr{C}_1 \times \mathscr{C}_2$ is a category.

Notice also that if a_1, a_2 have inverses a_1^{-1}, a_2^{-1} then (a_1, a_2) has inverse (a_1^{-1}, a_2^{-1}). It follows that if $\mathscr{C}_1, \mathscr{C}_2$ are groupoids then so also is $\mathscr{C}_1 \times \mathscr{C}_2$.

Let $p_1 : \mathscr{C}_1 \times \mathscr{C}_2 \to \mathscr{C}_1$, $p_2 : \mathscr{C}_1 \times \mathscr{C}_2 \to \mathscr{C}_2$ be the obvious projection functors. Then we have the universal property: *if $f_1 : \mathscr{D} \to \mathscr{C}_1$, $f_2 : \mathscr{D} \to \mathscr{C}_2$ are functors then there is a unique functor $f : \mathscr{D} \to \mathscr{C}_1 \times \mathscr{C}_2$ such that $p_1 f = f_1$, $p_2 f = f_2$.* The proof is easy and is left to the reader. As usual, this property characterizes the product up to isomorphism.

The functor $\pi : \mathscr{T}\mathrm{op} \to \mathscr{G}\mathrm{d}$ preserves products in the following sense.

6.4.4 *If $X = X_1 \times X_2$, then πX is isomorphic to $\pi X_1 \times \pi X_2$.*

Proof The projections $p_r : X \to X_r$ $(r = 1, 2)$ induce morphisms

$$\pi p_r : \pi X \to \pi X_r$$

which, by the universal property determine

$$f : \pi X \to \pi X_1 \times \pi X_2.$$

In fact f is the identity on objects and is defined on elements by

$$f(\mathrm{cls}\ a) = (\mathrm{cls}\ p_1 a, \mathrm{cls}\ p_2 a)$$

for any path a in X.

Let $x = (x_1, x_2)$, $y = (y_1, y_2)$ belong to X. We prove that f induces a bijection

$$\pi X(x, y) \to \pi X_1(x_1, y_1) \times \pi X_2(x_2, y_2).$$

Let a, b be paths in X from x to y and suppose first that there are homotopies $F^1 : p_1 a \sim p_1 b$, $F^2 : p_2 a \sim p_2 b$. Then it is easy to check that

$$F : [0, r] \times \mathbf{I} \to X_1 \times X_2$$
$$(s, t) \rightsquigarrow (F^1(s, t), F^2(s, t))$$

is a homotopy $F : a \sim b$. Suppose next only that

$$f(\text{cls } a) = f(\text{cls } b).$$

Then we know that $p_1 a$ is equivalent to $p_1 b$, $p_2 a$ is equivalent to $p_2 b$. Because $p_1 a$, $p_2 a$ have the same length, as do $p_1 b$, $p_2 b$, we can find real numbers $r, s \geqslant 0$ large enough so that both $r + p_1 a$ is homotopic to $s + p_1 b$ and $r + p_2 a$ is homotopic to $s + p_2 b$. It follows that $r + a$ is homotopic to $s + b$ and so cls a = cls b. Thus f is injective.

In order to prove that f is surjective, let cls $a_1 \in \pi X_1$, cls $a_2 \in \pi X_2$; we may suppose a_1, a_2 are both of length 1. The path $a = (a_1, a_2)$ then satisfies $p_1 a = a_1$, $p_2 a = a_2$, and so

$$f(\text{cls } a) = (\text{cls } a_1, \text{cls } a_2).$$

Therefore f is surjective. □

The proof of 6.4.4 shows a little more than stated—in fact the morphisms

$$\pi X_1 \xleftarrow{\pi p_1} \pi(X_1 \times X_2) \xrightarrow{\pi p_2} \pi X_2$$

are a product of groupoids in the sense of the universal property.

The *coproduct* $G = G_1 \sqcup G_2$ of groupoids G_1, G_2 is a simple construction. For simplicity, let us suppose that G_1, G_2 have no common elements or objects. Then we define

$$\text{Ob}(G) = \text{Ob}(G_1) \cup \text{Ob}(G_2)$$

and define

$$G(x, y) = \begin{cases} G_1(x, y) & \text{if } x, y \in \text{Ob}(G_1) \\ G_2(x, y) & \text{if } x, y \in \text{Ob}(G_2) \\ \varnothing & \text{otherwise.} \end{cases}$$

(The modification of this construction when G_1, G_2 are not disjoint is left to the reader.)

It is very easy to prove

6.4.5 *There is an isomorphism of groupoids*

$$\pi(X_1 \sqcup X_2) \to \pi X_1 \sqcup \pi X_2.$$

EXERCISES

1. Let $\mathscr{C}$ be a category and let $X \in \mathrm{Ob}(\mathscr{C})$. Define $\mathscr{C}_X : \mathscr{C} \to \mathscr{S}et$ by $\mathscr{C}_X C = \mathscr{C}(X, C)$ for $C \in \mathrm{Ob}(\mathscr{C})$, and if $f : C \to D$ is a morphism in $\mathscr{C}$, let

$$\mathscr{C}_X f : \mathscr{C}(X, C) \to \mathscr{C}(X, D)$$
$$g \rightsquigarrow fg.$$

Prove that $\mathscr{C}_X$ is a functor.

2. Let $\mathscr{C}$, $\mathscr{D}$ be categories. A *contravariant functor* $\Gamma : \mathscr{C} \to \mathscr{D}$ assigns to each object C of $\mathscr{C}$ on object ΓC of $\mathscr{D}$ and to each morphism $f : C \to D$ in $\mathscr{C}$ a morphism $\Gamma f : \Gamma D \to \Gamma C$ in $\mathscr{D}$ subject to the axioms (i) $\Gamma 1_C = 1_{\Gamma C}$, (ii) if $f : C \to D$, $g : D \to E$ in $\mathscr{C}$, then $\Gamma(gf) = \Gamma f \Gamma g$. Prove that the contravariant functors $\mathscr{C} \to \mathscr{D}$ are determined by the functors $\mathscr{C}^{\mathrm{op}} \to \mathscr{D}$ (where $\mathscr{C}^{\mathrm{op}}$ is the dual category of $\mathscr{C}$). [A functor as defined in the text is often called *covariant*.]

3. Prove that if $X \in \mathrm{Ob}(\mathscr{C})$, then there is a contravariant functor $\mathscr{C}^X : \mathscr{C} \to \mathscr{S}et$ such that $\mathscr{C}^X C = \mathscr{C}(C, X)$.

4. Let $\Gamma : \mathscr{C} \to \mathscr{S}et$ be a functor and let $C \in \mathrm{Ob}(\mathscr{C})$. An element u of ΓC is called *universal* (for Γ) if the function

$$\mathscr{C}(C, X) \to \Gamma X, \quad f \rightsquigarrow (\Gamma f)u$$

is bijective for all X in $\mathrm{Ob}(\mathscr{C})$. If such a u exists, we say Γ is *representable* and that (C, u)—or simply C—*represents* Γ. Prove that if (C, u), (C^1, u^1) represent Γ, then there is a unique morphism $f : C \to C^1$ such that $(\Gamma f)u = u^1$, and this f is an isomorphism.

5. The definition of the previous exercise is applied to a contravariant functor $\mathscr{C} \to \mathscr{S}et$ by considering the corresponding (covariant) functor $\mathscr{C}^{\mathrm{op}} \to \mathscr{S}et$. Write out the definition in detail.

6. Let $\mathscr{C}$ be a category and let $X \in \mathrm{Ob}(\mathscr{C})$. Which objects of $\mathscr{C}$ represent $\mathscr{C}_X$, $\mathscr{C}^X$?

7. Let R be an equivalence relation in a topological space C and for each space X let ΓX be the set of maps $f : C \to X$ such that $c \mathrel{R} c^1 \Rightarrow fc = fc^1$. If $g : X \to Y$ is a map, let $\Gamma g : \Gamma X \to \Gamma Y$ be defined by composition. Prove that Γ is a functor represented by (C, p) where $p : C \to C/R$ is the identification map.

8. For which functors are the product, sum of sets universal?

9. Let $\Gamma : \mathscr{G} \to \mathscr{S}et$ be the functor which assigns to each group G its underlying set, and to each morphism the corresponding function. Prove that Γ is representable.

10. Let $\Gamma : \mathscr{T}op \to \mathscr{S}et$ be the functor which assigns to each topological space its underlying set and to each map the corresponding function. Prove that Γ is representable.

11. Let $\mathscr{V}$ be the category of vector spaces over a given field and let Λ be a set. If V is a vector space, let ΓV be the set of all functions $\Lambda \to V$. Prove that this defines a representable functor $\Gamma : \mathscr{V} \to \mathscr{S}et$.

12. Let $\mathscr{P}^{\cdot} : \mathscr{S}et \to \mathscr{S}et$ be the contravariant functor assigning to each set X the set $\mathscr{P}(X)$ and to each function $f : X \to Y$ the function $\mathscr{P}(Y) \to \mathscr{P}(X)$, $A \rightsquigarrow f^{-1}[A]$. Prove that $\mathscr{P}^{\cdot}$ is representable. Is the similar (covariant) functor $\mathscr{P}_{\cdot}$, in which $\mathscr{P}_{\cdot}(f) : \mathscr{P}(X) \to \mathscr{P}(Y)$ sends $A \rightsquigarrow f[A]$, representable?

13. Let R be an integral domain (i.e., a commutative ring with identity and no

divisors of zero). Let $\mathscr{K}$ be the category of fields and morphisms of fields (i.e., functions such that $f(x + y) = fx + fy, f(xy) = fxfy$). Let $\Gamma : \mathscr{K} \to \mathscr{S}et$ assign to each field K the set of morphisms $R \to K$. Prove that Γ is representable.
14. The *product* of pointed spaces (X, x), (Y, y) is the pointed space

$$(X \times Y, (x, y)).$$

Prove that the fundamental group of the product of pointed spaces is isomorphic to the product of their fundamental groups.
15. Define and construct coproducts of categories.
16. Let X, Y be topological spaces. Are the following statements true: $P(X \times Y)$ is isomorphic to $PX \times PY$? $P(X \sqcup Y)$ is isomorphic to $PX \sqcup PY$?

6.5 Homotopies

In defining the fundamental groupoid we have used homotopies of paths. In describing the invariance properties of the fundamental groupoid we must use homotopies of maps.

Let X, Y be topological spaces. A map $F : X \times [0, q] \to Y$ will be called a *homotopy of length q*; for such F, the *initial map* and the *final map* of F are respectively the functions

$$\begin{aligned} f : X &\to Y \\ x &\rightsquigarrow F(x, 0) \end{aligned} \qquad \begin{aligned} g : X &\to Y \\ x &\rightsquigarrow F(x, q). \end{aligned}$$

We say F is a homotopy from f to g, and we write

$$F : f \simeq g.$$

If a homotopy (of some length) $F : f \simeq g$ exists then we say f, g are *homotopic* and write $f \simeq g$.

6.5.1 *The relation $f \simeq g$ is an equivalence relation on maps $X \to Y$.*

The proof is a simple generalization of the argument for homotopies of paths, and is left to the reader. It should be emphasized that the homotopies of 6.2 were more restricted since the end points of the paths had to be fixed during the homotopy (this kind of homotopy is subsumed under the notion of homotopy rel A which we use in chapter 7).

There is a continuous surjection $\lambda : \mathbf{I} \to [0, q]$; so if $F : X \times [0, q] \to Y$ is a homotopy of length q, then $G = F(1 \times \lambda) : X \times \mathbf{I} \to Y$ is a homotopy of length 1. Also F and G have the same initial and the same final maps. So in discussing homotopies of maps it is sufficient to restrict ourselves to homotopies of length 1, and this we shall do for the rest of this chapter. We also denote a homotopy of length 1 by $F_t : X \to Y$ (where $t \in \mathbf{I}$).

6.5.2 *Let* $f : W \to X, g_0, g_1 : X \to Y, h : Y \to Z$ *be maps. If* $g_0 \simeq g_1$, *then*

$$hg_0f \simeq hg_1f : W \to Z.$$

Proof Let $g_t : g_0 \simeq g_1$ be a homotopy. Then hg_tf is a homotopy $hg_0f \simeq hg_1f$. □

6.5.3 *Let* $f_0 \simeq f_1 : X \to Y, g_0 \simeq g_1 : Z \to W$. *Then* $f_0 \times g_0 \simeq f_1 \times g_1$.

Proof If $f_t : f_0 \simeq f_1, g_t : g_0 \simeq g_1$, then the required homotopy is $f_t \times g_t$. The detailed proof of continuity of $f_t \times g_t$ (as a function $X \times Z \times \mathbf{I} \to Y \times W$) is left to the reader [cf. Example 2, p. 33]. □

Let $f : X \to Y, g : Y \to X$ be maps. If

$$fg \simeq 1_Y$$

then we say g is a *right homotopy inverse* of f, that f is a *left homotopy inverse* of g, and that *X dominates Y*. If g is both a left and a right homotopy inverse of f, then g is called simply a *homotopy inverse* of f; further f is called a *homotopy equivalence* and we write $f : X \simeq Y$. If a homotopy equivalence $X \to Y$ exists, then we say X, Y are homotopy equivalent (or of the same *homotopy type*) we write $X \simeq Y$. (In much of the literature this relation is written $\equiv$). This relation is easily seen to be an equivalence relation on topological spaces.

EXAMPLES 1. Let Y be a convex subset of $\mathbf{R}^n$ and let $f_0, f_1 : X \to Y$ be maps. Then $f_t = (1 - t)f_0 + tf_1$ is a homotopy $f_0 \simeq f_1$.

2. A map is *inessential* if it is homotopic to a constant map—otherwise it is *essential*. For example, any map to a convex subset of $\mathbf{R}^n$ is inessential, by Example 1.

3. A space X is *contractible* if it is of the homotopy type of a space with only one point. In fact, X is *contractible if and only if the identity map* $1_X : X \to X$ *is inessential.*

Proof $\Rightarrow$ Suppose $f : X \simeq Y$ has homotopy inverse g, where Y is a single point space. Then $gf \simeq 1_X$ and gf is a constant map. Therefore 1_X is inessential.

$\Leftarrow$ Let $1_X \simeq f$ where $f : X \to X$ is a constant map with value x say. Let $Y = \{x\}$, let $i : Y \to X$ be the inclusion and let $f' : X \to Y$ be the unique map. Then $f'i = 1_Y$ and $if' = f \simeq 1_X$. □

It follows from these examples that any convex subset of $\mathbf{R}^n$ is contractible.

4. Let $f_t : X \to Y$ be a homotopy. Then for each x in X the map $t \rightsquigarrow f_tx$ is a path in Y from f_0x to f_1x_1 and so f_0x, f_1x lie in the same path-component of Y. It follows easily that, if X has more than one path-component, then $1 : X \to X$ is an essential map.

5. Let $y, z \in \mathbf{S}^1$ and let $g, h : \mathbf{S}^1 \to \mathbf{S}^1 \times \mathbf{S}^1$ be the maps $x \rightsquigarrow (x, y)$, $x \rightsquigarrow (x, z)$ respectively [Fig. 6.9]. Then it is easy to prove that $g \simeq h$—in fact if a is any path of length 1 in $\mathbf{S}^1$ from y to z, then $f_t : x \rightsquigarrow (x, at)$ is a homotopy $g \simeq h$. However, if k is the map $x \rightsquigarrow (y, x)$, then it is true that g is *not* homotopic to k, but the proof needs more theory than we have yet developed—in particular, we need to know that $\mathbf{S}^1$ is not simply-connected.

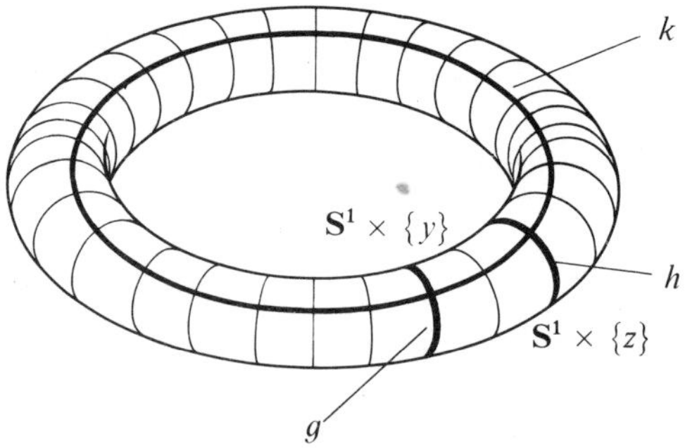

Fig. 6.9

We must now investigate the concept in the algebra of groupoids corresponding to homotopy of maps of spaces. For this it is convenient first to say something about functors on a product of categories.

Let $F : \mathscr{C} \times \mathscr{D} \to \mathscr{E}$ be a functor, where $\mathscr{C}, \mathscr{D}, \mathscr{E}$ are categories. If 1_x is the identity at x in $\mathscr{C}$, then let us write $F(x, b)$ for $F(1_x, b)$ where b is any morphism in $\mathscr{D}$. Similarly, let us write $F(a, y)$ for $F(a, 1_y)$ for any object y of $\mathscr{D}$ and morphism a of $\mathscr{C}$. Then, as is easily verified, $F(x, \ )$ is a functor $\mathscr{D} \to \mathscr{E}$ (called the *x-section* of F) and $F(\ , y)$ is a functor $\mathscr{C} \to \mathscr{E}$ (called the *y-section* of F). These two functors determine F. If $a : x \to x', b : y \to y'$ are morphisms in $\mathscr{C}, \mathscr{D}$ respectively then we have a commutative diagram

$$\begin{array}{ccc} F(x, y) & \xrightarrow{F(a, y)} & F(x', y) \\ {\scriptstyle F(x, b)}\downarrow & \searrow^{F(a, b)} & \downarrow{\scriptstyle F(x', b)} \\ F(x, y') & \xrightarrow[F(a, y')]{} & F(x', y') \end{array} \qquad (6.5.4)$$

since $F(1_{x'}a, b1_y) = F(a, b) = F(a1_x, 1_{y'}b)$.

6.5.5 *Suppose for each x in* Ob($\mathscr{C}$) *and y in* Ob($\mathscr{D}$) *we are given functors*

$$F(x, \ ) : \mathscr{D} \to \mathscr{E}, F(\ , y) : \mathscr{C} \to \mathscr{E}$$

such that $F(x, y)$ is a unique object of $\mathscr{E}$. Suppose for each $a : x \to x'$ in $\mathscr{C}$ and $b : y \to y'$ in $\mathscr{D}$ the outer square of (6.5.4) *commutes. Then the diagonal composite $F(a, b)$ makes F a functor $\mathscr{C} \times \mathscr{D} \to \mathscr{E}$. All functors $\mathscr{C} \times \mathscr{D} \to \mathscr{E}$ arise in this way.*

Proof The verification of FUN 1 for F is easy since

$$\begin{aligned} F(1_x, 1_y) &= F(1_x, y)F(x, 1_y) \\ &= 1_{F(x, y)}1_{F(x, y)} \\ &= 1_{F(x, y)}. \end{aligned}$$

The verification of FUN 2 involves a diagram of four commutative squares, and is left to the reader. The last statement is clear from the discussion preceding 6.5.5. □

In order to model the notion of homotopy we need a model for categories of the unit interval. This is provided by the tree groupoid I.

Let $\mathscr{C}$ and $\mathscr{E}$ be categories. A *homotopy* (or *natural equivalence*) of functors from $\mathscr{C}$ to $\mathscr{E}$ is a functor

$$F : \mathscr{C} \times I \to \mathscr{E}.$$

The *initial functor* of F is then $f = F(\ , 0)$ and the *final functor* of F is $g = F(\ , 1)$; we say F is a homotopy from f to g and write $F : f \simeq g$. If such a homotopy from f to g exists then we say f, g are *homotopic* and write $f \simeq g$.†

According to 6.5.5, in order to specify a homotopy F it is sufficient to give the initial and final functors f and g of F, and also for each object x of $\mathscr{C}$ a functor $F(x,\ ) : I \to \mathscr{E}$ in such a way that the outside of (6.5.4) commutes. However, the functors $F(x,\ ) : I \to \mathscr{E}$ are entirely specified by invertible elements θx of $\mathscr{E}$ where $\theta x = F(x, \iota)$. In these terms, (6.5.4) becomes (with $b = \iota$)

$$\begin{array}{ccc} fx & \xrightarrow{fa} & fy \\ {\scriptstyle \theta x}\downarrow & & \downarrow{\scriptstyle \theta y} \\ gx & \xrightarrow[ga]{} & gy \end{array}$$

the commutativity of which asserts

$$(ga)(\theta x) = (\theta y)(fa). \tag{6.5.6}$$

† This definition of homotopy was pointed out to me by P. J. Higgins. I am grateful to W. F. Newns for suggesting that the emphasis be placed on this definition (rather than that by the function θ as below) and for other helpful comments on this section and section 6.7.

Since θx is invertible this shows that g is determined by f and θ. Thus, given any functor $f : \mathscr{C} \to \mathscr{E}$ and for each object x of $\mathscr{C}$ an invertible element θx of $\mathscr{E}$ with initial point fx, then there is a homotopy $f \simeq g$ where g is defined by (6.5.6). We call θ a *homotopy function* from f to g and write also $\theta : f \simeq g$.

The function θ also gives rise to the useful diagram

$$\begin{array}{ccc} \mathscr{C}(x, y) & \xrightarrow{\ f\ } & \mathscr{E}(fx, fy) \\ & \searrow^{g} & \big\downarrow \theta(x, y) \\ & & \mathscr{E}(gx, gy) \end{array} \qquad (6.5.7)$$

in which $\theta(x, y)$ is the bijection $a' \rightsquigarrow (\theta y)a'(\theta x)^{-1}$. The commutativity of (6.5.7) is immediate from (6.5.7) with $a' = fa$.

6.5.8 *Homotopy of functors is an equivalence relation.*

Proof (*a*) The function $x \rightsquigarrow 1_{fx}$ is a homotopy function $f \simeq f$.
(*b*) If θ is a homotopy function $f \simeq g$, then $x \rightsquigarrow (\theta x)^{-1}$ is a homotopy function $g \simeq f$.
(*c*) If θ, φ are homotopy functions $f \simeq g$, $g \simeq h$ respectively then $x \rightsquigarrow \varphi x\, \theta x$ is a homotopy function $f \simeq h$. □
The proof of the following proposition is left as an exercise.

6.5.9 *Let $f : \mathscr{C} \to \mathscr{D}$, $g : \mathscr{D} \to \mathscr{E}$, $h : \mathscr{E} \to \mathscr{F}$ be functors and suppose $g \simeq g'$. Then $hgf \simeq hg'f$.*

Exactly as for topological spaces we have the notions of *homotopy inverse* of a functor, and homotopy equivalence of categories. Further, all these notions apply *a fortiori* to groupoids and morphisms of groupoids.

The utility for topology of these notions is shown by

6.5.10 *If $f, g : X \to Y$ are homotopic maps of spaces, then the induced morphisms $\pi f, \pi g : \pi X \to \pi Y$ are homotopic.*

Proof Let $F : X \times \mathbf{I} \to Y$ be a homotopy from f to g. Consider the composite

$$\pi X \times I \to \pi X \times \pi \mathbf{I} \to \pi(X \times \mathbf{I}) \xrightarrow{\pi F} \pi Y$$

in which the first morphism is inclusion (since $I \subset \pi\mathbf{I}$) and the second morphism is the isomorphism constructed in 6.4.4. We prove that the composite G is a homotopy $\pi f \simeq \pi g$.

Let $\alpha \in \pi X(x, y)$, let $a \in \alpha$ be of length 1, and let c_ε be a constant path at

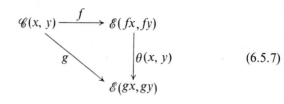

ε of length 1 for $\varepsilon = 0, 1$. Then

$$\begin{aligned} G(\alpha, \varepsilon) &= (\pi F)\, \varphi\, (\alpha, \varepsilon) \\ &= (\pi F)\,(\text{cls}\,(a, c_\varepsilon)) \\ &= \text{cls}\, F(a, c_\varepsilon) \\ &= \begin{cases} \text{cls}\, fa, & \varepsilon = 0 \\ \text{cls}\, ga, & \varepsilon = 1. \end{cases} \end{aligned} \quad \square$$

6.5.10 (*Corollary 1*) *If $f : X \to Y$ is a homotopy equivalence of spaces, then $\pi f : \pi X \to \pi Y$ is a homotopy equivalence of groupoids.*

6.5.11 *Let $f : \mathscr{C} \to \mathscr{C}$ be a functor such that $f \simeq 1$. Then for all objects x, y of $\mathscr{C}$*

$$f : \mathscr{C}(x, y) \to \mathscr{C}(fx, fy)$$

is a bijection.

Proof This is immediate from the commutativity of (6.5.7) and the fact that $\theta(x, y)$ is a bijection. $\square$

6.5.12 *Let $f : \mathscr{C} \to \mathscr{E}$ be a homotopy equivalence of categories. Then for all objects x, y of $\mathscr{C}$, $f : \mathscr{C}(x, y) \to \mathscr{E}(fx, fy)$ is a bijection. Hence, if f is bijective on objects then f is an isomorphism.*

Proof Let $g : \mathscr{E} \to \mathscr{C}$ be a homotopy inverse of f, so that $gf \simeq 1, fg \simeq 1$. Consider the functions,

$$\mathscr{C}(x, y) \xrightarrow{f} \mathscr{E}(fx, fy) \xrightarrow{g} \mathscr{C}(gfx, gfy) \xrightarrow{f} \mathscr{E}(fgfx, fgfy).$$

By 6.5.11, the composites of the first two, and of the last two, functions are bijections. It follows easily that each function is a bijection. The last part follows from this and 6.4.3. $\square$

6.5.12 (*Corollary 1*) *If $f : X \to Y$ is a homotopy equivalence of spaces, then for each x in X, $\pi f : \pi(X, x) \to \pi(Y, fx)$ is an isomorphism of fundamental groups.*

The application of this result to spaces must wait until we can compute more readily the fundamental groupoid. However 6.5.10 (corollary 1) focuses attention on the homotopy type of groupoids and so, more generally, of categories. The rest of this section is devoted to a simple result which enables us to replace a given category by a simpler but homotopically equivalent category. This process is especially useful for computations of the fundamental groupoid—in fact πX has an embarrassingly large number of objects, and so any simplification of πX is welcome.

Let $F : \mathscr{C} \times I \to \mathscr{E}$ be a homotopy, where $\mathscr{C}$ and $\mathscr{E}$ are categories. The homotopy is called *constant* if

$$F(a, 0) = F(a, \iota) = F(a, 1)$$

for all elements a of $\mathscr{C}$. Let θ be the homotopy function defined by F (so that $\theta x = F(x, \iota)$, $x \in \text{Ob}(\mathscr{C})$). Then clearly F is constant if and only if θx is an identity of $\mathscr{E}$ for all x.

Suppose that $\mathscr{D}$ is a subcategory of $\mathscr{C}$. Then we say F is a *homotopy* rel $\mathscr{D}$ if $F \mid \mathscr{D} \times I$ is a constant homotopy, or, equivalently, if $\theta x = 1$ for all x in $\text{Ob}(\mathscr{D})$. In such case, the initial and final functors f, g of F are said to be *homotopic* rel $\mathscr{D}$, written $f \simeq g$ rel $\mathscr{D}$; these functors must of course agree on $\mathscr{D}$, but this alone, or even with the existence of a homotopy $f \simeq g$, is not enough to ensure $f \simeq g$ rel $\mathscr{D}$.

A subcategory $\mathscr{D}$ of $\mathscr{C}$ is a *deformation retract* of $\mathscr{C}$ if there is a functor $r : \mathscr{C} \to \mathscr{D}$ such that $ir \simeq 1_{\mathscr{C}}$ rel $\mathscr{D}$, where $i : \mathscr{D} \to \mathscr{C}$ is the inclusion. Such a functor r is called a *deformation retraction*. It is a retraction, since $ir \mid \mathscr{D} = 1_{\mathscr{C}} \mid \mathscr{D} = i$ and so $ri = 1_{\mathscr{D}}$. Further, if F is the homotopy $ir \simeq 1_{\mathscr{C}}$ and $a : x \to x'$ is an element of $\mathscr{C}$, then by 6.5.4 with $b = \iota : 0 \to 1$

$$F(x', \iota)ir(a) = aF(x, \iota).$$

In particular, let $a = F(x', \iota) : rx' \to x'$; then $F(x, \iota) = 1$ since F is rel $\mathscr{D}$. So we obtain

$$F(x', \iota)r(a) = a$$

whence $rF(x', \iota) = 1$. This shows that rF is the constant homotopy $r \simeq r$.

We note also that if $\mathscr{D}$ is a deformation retract of $\mathscr{C}$ then $i : \mathscr{D} \to \mathscr{C}$ is a homotopy equivalence.

To characterize deformation retracts we need a further definition. A subcategory $\mathscr{D}$ of $\mathscr{C}$ is *representative* in $\mathscr{C}$ if each object of $\mathscr{C}$ is isomorphic to an object of $\mathscr{D}$.

6.5.13 *A subcategory $\mathscr{D}$ of $\mathscr{C}$ is a deformation retract of $\mathscr{C}$ if and only if $\mathscr{D}$ is a full, representative subcategory of $\mathscr{C}$. In fact, if $\mathscr{D}$ is a full representative subcategory of $\mathscr{C}$ and if we define $\theta x = 1_x$ for each object x of $\mathscr{D}$, and choose θy for any other object y of $\mathscr{C}$ to be an isomorphism of some x in* $\text{Ob}(\mathscr{D})$ *with y, then θ determines a deformation retraction $r : \mathscr{C} \to \mathscr{D}$ and a homotopy $ir \simeq 1$* rel $\mathscr{D}$, *where $i : \mathscr{D} \to \mathscr{C}$ is the inclusion.*

Proof We leave the proof of 'only if' as an exercise. That θ can be chosen follows from the fact that $\mathscr{D}$ is representative in $\mathscr{C}$. By the remarks following (6.5.6), θ^{-1} determines a homotopy $1_{\mathscr{C}} \simeq g$ for some functor g, and this homotopy is rel $\mathscr{D}$ by construction. Now gy is an object of $\mathscr{D}$ for each object y of $\mathscr{C}$; since $\mathscr{D}$ is full it follows that $ga \in \mathscr{D}$ for each element a of $\mathscr{C}$. Therefore, we can write $g = ir$ for some functor $r : \mathscr{C} \to \mathscr{D}$. □

6.5.13 (Corollary 1) *Any groupoid is of the homotopy type of a totally disconnected groupoid.*

Proof Let G be a groupoid and H a full subgroupoid of G consisting of one object group in each component of G. By 6.5.13, the inclusion $H \to G$ is a homotopy equivalence. □

This shows that the interesting invariant of homotopy type of a space X is a groupoid consisting of one object group in each path-component of X. It would thus seem likely that groupoids which are not totally disconnected are of little interest.

We shall see later that this view is not valid. In fact, groupoids and groupoid morphisms carry information (essentially of a graph theoretic character) which it is more difficult to describe in terms of groups alone.

EXERCISES

1. Prove that a category $\mathscr{HT}$op is defined as follows: the objects of $\mathscr{HT}$op are the topological spaces; the maps $X \to Y$ are the homotopy classes of maps $X \to Y$; composition is defined by

$$(\text{cls } g)(\text{cls } f) = \text{cls}(gf).$$

Define $\mathscr{HG}$d similarly and prove that $\pi : \mathscr{T}\text{op} \to \mathscr{G}\text{d}$ determines a functor $\mathscr{HT}\text{op} \to \mathscr{HG}\text{d}$.

2. For any space X, let $\pi_0 X$ be the set of path components of X. Prove that π_0 determines a functor $\mathscr{HT}\text{op} \to \mathscr{S}\text{et}$. Deduce that if $X \simeq Y$, then ※$(\pi_0 X) =$ ※$(\pi_0 Y)$.

3. Prove that if X, X', Y, Y' are spaces and $X \simeq X'$, $Y \simeq Y'$ then $X \times Y \simeq X' \times Y'$, $X \sqcup Y \simeq X' \sqcup Y'$.

4. Prove that if Y is contractible then any maps $X \to Y$ are homotopic. Prove that if X and Y are contractible then any map $X \to Y$ is a homotopy equivalence. Prove also that a retract of a contractible space is contractible. Prove results for groupoids similar to those of this and the preceding two exercises.

5. Let $\mathscr{V}_K$ be the category of finite dimensional vector spaces over a given field K. Let $\mathscr{K}$ be the full subcategory of $\mathscr{V}_K$ on the vector spaces K^n, $n \geqslant 0$. Prove that $\mathscr{K}$ is a deformation retract of $\mathscr{V}_K$.

6. Let $I^{\to}$ be the category with two objects 0, 1 and only one non-identity element $\iota : 0 \to 1$. If $F : \mathscr{C} \times I^{\to} \to \mathscr{E}$ is a functor, then $f = F(\ ,\ 0)$ and $g = F(\ ,1)$ are called the *initial* and *final* functors of F, and F is called a *morphism*, or *natural transformation*, from f to g. Prove that there is a category $\mathscr{F}\text{un}(\mathscr{C}, \mathscr{E})$, whose objects are the functors $\mathscr{C} \to \mathscr{E}$, whose morphisms are the morphisms of functors and such that the invertible elements of this category are the homotopies.

7. Let $\mathscr{V}$ be the category of vector spaces over a (commutative) field K. For each object V of $\mathscr{V}$ let $V^* = \mathscr{V}(V, K)$, considered again as a vector space over K, and let $gV = V^{**}$; let $\theta V : V \to V^{**}$ be the function $v \rightsquigarrow (\lambda \rightsquigarrow \lambda v)$. Prove that g is a functor $\mathscr{V} \to \mathscr{V}$ and that θ determines a morphism $1_{\mathscr{V}} \to g$ such that θV is invertible if V is of finite dimension.

8. Let $\Gamma : \mathscr{C} \to \mathscr{S}et$ be a functor represented by (C, u) and let $\Delta : \mathscr{C} \to \mathscr{S}et$ be any functor. Prove that the function

$$\begin{aligned} \mathscr{F}un\,(\Gamma, \Delta) &\to \Delta C, \\ F &\rightsquigarrow F(C, 1)(u) \end{aligned}$$

is a bijection. (This says that the natural transformations of a representable functor are entirely determined by their values on the universal element.)

9. Prove that, if C, D are objects of $\mathscr{C}$, then there is a bijection

$$\mathscr{F}un\,(\mathscr{C}_C, \mathscr{C}_D) \to \mathscr{C}(C, D).$$

10. Prove that a functor $\Gamma : \mathscr{C} \to \mathscr{S}et$ is representable by C if and only if there is a homotopy $\Gamma \simeq \mathscr{C}_C$.

11. Let $\mathscr{D}$ be a category. Prove that if $\hat{\mathscr{D}}$ assigns to each pair of x, y of $\mathscr{D}$ the set $\mathscr{D}(x, y)$ and to each pair of morphisms $a : x \to x'$, $b : y \to y'$ the morphism

$$\begin{aligned} \mathscr{D}(a, b) : \mathscr{D}(x', y) &\to \mathscr{D}(x, y'), \\ c &\rightsquigarrow bca \end{aligned}$$

then $\hat{\mathscr{D}}$ is a functor $\mathscr{D}^{op} \times \mathscr{D} \to \mathscr{S}et$.

12. Let $\mathscr{C}$, $\mathscr{D}$ be categories and $\Gamma : \mathscr{C}^{op} \times \mathscr{D} \to \mathscr{S}et$ a functor such that for each object C of $\mathscr{C}$ the functor $\Gamma(C, \ )$ is representable by $(\Delta C, uC)$. Prove that the function $C \rightsquigarrow \Delta C$ of objects extends uniquely to a functor $\Delta : \mathscr{C} \to \mathscr{D}$ such that the bijection

$$\begin{aligned} \mathscr{D}(\Delta C, D) &\to \Gamma(C, D), \\ f &\rightsquigarrow \Gamma(C, f)(uC) \end{aligned}$$

is a homotopy $\hat{\mathscr{D}}(\Delta \times 1) \simeq \Gamma$(where $\hat{\mathscr{D}}$ is as in Exercise 11).

13. Let $\mathscr{C}$, $\mathscr{D}$, $\mathscr{E}$ be categories. Prove that there is an isomorphism of categories

$$\mathscr{F}un\,(\mathscr{C}, \mathscr{F}un\,(\mathscr{D}, \mathscr{E})) \to \mathscr{F}un\,(\mathscr{C} \times \mathscr{D}, \mathscr{E}).$$

14. A functor $f : \mathscr{C} \to \mathscr{D}$ induces for all objects x, y of $\mathscr{C}$ a function $f : \mathscr{C}(x, y) \to \mathscr{D}(fx, fy)$. If this function is injective for all x, y then f is called *faithful*; if it is surjective for all x, y then f is called *full*; finally if each object z of $\mathscr{D}$ is isomorphic to some object fx, then f is called *representative*. Prove that f is a homotopy equivalence of categories if and only if f is full, faithful and representative. Prove also that if g is a homotopy inverse of f, and if θ is a homotopy function $fg \simeq 1_{\mathscr{D}}$, then we can choose a homotopy function $\varphi : gf \simeq 1_{\mathscr{C}}$ such that $f\varphi = \theta f$, $g\theta = \varphi g$.

15. Let both the path functor P and the fundamental groupoid functor π be considered as functors $\mathscr{T}op \to \mathscr{C}at$. Prove that the assignment to each topological space X of the projection $p_. : PX \to \pi X$ defines a natural transformation $P \to \pi$.

6.6 Coproducts and pushouts

We have already used in several categories the idea of a coproduct. It seems reasonable to formulate now the general definition. We shall also define pushouts in a general category—there is a close relation between coproducts and pushouts which is presented briefly in the Exercises.

Let $\mathscr{C}$ be a category. A *coproduct* (or *sum*) of two objects C_1, C_2 of $\mathscr{C}$ is a diagram

$$C_1 \xrightarrow{i_1} C \xleftarrow{i_2} C_2 \tag{6.6.1}$$

of morphisms of $\mathscr{C}$ with the following φ-universal property: for any diagram

$$C_1 \xrightarrow{v_1} C' \xleftarrow{v_2} C_2$$

of morphisms of $\mathscr{C}$, there is a unique morphism $v : C \to C'$ such that

$$vi_1 = v_1, \qquad vi_2 = v_2.$$

The usual universal argument shows that this property characterizes coproducts up to isomorphism. If (6.6.1) is a coproduct in $\mathscr{C}$ it is usual to write

$$C = C_1 \sqcup C_2$$

and by an abuse of language, to refer to C (rather than i_1, i_2) as the coproduct of C_1 and C_2.

This definition is of course a simple extension to arbitrary categories of definitions encountered already in the categories $\mathscr{S}$et, $\mathscr{T}$op, and $\mathscr{G}$d—in each of these categories, any two objects have a coproduct. This is also true in the category $\mathscr{G}$, but the proof, which will be given in chapter 8, is non-trivial.

We now discuss pushouts. A diagram

$$\begin{array}{ccc} C_0 & \xrightarrow{i_1} & C_1 \\ {\scriptstyle i_2}\downarrow & & \downarrow{\scriptstyle u_1} \\ C_2 & \xrightarrow[u_2]{} & C \end{array} \tag{6.6.2}$$

of morphisms of $\mathscr{C}$ is called a *pushout* (of i_1, i_2) if

(*a*) The diagram is commutative: that is $u_1 i_1 = u_2 i_2$.

(*b*) u_1, u_2 are φ-universal for property (*a*); that is, if the diagram

$$\begin{array}{ccc} C_0 & \xrightarrow{i_1} & C_1 \\ {\scriptstyle i_2}\downarrow & & \downarrow{\scriptstyle v_1} \\ C_2 & \xrightarrow[v_2]{} & C' \end{array} \tag{6.6.3}$$

of morphisms of $\mathscr{C}$ is commutative, then there is a unique morphism $v : C \to C'$ such that $vu_1 = v_1$, $vu_2 = v_2$.

The usual universal argument shows that if (6.6.2) is a pushout, then (6.6.3) is a pushout if and only if there is an isomorphism $v : C \to C'$ such that $vu_\alpha = v_\alpha$, $\alpha = 1, 2$. Thus a pushout is determined up to isomorphism by i_1, i_2.

If (6.6.2) is a pushout it is usual to write

$$C = C_1 \; {}_{i_1}{\sqcup}_{i_2} \; C_2$$

and, by an abuse of language, to refer to C itself as the pushout of i_1, i_2.

It is important to note that we have not asserted that pushouts exist for any $\mathscr{C}$ and i_1, i_2. The existence of arbitrary pushouts is to be regarded as a good property of $\mathscr{C}$. In the exercises we give a condition for pushouts to exist—here we are more concerned with giving useful lemmas for proving that a particular diagram is a pushout.

Suppose given in $\mathscr{C}$ a commutative diagram

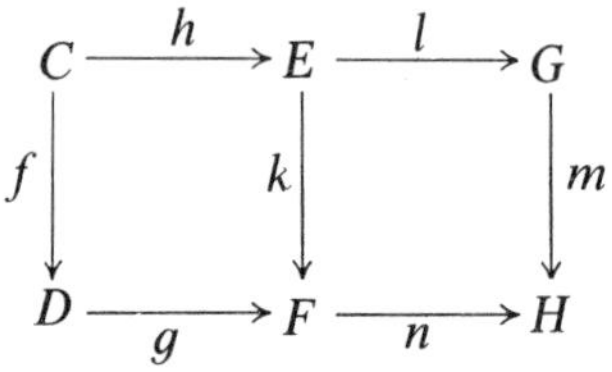

(6.6.4)

Then the outer part of this diagram is also a commutative square which we call the *composite* of the two individual squares. The abstract algebra of such compositions is a 'double category', since there are possible compositions both horizontally and vertically [cf. Ehresmann [1]]. But we shall need only the horizontal composition.

6.6.5 *In any category $\mathscr{C}$, the composite of two pushouts is a pushout.*

Proof We use the notation of (6.6.4). Suppose given a commutative diagram

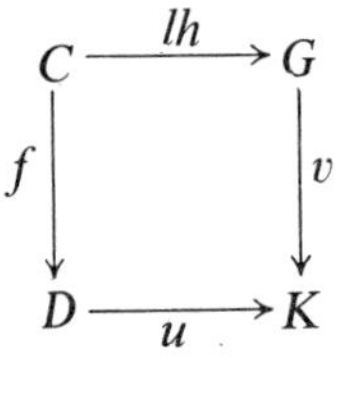

By the pushout property for the first square, there is a unique morphism $w : F \to K$ such that

$$wg = u, \qquad wk = vl.$$

The two maps w and v determine, by the pushout property of the second square, a unique map $x : H \to K$ such that

$$xn = w, \qquad xm = v;$$

it follows that

$$xng = u, \qquad xm = v.$$

To complete the proof we must show that x is the only morphism satisfying these last equations.

Suppose that x' satisfies

$$x'ng = u, \qquad x'm = v.$$

Then $x'ng = u, x'nk = x'ml = vl$. By uniqueness of the construction of w, $x'n = w$. Further, by uniqueness of the construction of x, $x' = x$. □

Suppose, in particular, that $\mathscr{C}$ is the category $\mathscr{T}op$ and that $f : C \to D$ is the inclusion of the closed sub-space C of D. Then F, H are adjunction spaces and 6.6.5 can be stated as

$$G \,{}_{l}\!\sqcup (E \,{}_{h}\!\sqcup D) = G \,{}_{lh}\!\sqcup D.$$

Other applications of 6.6.5 (mainly to groupoids) will occur later.

Our next result says, roughly, that a retract of a pushout is a pushout. Now the term retract is meaningful in any category, so in order to give meaning to the last sentence it is enough to define morphisms of commutative squares.

Consider the following diagram of morphisms of $\mathscr{C}$

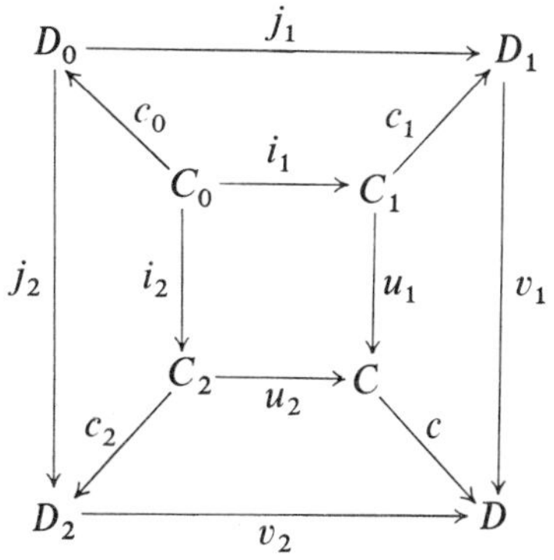

(6.6.6)

Let us write C for the inside square and D for the outer square, both of which we suppose commutative. If the whole diagram is commutative, then it is called a *morphism* $\mathsf{c} : \mathsf{C} \to \mathsf{D}$. Clearly we have an identity morphism $\mathsf{C} \to \mathsf{C}$, and the composite of morphisms is again a morphism.

So we have a category $\mathscr{C}_\square$ of commutative squares and morphisms of squares and in this category the notion of retraction is well-defined.

6.6.7 *Let* C, D *be commutative squares in* $\mathscr{C}$ *such that* D *is a pushout. If there is a retraction* D → C *then* C *is a pushout.*

Proof Let c : C → D, d : D → C be morphisms such that d c = 1_{C}. Suppose also (referring to the diagram (6.6.6)) that we are given morphisms

$$C_1 \xrightarrow{w_1} C' \xleftarrow{w_2} C_2$$

such that $w_1 i_1 = w_2 i_2$. Consider the morphisms

$$D_1 \xrightarrow{w_1 d_1} C' \xleftarrow{w_2 d_2} D_2.$$

Since d is a map of squares, $d_1 f_1 = i_1 c_0$, $d_2 j_2 = i_2 c_0$. Hence

$$(w_1 d_1) j_1 = (w_2 d_2) j_2.$$

Since D is a pushout, there is a unique morphism $x : D \to C'$ such that

$$x v_1 = w_1 d_1, \qquad x v_2 = w_2 d_2.$$

Let $w = xc : C \to C'$. Then for $\alpha = 1, 2$.

$$w u_\alpha = x c u_\alpha = x v_\alpha c_\alpha = w_\alpha d_\alpha c_\alpha = w_\alpha$$

as we required.

Suppose $w' : C \to C'$ also satisfied $w'_\alpha u_\alpha = w_\alpha$, $\alpha = 1, 2$. Then

$$w' d v_\alpha = w' u_\alpha d_\alpha = w_\alpha d_\alpha = x v_\alpha, \quad \alpha = 1, 2.$$

Hence $w'\mathrm{d} = x$ and so $w' = w'\mathrm{dc} = xc = w$. □

EXERCISES

1. Let $\mathscr{C}$ be a category and let C_1, C_2 be objects of $\mathscr{C}$. Let $\mathscr{C}$ be the category whose objects are diagrams $C_1 \xrightarrow{i_1} C \xleftarrow{i_2} C_2$ (written (i_1, i_2, C)) and whose morphisms (i'_1, i'_2, d') are morphisms $f : C \to C$ of $\mathscr{C}$ such that $f i_\alpha = i'_\alpha$, $\alpha = 1, 2$. Prove that a copoint of $\mathscr{C}'$ [Exercise 7 of 6.1] is a coproduct diagram $C_1 \to C_1 \sqcup C_2 \leftarrow C_2$.
2. Let $\mathscr{C}$ be a category and let C_1, C_2 be objects of $\mathscr{C}$. A *product* of C_1, C_2 is a point in the category $\mathscr{C}''$ of diagrams (p_1, p_2, C) (where $p_1 : C \to C_1$, $p_2 : C \to C_2$) and morphisms of such diagrams. Prove that the category of groups admits products.
3. Prove that if a category has a copoint and arbitrary pushouts, then it has coproducts.
4. Let $f, g : A \to B$ be morphisms in $\mathscr{C}$. A *difference cokernel* of f, g is a morphism $c : B \to C$ for some C such that c is φ-universal for the property $cf = cg$. Prove that the categories $\mathscr{S}$et, $\mathscr{T}$op, $\mathscr{G}$ all admit arbitrary difference cokernels.

5. Prove that if a category $\mathscr{C}$ has coproducts and difference cokernels, then it has arbitrary pushouts.
6. Let $f : A \to B, g : B \to C$ be morphisms in $\mathscr{C}$. Prove that the square

$$\begin{array}{ccc} A & \xrightarrow{f} & B \\ {\scriptstyle gf}\downarrow & & \downarrow{\scriptstyle g} \\ C & \xrightarrow[1]{} & C \end{array}$$

is a pushout if f is epic.
7. Prove that if in diagram (6.6.4) the composite square is a pushout, and g is epic, then the right-hand square is a pushout.
8. Prove that the category of pointed spaces has a coproduct. [This coproduct is called the *wedge* and is written $X \vee Y$.]
9. The *square category* $\square$ has four objects 0, 1, 2, 3 and it has non-identity morphisms $i_1 : 0 \to 1, i_2 : 0 \to 2, u_1 : 1 \to 3, u_2 : 2 \to 3, u : 0 \to 3$ with the rule $u_1 i_1 = u_2 i_2 = u$. Prove that this does specify a category, that an object of $\mathscr{C}_\square$ corresponds to a functor $C : \square \to \mathscr{C}$, and that a morphism of squares corresponds to a morphism of functors $\square \to \mathscr{C}$.
10. The *wedge category* $\vee$ has three objects 0, 1, 2 and two non-identity morphisms $i_1 : 0 \to 1, i_2 : 0 \to 2$. Prove that a commutative square in a category $\mathscr{C}$ can be regarded as a morphism $\Gamma \to \Gamma'$ of functors $\vee \to \mathscr{C}$ such that Γ' is a constant functor.
11. Let $\Gamma : \mathscr{C} \to \mathscr{D}$ be a functor. A morphism $\eta : \Gamma \to \Gamma'$ of functors is said to be a *right-root* of Γ if (i) $\Gamma' : \mathscr{C} \to \mathscr{D}$ is a constant functor, (ii) η is φ-universal for morphisms of Γ to a constant functor: that is, if $\delta : \Gamma \to \Delta$ is a morphism such that Δ is constant then there is a unique morphism $\delta' : \Gamma' \to \Delta$ such that $\delta' \eta = \delta$. Prove that a pushout square in $\mathscr{C}$ can be regarded as a right-root of a functor $\vee \to \mathscr{C}$. [Mitchell [1] uses *limit* for right-root.]
12. Let $\mathscr{U}$ be an open cover of a space X. Regard $\mathscr{U}$ as a subcategory of $\mathscr{T}op$ with objects the elements of $\mathscr{U}$ and morphisms all inclusions $U \to V$ such that $U, V \in \mathscr{U}$. Prove that the inclusion functor $\mathscr{U} \to \mathscr{T}op$ has as right-root a morphism to the constant functor with value X.
13. Given categories $\mathscr{C}$ and $\mathscr{D}$ there is a category whose objects are the morphisms of functors $\mathscr{C} \to \mathscr{D}$—this category is essentially $\mathscr{F}un\ (\mathrm{I}^{\to}, \mathscr{F}un\ (\mathscr{C}, \mathscr{D}))$ and it is isomorphic to $\mathscr{F}un\ (\mathscr{C} \times I^{\to}, \mathscr{D})$. In any case, the notion of a retract of a morphism of functors is well defined. Prove that a retract of a right-root is again a right-root. [This generalizes 6.6.7.]
14. The *dual* of a construction for categories is obtained in a category by carrying out this construction in the opposite category $\mathscr{C}^{op}$ and then transferring the construction to $\mathscr{C}$ by means of the obvious contravariant functor $D : \mathscr{C}^{op} \to \mathscr{C}$. In this way a coproduct $C_1 \sqcup C_2$ in $\mathscr{C}^{op}$ becomes a *product* $C_1 \sqcap C_2$ in $\mathscr{C}$; a pushout in $\mathscr{C}^{op}$ becomes a *pullback* in $\mathscr{C}$; difference cokernel becomes *difference kernel*;

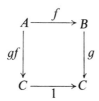

point becomes copoint and copoint becomes point. Write out the definitions in $\mathscr{C}$ of these constructions and discuss their properties.

6.7 The fundamental groupoid of a union of spaces

Throughout this section let X be a topological space and let X_0, X_1, X_2 be subspaces of X such that $X_0 = X_1 \cap X_2$ and the interiors of X_1, X_2 cover X. Our object is to determine the groupoid πX, and also certain of its full subgroupoids, in terms of the groupoids πX_i, $i = 0, 1, 2$ and the morphisms induced by inclusions. The general interpretation of the theorem which we prove must, however, wait until chapter 8 when the necessary algebraic theory of groupoids will be developed. It is hoped that the reader will by now be so familiar with the concept of a pushout that the theorem will have its appeal even with only one example of its use in computations.

By 2.5.11, the following square of inclusions

$$\begin{array}{ccc} X_0 & \xrightarrow{i_1} & X_1 \\ {\scriptstyle i_2}\downarrow & & \downarrow{\scriptstyle u_1} \\ X_2 & \xrightarrow[u_2]{} & X \end{array} \qquad (6.7.1)$$

is a pushout in the category of topological spaces. We denote this square by X.

Let A be any set. If $i : Y \to X$ is the inclusion of a subspace Y of X, then i induces a morphism

$$\pi YA \to \pi XA$$

which should be denoted by πiA, but which we shall denote simply by i. We shall also denote $Pi : PY \to PX$ by i'.

A set A is called *representative* in X if A meets each path-component of X. Our main result is the following.

6.7.2 *If A is representative in X_0, X_1, X_2, then the square*

$$\begin{array}{ccc} \pi X_0 A & \xrightarrow{i_1} & \pi X_1 A \\ {\scriptstyle i_2}\downarrow & & \downarrow{\scriptstyle u_1} \\ \pi X_2 A & \xrightarrow[u_2]{} & \pi XA \end{array}$$

is a pushout in the category of groupoids.

The above square of groupoid morphisms we write $\pi\mathsf{X}A$.

In using this theorem, it is usual to take X to be path-connected—otherwise the result is used on each path-component of X at a time. Also, we shall assume that $A \subset X$, since any points of $A \setminus X$ are irrelevant to the theorem. The sort of picture for X_1, X_2 is the following, in which shading is used to distinguish X_1, X_2; points of A are denoted by dots; and the components of X_1, X_2 should be thought of as complicated spaces (for example, a real projective space) rather than the simple subsets of $\mathbf{R}^2$ shown.

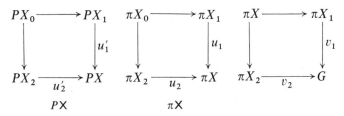

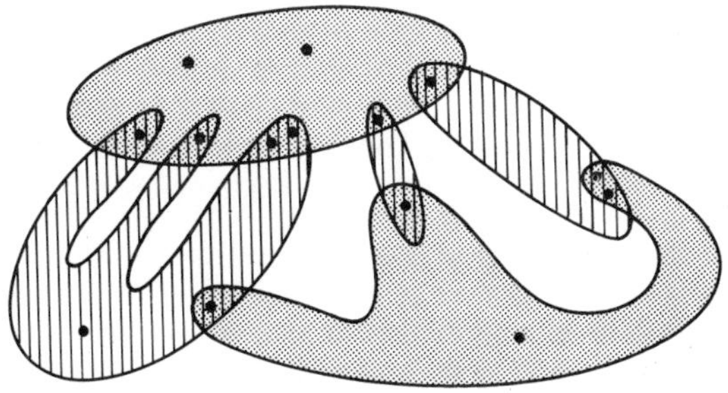

Fig. 6.10

The proof of 6.7.2 is in two parts—the case $A = X$, and the general case. The first case is the only one involving topology.

Proof of 6.7.2—*the case* $A = X$

Consider the following three diagrams

$$\begin{array}{ccc} PX_0 & \longrightarrow & PX_1 \\ \downarrow & & \downarrow{\scriptstyle u_1'} \\ PX_2 & \xrightarrow[u_2']{} & PX \end{array} \qquad \begin{array}{ccc} \pi X_0 & \longrightarrow & \pi X_1 \\ \downarrow & & \downarrow{\scriptstyle u_1} \\ \pi X_2 & \xrightarrow[u_2]{} & \pi X \end{array} \qquad \begin{array}{ccc} \pi X & \longrightarrow & \pi X_1 \\ \downarrow & & \downarrow{\scriptstyle v_1} \\ \pi X_2 & \xrightarrow[v_2]{} & G \end{array}$$

$\mathsf{P}X$ $\qquad\qquad$ $\pi\mathsf{X}$

and suppose that the third diagram is commutative. We wish to prove that there is a unique morphism $v : \pi X \to G$ such that $vu_1 = v_1$, $vu_2 = u_2$.

Step 1. We know that the projections $p_\lambda : PX_\lambda \to \pi X_\lambda$ ($\lambda = 0, 1, 2$ or $\;$) are morphisms (a word we use here for functor). Let $w_\lambda = v_\lambda p_\lambda$ ($\lambda = 1, 2$). We use w_1, w_2 to construct a morphism $w : PX \to G$.

Let a be a path of X and suppose first of all that Im a is contained in one or other of X_1, X_2. Then a is $u_1'b_1$ or $u_2'b_2$ for a unique path b_λ in X_λ and we define

$$wa = w_\lambda b_\lambda.$$

This definition is sensible because if Im a is contained in $X_1 \cap X_2 = X_0$, then $b_1 = i_1 b$, $b_2 = i_2 b$ for a path b in X_0 and

$$w_1 b_1 = w_1 i_1 b = w_2 i_2 b = w_2 b_2.$$

Next, suppose a is any path in X. By a corollary of the Lebesgue covering theorem [3.6.4 (corollary 1)] there is a subdivision

$$a = a_n + \cdots + a_1$$

such that Im a_i is contained in one or other of X_1, X_2 for each $i = 1, \ldots, n$. Then wa_i is well-defined and we set

$$wa = wa_n + \cdots + wa_1.$$

The usual arguments of superimposing subdivisions show that wa is independent of the subdivision, and that if a, b are paths such that $b + a$ is defined then

$$w(b + a) = wb + wa.$$

Clearly $w : PX \to G$ is a morphism, in fact, the only morphism such that

$$wu_1' = w_1, \qquad wu_2' = w_2.$$

It is also clear that w maps constant paths to zeros of G, since both w_1 and w_2 do so.

(This part of the proof does not use the inverses of G. In effect we have proved that $P\mathsf{X}$ is a pushout in $\mathscr{C}$at.)

We next show that w maps equivalent paths to the same element of G. We know that this is true for w_1, w_2.

Step 2. Consider a rectangle R in $\mathbf{R}^2$ and a map $F : R \to X$

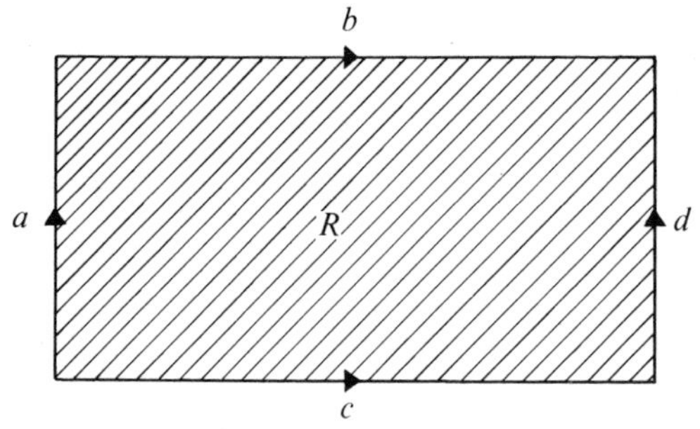

Fig. 6.11

such that Im F is contained in X_λ ($\lambda = 1$ or 2). The map F determines on the sides of R paths a, b, c, d [Fig. 6.11], such that $a = u_\lambda' a_\lambda, \ldots, d = u_\lambda' d_\lambda$.

Since R is convex

$$b_\lambda + a_\lambda \sim d_\lambda + c_\lambda$$

whence

$$wb + wa = w_\lambda(b_\lambda + a_\lambda) = w_\lambda(d_\lambda + c_\lambda) = wd + wc.$$

Step 3. This step is a quite simple cancellation argument.

Let a, b be paths in X and let $F : [0, r] \times \mathbf{I} \to X$ be a homotopy $a \sim b$. The Lebesgue covering lemma shows that we can, by a grid composed of lines

$$\{(ri/n, t) : t \in \mathbf{I}\}, \qquad i = 0, \ldots, n$$

$$\{(s, j/n) : s \in [0, r]\}, \quad j = 0, \ldots, n$$

subdivide $[0, r] \times \mathbf{I}$ into rectangles so small that each is mapped by F into X_1 or X_2. Let a_j be the path $s \rightsquigarrow F(s, j/n)$

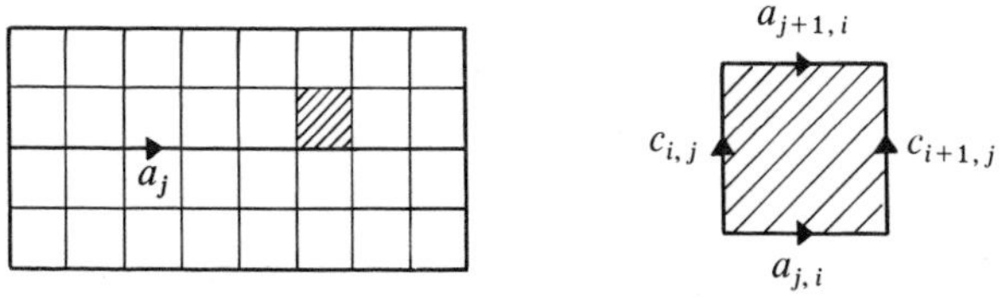

Fig. 6.12

so that $a_0 = a$, $a_n = b$. The vertical lines determine a subdivision

$$a_j = a_{j,n-1} + \cdots + a_{j,0}$$

and also paths $c_{i,j} : t \rightsquigarrow F(\mathrm{ri}/n, t + j/n)$ such that

$$a_{j+1,i} + c_{i,j} \sim c_{i+1,j} + a_{j,i}.$$

Therefore

$$\begin{aligned} wa_j &= \sum_{i=0}^{n-1} wa_{j,i} \\ &= \sum_{i=0}^{n-1} \{-wc_{i+1,j} + wa_{j+1,i} + wc_{i,j}\} \quad \text{by Step 2} \\ &= -wc_{n,j} + \sum_{i=0}^{n-1} wa_{j+1,i} + wc_{0,j} \\ &= wa_{j+1} \end{aligned}$$

the first and last terms being zero since $c_{0,j}$, $c_{n,j}$ are constant paths. Hence by induction $wa = wb$.

Step 4. Let a, b be equivalent paths in X. Then there are constant paths r s such that $r + a$ is homotopic to $s + b$. It follows that

$$wa = w(r + a) = w(s + b) = wb.$$

Thus w maps equivalent paths to the same element of G and so defines a morphism $v : \pi X \to G$ such that $vp = w$. Thus for any path a in X_λ $(\lambda = 1, 2)$

$$vu_\lambda\,(\text{cls } a) = v \text{ cls } u'_\lambda a = wu'_\lambda a = w_\lambda a$$

$$= v_\lambda\,(\text{cls } a)$$

whence $$vu_\lambda = v_\lambda.$$

Step 5. Let $v' : \pi X \to G$ be any morphism such that

$$v'u_\lambda = v_\lambda, \quad \lambda = 1, 2.$$

Then
$$v'pu'_\lambda = v'u_\lambda p_\lambda$$
$$= v_\lambda p_\lambda$$
$$= w_\lambda.$$

Since $v'p$ is a morphism it follows that

$$v'p = w = vp.$$

Since $p : PX(x, y) \to \pi X(x, y)$ is surjective for each x, y in X, it follows that

$$v' = v. \quad \square$$

Proof of 6.7.2—*the general case*

For this it is sufficient, by 6.6.7, to prove that the square $\pi\mathbf{X}A$ is a retract of $\pi\mathbf{X}$. We do this by constructing retractions $r_\lambda : \pi X_\lambda \to \pi X_\lambda A$ $(\lambda = 0, 1, 2$ or $\ )$ consistent with the various morphisms induced by inclusion.

Since A is representative in X_0 we can choose for each x in X_0 a road $\theta_0 x$ in πX_0 from some point of $A \cap X_0$ to x. Since A is representative in X_1 and in X_2 we can choose for each x in $X_\lambda \setminus X_0$ $(\lambda = 1, 2)$ a road $\theta_\lambda x$ in πX_λ from some point of $A \cap X_\lambda$ to x, and we extend θ_λ over all of X_λ by setting $\theta_\lambda x = i_\lambda \theta_0 x$ for x in X_0. Finally, for any x in X we define

$$\theta x = \begin{cases} u_1\theta_1 x, & x \in X_1 \\ u_2\theta_2 x, & x \in X_2. \end{cases}$$

Then θ is well-defined. Further, θ_λ ($\lambda = 0, 1, 2$ or   ) defines, by 6.5.14, a retraction $r_\lambda : \pi X_\lambda \to \pi X_\lambda A$; these morphisms define a retraction $\mathsf{r} : \pi \mathsf{X} \to \pi \mathsf{X} A$. □

As will be clear from chapter 8, 6.7.2 can often be interpreted to give information on the fundamental groups themselves. For example, if X_0, X_1, X_2 are path-connected, then we can take $A = \{x_0\}$ where x_0 is a point of X_0, and then 6.7.2 determines completely the fundamental group $\pi(X, x_0)$. But we shall want to use 6.7.2 when X_0 at least is not path-connected, and in this case it is useful to carry out a further retraction. The main lemma for this purpose is the following.

6.7.3 *Let $\mathscr{C}, \mathscr{D}$ be categories and $f : \mathscr{C} \to \mathscr{D}$ a functor such that* $\mathsf{Ob}(f)$ *is injective. Then any full representative subcategory $\mathscr{C}'$ of $\mathscr{C}$ gives rise to a pushout square*

$$\begin{array}{ccc} \mathscr{C} & \xrightarrow{r} & \mathscr{C}' \\ {\scriptstyle f}\downarrow & & \downarrow{\scriptstyle f'} \\ \mathscr{D} & \xrightarrow[s]{} & \mathscr{D}' \end{array} \qquad (*)$$

in which f' is a restriction of f and r, s are deformation retractions.

Proof Let $\mathscr{D}'$ be the full subcategory of $\mathscr{D}$ whose objects are those equal to fx for x in $\mathsf{Ob}(\mathscr{C}')$, and those not equal to fx for any x in $\mathsf{Ob}(\mathscr{C})$; that is,

$$\mathsf{Ob}(\mathscr{D}') = f[\mathsf{Ob}(\mathscr{C}')] \cup (\mathsf{Ob}(\mathscr{D}) \setminus f[\mathsf{Ob}(\mathscr{C})])$$

Then we have a commutative diagram

$$\begin{array}{ccc} \mathscr{C}' & \xrightarrow{i} & \mathscr{C} \\ {\scriptstyle f'}\downarrow & & \downarrow{\scriptstyle f} \\ \mathscr{D}' & \xrightarrow[j]{} & \mathscr{D} \end{array}$$

in which i, j are inclusions and f' is the restriction of f. We choose a deformation retraction $r : \mathscr{C} \to \mathscr{C}'$ and a homotopy function $\theta : ir \simeq 1$ rel $\mathscr{C}'$ as is possible by 6.5.13.

For each y in $\mathsf{Ob}(\mathscr{D})$ let $\varphi y = f\theta x$ if $y = fx$ and otherwise let $\varphi y = 1_y$. Thus $\varphi f = f\theta$; φy is invertible, has initial point in $\mathsf{Ob}(\mathscr{D}')$ and has final point y; further, $\varphi y = 1_y$ if $y \in \mathsf{Ob}(\mathscr{D}')$. It follows that φ is a homotopy function $js \simeq 1$ rel $\mathscr{D}'$ where $s : \mathscr{D} \to \mathscr{D}'$ is a deformation retraction.

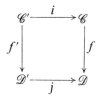

If $a \in \mathscr{C}(x, x')$ then

$$\begin{aligned} f'r(a) &= f((\theta x')^{-1}a(\theta x)) \\ &= (\varphi f x')^{-1} fa(\varphi f x) \\ &= sf(a). \end{aligned}$$

Therefore $f'r = sf$ and the square (*) commutes.

To prove (*) to be a pushout suppose there is given a commutative diagram

$$\begin{array}{ccc} \mathscr{C} & \xrightarrow{r} & \mathscr{C}' \\ {\scriptstyle f}\downarrow & & \downarrow{\scriptstyle u} \\ \mathscr{D} & \xrightarrow[v]{} & \mathscr{E} \end{array}$$

If there is a functor $w : \mathscr{D}' \to \mathscr{E}$ such that $ws = v$ then

$$w = wsj = vj$$

and so there is at most one such w. On the other hand, let $w = vj$. Then

$$wf' = vjf' = vfi = uri = u.$$

Further φy, and hence also $v\varphi y$, is the identity if $y \neq fx$ for any x, while if $y = fx$ then

$$v\varphi f(x) = vf\theta(x) = ur\theta(x) = u(1) = 1.$$

It follows that $v\varphi$ is the constant homotopy function and so

$$ws = vjs = v.$$

This proves that the square is a pushout. □

We apply this result to the situation at the beginning of this section. Thus we are given subspaces X_1, X_2 of X whose interiors cover X. Further $X_0 = X_1 \cap X_2$, and A is a subset of X representative in X_0, X_1, X_2: if X is path-connected then each path component of one of the spaces X_1, X_2 meets the other space [Exercise 8 of 3.4]—it is then common and convenient to take A as one point in each path-component of X_0, but this assumption is not essential.

6.7.4 *Let A' be a subset of $A \cap X_1$ representative in X_1 and let $A_1 = A' \cup (A \setminus X_1)$. Then there is a pushout diagram*

$$\begin{array}{ccccc} \pi X_0 A & \xrightarrow{i_1} & \pi X_1 A & \xrightarrow{r} & \pi X_1 A_1 \\ {\scriptstyle i_2}\downarrow & & & & \downarrow{\scriptstyle u_1} \\ \pi X_2 A & \xrightarrow[u_2]{} & \pi X A & \xrightarrow[s]{} & \pi X A_1 \end{array} \qquad (*)$$

in which r, s are deformation retractions.

Proof If we insert the morphism $u_1 : \pi X_1 A \to \pi XA$ in (*) we obtain two squares. The left-hand square is a pushout by 6.7.2. To prove the right-hand square a pushout we use 6.7.3 with the substitutions $\mathscr{C} = \pi X_1 A$, $\mathscr{D} = \pi XA$, $\mathscr{D}' = \pi X_1 A' = \pi X_1 A_1$. Then we have

$$\text{Ob}\,(\mathscr{D}') = A' \cup (A \setminus (A \cap X_1)) = A' \cup (A \setminus X_1) = A_1.$$

Finally, (*) itself is a pushout, since it is the composite of two pushouts. □

This result will be used in chapter 8 to prove the van Kampen theorem on the fundamental group of an adjunction space. The interpretation of that theorem in general requires a lot of preliminary algebra. Here we finish this chapter by showing that 6.7.4 enables us to determine the fundamental group of a circle—this gives us our first simple example of a path-connected but not simply-connected space.

6.7.5 *There is an isomorphism*

$$\pi(\mathbf{S}^1, 1) \cong \mathbf{Z}$$

Proof

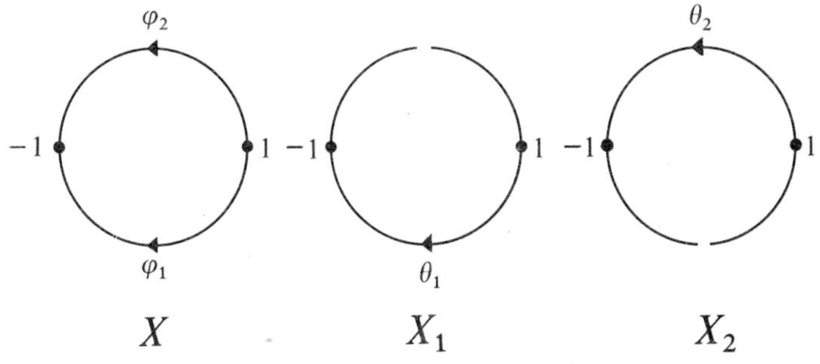

Fig. 6.13

We use complex number notation. Let $X_1 = \mathbf{S}^1 \setminus \{i\}$, $X_2 = \mathbf{S}^1 \setminus \{-i\}$, $A = \{-1, 1\}$, $A_1 = \{1\}$. Then X_1, X_2 are simply connected (they are both homeomorphic to $]0, 1[$) while X_0 is the topological sum of two simply-connected components. Therefore $\pi X_2 A$ is isomorphic to the groupoid I while $\pi X_0 A$ is isomorphic to the discrete groupoid $\{0, 1\}$. Thus by 6.7.4 we have a pushout

$$\begin{array}{ccc} \{0, 1\} & \longrightarrow & 0 \\ \downarrow & & \downarrow \\ I & \xrightarrow[g]{} & \pi(\mathbf{S}^1, 1) \end{array}$$

in which 0 is the trivial group and g is specified completely as a morphism by the fact that

$$g(i) = \varphi, \quad \text{say.}$$

The above pushout implies the following: *if $f : I \to K$ is any morphism to a group K, then there is a unique morphism $h : \pi(\mathbf{S}^1, 1) \to K$ such that $hg = f$.* In particular, letting $f : I \to \mathbf{Z}$ be the morphism such that $f(\iota)$ is the element 1 of $\mathbf{Z}$, there is a unique morphism $h : \pi(\mathbf{S}^1, 1) \to \mathbf{Z}$ such that $h(\varphi) = 1$.

Let $k : \mathbf{Z} \to \pi(\mathbf{S}^1, 1)$ be the morphism $n \rightsquigarrow n\varphi$. Clearly $hk(1) = 1$ and so $hk = 1 : \mathbf{Z} \to \mathbf{Z}$. On the other hand,

$$khg(\iota) = kf(\iota) = k(1) = \varphi = g(\iota).$$

Therefore $khg = g$ and so, by the uniqueness part of the universal property, $kh = 1 : \pi(\mathbf{S}^1, 1) \to \pi(\mathbf{S}^1, 1)$. □

The proof also shows that φ is a generator of $\pi(\mathbf{S}^1, 1)$. In order to determine φ, let θ_i be the unique element of $\pi X_i(1, -1)$ $(i = 1, 2)$ and let $\varphi_i = u_i\theta_i$ in $\pi\mathbf{S}^1(1, -1)$. The retraction s satisfies

$$s\varphi_2 = -\varphi_1 + \varphi_2$$

and so if we take the isomorphism $\pi X_2 A \to I$ to be that which sends φ_2 to ι we deduce that

$$\varphi = -\varphi_1 + \varphi_2.$$

Clearly φ is the class of the path $t \rightsquigarrow e^{2\pi i t}$ of length 1.

6.7.5 (*Corollary 1*) $\mathbf{S}^1$ *is not a retract of* $\mathbf{E}^2$.

Proof If $r : \mathbf{E}^2 \to \mathbf{S}^1$ were a retraction, then so also would be $r : \pi\mathbf{E}^2\{1\} \to \pi\mathbf{S}^1\{1\}$. Since a retraction is surjective this is clearly impossible. □

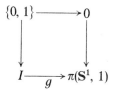

EXERCISES

1. Prove that the fundamental group of the torus $\mathbf{S}^1 \times \mathbf{S}^1$ is isomorphic to $\mathbf{Z} \times \mathbf{Z}$. Prove also that the two maps $\mathbf{S}^1 \to \mathbf{S}^1 \times \mathbf{S}^1$ which send $x \rightsquigarrow (x, a)$, $x \rightsquigarrow (a, x)$ respectively are not homotopic.

2. The coproduct in the category of groups exists [cf. chapter 8]—it is called the free product and is written $*$. Prove that the fundamental group of $\mathbf{S}^1 \vee \mathbf{S}^1$ is isomorphic to $\mathbf{Z}*\mathbf{Z}$.

3. Suppose there is given the commutative square of 6.7.3 in which r, s are de-deformation retractions and f' is the restriction of f. Suppose also that θ, φ are homotopy functions $ir \simeq 1$ rel $\mathscr{C}'$, $js \simeq 1$ rel $\mathscr{D}'$ respectively. Prove that the square is a pushout if the following condition holds: if $y \in \mathrm{Ob}(\mathscr{D})$ and $\varphi y \neq 1$, then there are objects $x_1, \ldots, x_n$ of $\mathscr{C}$ such that $\varphi y = (f\theta x_n)\cdots(f\theta x_1)$.

4. Suppose that in the situation of 6.7.2, A_1, A_2 are subsets of A representative in X_1, X_2 respectively, and that $A \cap X_0 = A_1 \cap A_2$. Prove that there is a pushout square

$$\begin{array}{ccc} \pi X_0 A & \longrightarrow & \pi X_1 A_1 \\ \downarrow & & \downarrow \\ \pi X_2 A_2 & \longrightarrow & \pi X A_0 \end{array}$$

5. Prove that $\mathbf{S}^{n-1}$ is a retract of $\mathbf{E}^n$ if and only if there is a map $\mathbf{E}^n \to \mathbf{E}^n$ without fixed points.

6. Let the open cover $\mathscr{U}$ of the space X be regarded as a category as in Exercise 12 of 6.6. The fundamental groupoid functor restricts to a functor $\pi \mid \mathscr{U} : \mathscr{U} \to \mathscr{G}d$. Prove that if $\mathscr{U}$ has a subcover $\mathscr{V}$ such that $\mathscr{U}$ consists of all intersections $V \cap V'$ for V, V' in $\mathscr{V}$, then the functor $\pi \mid \mathscr{U}$ has as right-root a morphism to the constant functor with value πX.

7. [This exercise requires knowledge of ordinals and transfinite induction.] Let $\mathscr{U}$ be an open cover of X such that the intersection of two elements of $\mathscr{U}$ belongs to $\mathscr{U}$. We say $\mathscr{U}$ is *stratified* if there is a function f from $\mathscr{U}$ to the ordinals such that $fU < fU'$ whenever U is a proper subset of U'. Prove $\mathscr{U}$ is stratified if $\mathscr{U}$ is finite or is well-ordered by inclusion. Let $\mathscr{U}$ be stratified, and let A be a subset of X representative in each U of $\mathscr{U}$. Prove that the functor $\mathscr{U} \to \mathscr{G}\mathrm{d}$, $U \rightsquigarrow \pi UA$, has as right-root a morphism to the constant functor with value πXA. [Use the previous exercise and Exercises 12, 13, of 6.6.]

8. Let σX be the semi-fundamental groupoid of X as constructed in Exercise 7 of 6.2. Investigate whether or not theorems corresponding to 6.7.2, 6.7.4 hold for σ.

9. Prove that the coproduct of trivial groups is trivial. Deduce that in the situation of 6.7.2, if X_0, X_1, X_2 are 1-connected, then so also is $X = X_1 \cup X_2$.

NOTES

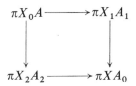

The literature of category theory up to 1965 is surveyed in Maclane [2]. The *abelian categories* model closely the categories of modules—an account of these is given in Freyd [1] and Mitchell [1]. Many general results on categories are given in the book of Ehresmann [1] which, unfortunately, contains little motivation, examples or applications. Groupoids have been known since Brandt [1]; their applications to group theory were spotlighted by Higgins [1].

The fundamental group of a union $K_1 \cup K_2$ of simplicial complexes K_1, K_2 such that K_1, K_2 and $K_1 \cap K_2$ are all path-connected was described by Seifert [1]. The fundamental group of a certain kind of adjunction space was determined by van Kampen [1] (apparently not knowing of Seifert's paper); the theorems given in this paper are more general than Seifert's, but the name van Kampen has usually been applied to any theorem (including Seifert's theorem) on the fundamental group of a union of spaces. Other results on the fundamental group of a union have been given by Olum [1], Crowell [1], Brown [3], Weinzweig [1], and Brown [4]: the results of 6.7, and the general van Kampen theorem given in 8.4, are taken from the last paper.

7. The homotopy extension property

In the last chapter, we introduced the notion of homotopy type of spaces. An obvious question is: how do we determine whether or not two spaces X, Y are of the same homotopy type? This is a difficult question in general (and even for cell complexes), but there are two useful techniques.

First, to prove X, Y are not of the same homotopy type, we construct homotopy type invariants. For example, $\mathbf{S}^1$ is not of the homotopy type of a point because $\mathbf{S}^1$ has fundamental group $\mathbf{Z}$, which is not isomorphic to the trivial group.

Second, to prove X, Y are of the same homotopy type, we must construct a homotopy equivalence $X \to Y$. For this, we need to know something about X and Y; a common case is that X and Y are given as repeated adjunction spaces, and because of this we take as our main aim the discussion of the homotopy type of adjunction spaces. The main tool used is a glueing theorem for homotopy equivalences. To prove this theorem, we need the notion of the *homotopy extension property*, and to conceptualize the proof it is convenient to use the *track groupoid*, and the *operations* of this groupoid. These operations are important in their own right.

7.1 The track groupoid

In this section, we generalize the fundamental groupoid to the track groupoid by replacing paths by homotopies. (There is an alternative exposition of this subject in which homotopies are defined by means of paths in spaces of functions, but we shall give here only the direct approach.) In order to link this idea with other developments in algebraic topology, particularly the homotopy groups, we define the track groupoid in greater generality than is strictly necessary for our purposes.

Let A be a subspace of X. A map $F : X \times [0, q] \to Y$ is called a *homotopy* rel A (of length q) if F is on A the constant homotopy, that is, if

$$F(a, t) = F(a, 0), \quad a \in A, t \in [0, q].$$

In this case, we say that the initial and final maps, f and g, of F are *homotopic* rel A, and we write $f \simeq g$ rel A. Note that, if $f \simeq g$ rel A then f and g agree on A.

If F and G are homotopies rel A such that the final map of F is the same as the initial map of G, then we can define a homotopy $G + F$ of length $q + r$ (where q, r are the lengths of F, G) by

$$(G + F)(x, t) = \begin{cases} F(x, t), & t \leqslant q \\ G(x, t - q), & q \leqslant t \leqslant q + r \end{cases}$$

and it is obvious that $G + F$ is a homotopy rel A. Thus, as in the case of paths, these homotopies rel A form a category whose objects are the maps $X \to Y$ and whose morphisms from f to g are the homotopies rel A with initial map f and final map g. This category will be written

$$P(Y^X \text{ rel } A).$$

We now construct from this category a groupoid.

Let F, G be homotopies rel A with initial map f and final map g. Suppose first that F and G have the same length q. Then F and G are said to be *A-homotopic rel end points* if there is a map

$$\mathscr{H} : X \times [0, q] \times \mathbf{I} \to Y$$

such that for x in X, s in $[0, q]$, t in $\mathbf{I}$ and a in A

$$\mathscr{H}(x, s, 0) = F(x, s) \qquad \mathscr{H}(x, s, 1) = G(x, s)$$
$$\mathscr{H}(x, 0, t) = fx \qquad \mathscr{H}(x, q, t) = gx$$
$$\mathscr{H}(a, s, t) = fa.$$

This amounts to saying that if $\mathscr{H}_t$ denotes the map $(x, s) \rightsquigarrow \mathscr{H}(x, s, t)$, then $\mathscr{H}_0 = F$, $\mathscr{H}_1 = G$ and each $\mathscr{H}_t$ is a homotopy $f \simeq g$ rel A; or alternatively, that $\mathscr{H}$ is a homotopy $F \simeq G$ rel $(X \times \{0, q\}) \cup (A \times [0, q])$.

We now discuss the general case where F and G need not have the same length. A *constant homotopy of length* r is a map $F : X \times [0, r] \to Y$ such that $F(x, 0) = F(x, t)$, all x in X, t in $[0, r]$. Such a homotopy is denoted ambiguously by r. Then F and G are *equivalent* if there are real numbers $r, s \geqslant 0$ such that $r + F$, $s + G$ are A-homotopic rel end points. This is an equivalence relation and it is shown by methods only formally different from those of the previous chapter that the equivalence classes form a groupoid. This groupoid we call a *track groupoid* and we write it

$$\pi(Y^X \text{ rel } A).$$

The elements of this groupoid will be called *tracks*.

Two maps $f, g : X \to Y$ are homotopic rel A if and only if f and g are objects in the same component of this track groupoid. It follows (as is easy to verify anyway) that the relation $f \simeq g$ rel A is an equivalence relation on the set of maps $X \to Y$.

Let $u : A \to Y$ be a given map. If u is the restriction of a map $f : X \to Y$ then we say u *extends* over X. Let

$$X /\!/ u : A \to Y, \quad \text{or, simply, } X/\!/u$$

denote the set of homotopy classes rel A of maps $X \to Y$ extending u—thus $X/\!/u$ is non-empty if and only if u extends over X. Notice also that if f extends u and $f \simeq g$ rel A then g extends u.

The track groupoid $\pi(Y^X \text{ rel } A)$ can also be described in terms of sets such as those just introduced. In fact, any element of this track groupoid is represented by a homotopy $X \times \mathbf{I} \to Y$ of length 1, and two homotopies of length 1 are equivalent if and only if they are homotopic rel $(X \times \dot{\mathbf{I}} \cup A \times \mathbf{I})$ (where $\dot{\mathbf{I}} = \{0, 1\}$). Thus, the set of elements of $\pi(Y^X \text{ rel } A)$ can be identified with the union of the sets

$$X \times \mathbf{I} /\!/ U$$

for all $U : X \times \dot{\mathbf{I}} \cup A \times \mathbf{I} \to Y$ such that $U \mid A \times \mathbf{I}$ is a constant homotopy.

We are interested mainly in two special cases of the track groupoid. First, when $A = \varnothing$ the track groupoid is written

$$\pi Y^X.$$

Second, in the category of pointed spaces we take $A = \cdot$, the base point of X, and consider the full subgroupoid of $\pi(Y^X \text{ rel } \cdot)$ on the set of pointed maps $X \to Y$—this subgroupoid is written

$$\pi Y^X_\cdot.$$

For each pointed map $f : X \to Y$ we have a group

$$\pi Y^X_\cdot\{f\}$$

which is the object group of the track groupoid at f. In particular, let $\cdot : Z \to Y$ denote for any Z the constant pointed map; then we can if $f = \cdot$ identify the underlying set of the above groupoid with

$$(X \times \mathbf{I})/\!/ \cdot$$

where in this case $\cdot$ is the constant pointed map $X \times \dot{\mathbf{I}} \cup \cdot \times \mathbf{I} \to Y$.

7.1.1 *Let A be a subspace of X and let $p : X \to X/A$ denote the projection, where X/A has base point the set A. Then p induces a bijection between the sets*

$$X/\!/ \cdot : A \to Y \quad \text{and} \quad (X/A)/\!/ \cdot : \cdot \to Y.$$

Proof The latter sets we abbreviate to $X/\!/\cdot$ and $(X/A)/\!/\cdot$.

Certainly, p induces by the rule $f \rightsquigarrow fp$ a bijection between the set of pointed maps $X/A \to Y$ and the set of maps $X \to Y$ which agree with $\cdot$ on A. Also, if $f, g : X/A \to Y$ are pointed maps and $F : f \simeq g$ rel $\cdot$, then the composite

$$X \times \mathbf{I} \xrightarrow{p \times 1} (X/A) \times \mathbf{I} \xrightarrow{F} Y$$

is a homotopy $fp \simeq gp$ rel A. Conversely, if G is a homotopy $fp \simeq gp$ rel A then for each t the map $x \rightsquigarrow G(x, t)$ is constant on A and so defines a function $F : (X/A) \times \mathbf{I} \to Y$ such that $F(p \times 1) = G$. But by 4.3.2, $p \times 1$ is an identification map. Therefore, F is continuous and is a homotopy $f \simeq g$ rel $\cdot$. □

If X, Y are pointed spaces then we denote by $[X, Y]_{\cdot}$ the set of homotopy classes rel $\cdot$ of pointed maps $X \to Y$.

7.1.1 (*Corollary 1*) *There is a bijection*

$$\pi Y^X_{\cdot}\{\cdot\} \to [\Sigma X, Y]_{\cdot}$$

where ΣX is the reduced suspension of X.

Proof The former set is bijective with

$$(X \times \mathbf{I})/\!/\cdot : X \times \dot{\mathbf{I}} \cup \cdot \times \mathbf{I} \to Y.$$

Since $\Sigma X = (X \times \mathbf{I})/(X \times \dot{\mathbf{I}} \cup \cdot \times \mathbf{I})$ the result follows from 7.1.1. □

It is convenient to identify these sets by means of this bijection. Then $[\Sigma X, Y]_{\cdot}$ is identified with an object group of $\pi Y^X_{\cdot}$ and so has a group structure. In particular, if $X = \mathbf{S}^{n-1}$, then we can identify $\Sigma\mathbf{S}^{n-1}$ with $\mathbf{S}^n$, and we obtain a group structure on the set

$$[\mathbf{S}^n, Y]_{\cdot}.$$

This group is called the *nth homotopy group* of Y and is written $\pi_n(Y, \cdot)$ (so that $\pi_1(Y, \cdot)$ can be identified with the fundamental group $\pi(Y, \cdot)$).

The homotopy groups will be used in this book only in the Exercises in the last section of this chapter. For further information the reader should consult the books of Hu [1] and particularly Spanier [3], Toda [1].

EXERCISES

1. Let the pointed space X be identified with the subset $X \times 1$ of the reduced cone ΓX. Let $u : X \to Y$ be a pointed map. Prove that the set $\Gamma X/\!/u$ may be identified with $\pi Y^X_{\cdot}(\cdot, u)$. Let β'', β', β be elements of $\Gamma X/\!/u$. By using this identification and also that given in 7.1.1 (corollary 1), the element $-\beta' + \beta$ becomes the *difference element* $\mathrm{d}(\beta', \beta)$ of $[\Sigma X, Y]_{\cdot}$. Prove that $\mathrm{d}(\beta'', \beta') + \mathrm{d}(\beta', \beta) = \mathrm{d}(\beta'', \beta)$.
2. Suppose further to the previous exercise that $\alpha', \alpha \in [\Sigma X, Y]_{\cdot}$. Then $\beta + \alpha$ can be regarded as an element of $\Gamma X/\!/u$ and is then written $\beta \perp \alpha$. Prove that (i) $(\beta \perp \alpha) \perp \alpha' = \beta \perp (\alpha + \alpha')$, (ii) $\beta \perp 0 = \beta$, (iii) $\mathrm{d}(\beta, \beta \perp \alpha) = \alpha$.

3. Let $\mathscr{C}$ be a category which admits coproducts. Prove that for any objects C_1, C_2, C_3 of $\mathscr{C}$ there are isomorphisms

$$C_1 \sqcup C_2 \to C_2 \sqcup C_1, \qquad C_1 \sqcup (C_2 \sqcup C_3) \to (C_1 \sqcup C_2) \sqcup C_3.$$

Prove also that if 0 is a copoint of $\mathscr{C}$ then there are isomorphisms

$$C_1 \to C_1 \sqcup 0, \qquad C_1 \to 0 \sqcup C_1.$$

4. Suppose further to the previous exercise that for each pair (C_1, C_2) of $\mathscr{C}$ we choose a specific coproduct and write this coproduct $C_1 \sqcup C_2$. Prove that $\sqcup$ is a functor $\mathscr{C} \times \mathscr{C} \to \mathscr{C}$, and interpret the isomorphisms of the previous exercises as natural equivalences of functors.

5. Let A, B, C be objects of the category $\mathscr{C}$. Let $\mathscr{C}^{A,B}$ denote the functor $X \rightsquigarrow \mathscr{C}(A, X) \times \mathscr{C}(B, X)$ from $\mathscr{C}$ to $\mathscr{S}$et, and let $\mathscr{C}^C$ denote as usual the functor $X \rightsquigarrow \mathscr{C}(C, X)$. Prove that C is a coproduct of A and B if and only if there is a natural equivalence of functors $\mathscr{C}^C \simeq \mathscr{C}^{A,B}$. Use this and similar results to give another solution to Exercise 3.

6. A *comultiplication* on an object A of $\mathscr{C}$ is a morphism $c : A \to A \sqcup A$. Prove that a comultiplication on A induces a natural transformation $\mathscr{C}^{A,A} \to \mathscr{C}^A$. On the other hand, any such natural transformation is called a *natural multiplication* on the sets $\mathscr{C}(A, X)$, $X \in \mathrm{Ob}(\mathscr{C})$. Prove, conversely, that any natural multiplication on these sets is induced by a comultiplication $A \to A \sqcup A$.

7. Let $c : A \to A \sqcup A$ be a comultiplication on the object A of $\mathscr{C}$. We say c is *associative* if the following diagram commutes

$$\begin{array}{ccccc} & \overset{c}{\nearrow} & A \sqcup A & \xrightarrow{1 \sqcup c} & A \sqcup (A \sqcup A) \\ A & & & & \downarrow a \\ & \underset{c}{\searrow} & A \sqcup A & \xrightarrow[c \sqcup 1]{} & (A \sqcup A) \sqcup A \end{array}$$

where a is the isomorphism given by Exercise 3. Prove that c is associative if and only if the multiplication induced on each $\mathscr{C}(A, X)$, for $X \in \mathrm{Ob}(\mathscr{C})$, is associative.

8. Let $c : A \to A \sqcup A$ be a comultiplication on the object A of $\mathscr{C}$, and for each X in $\mathscr{C}$ let $c_X : \mathscr{C}(A, X) \times \mathscr{C}(A, X) \to \mathscr{C}(A, X)$ be the induced multiplication. A *natural identity* for c_X is for each X an element e_X of $\mathscr{C}(A, X)$ which is an identity for the multiplication c_X and which is natural, i.e., for any morphism $f : X \to Y$ the induced function $f_* : \mathscr{C}(A, X) \to \mathscr{C}(A, Y)$ maps e_X to e_Y. Find necessary and sufficient conditions on c for the induced natural multiplications to have a natural identity. Further, supposing that c_X has a natural identity, define a *natural inverse* on $\mathscr{C}(A, X)$, $X \in \mathrm{Ob}(\mathscr{C})$ and find necessary and sufficient conditions on c for c_X to have a natural inverse.

9. Prove that, in the category $\mathscr{H}\mathscr{T}$op. of pointed spaces and homotopy classes of pointed maps, the set $[\Sigma A, X]_.$ for all pointed spaces X, has a natural group structure (that is, a natural multiplication which is associative and has natural identity and inverse). Prove that the wedge of pointed spaces is a coproduct in $\mathscr{H}\,\mathscr{T}$op.. Give an explicit formula for a comultiplication $\Sigma A \to \Sigma A \vee \Sigma A$ which induces the above natural group structure.

10. Dualize (in the sense of Exercise 14 of 6.6) Exercises 3–8. [Here the dual of a comultiplication $A \to A \sqcup A$ is a multiplication $A \sqcap A \to A$.]

11. Let S be a set with two functions $m_0, m_1 : S \times S \to S$; we write $m_i(x, y) = x +_i y$, $x, y \in S$. These additions induce additions in $S \times S$ by the usual rule

$$(x, y) +_i (x', y') = (x +_i x', y +_i y').$$

We say that m_i is a *morphism* for m_{1-i} if $m_i(z +_{1-i} w) = m_i(z) +_{1-i} m_i(w)$ for all $z, w \in S \times S$. Prove that m_0 is a morphism for m_1 if and only if

$$(x +_1 x') +_0 (y +_1 y') = (x +_0 y) +_1 (x' +_0 y')$$

for all $x, x', y, y' \in S$ and that this is equivalent to m_1 is a morphism for m_0. Suppose further that there is an element 0 of S which is a zero for both $+_0$, $+_1$. Prove that $m_1 = m_2$, and that both additions are associative and commutative.

12. Prove that the group structure on $[\Sigma A, X]_.$ is natural with respect to maps of A in the sense that a pointed map $f : A \to B$ induces a morphism of groups $f^* : [\Sigma B, X]_. \to [\Sigma A, X]_.$, cls $g \rightsquigarrow$ cls $(g(\Sigma f))$. Deduce that $[\Sigma^2 A, X]_.$ has two multiplications each of which is a morphism for the other; hence, show that these two multiplications coincide and are commutative. Express these facts as statements about $\Sigma^2 A$ and its coproducts.

7.2 Operations of the track groupoid

Let A be a subspace of X and let Y be a space. An obvious question is: if u, v are homotopic maps $A \to Y$, are the sets $X//u$, $X//v$ related? We shall answer this by constructing a functor

$$X// : \pi Y^A \to \mathscr{S}\text{et}$$

which assigns to each map $u : A \to Y$ the set $X//u$. Thus $X//\gamma$ assigns to an element γ of $\pi Y^A(u, v)$ a function

$$X//\gamma : X//u \to X//v$$

which, because $X//$ is a functor from a groupoid, must be a bijection. This gives the required relation between these sets.

The existence of this functor will be described by saying that πY^A *operates* on the family of sets $X//u$. However, for the construction of this operation we need an extra condition on the subspace A of X.

In order to motivate this condition, suppose that $f : X \to Y$ extends $u : A \to Y$. If F is a homotopy of f, then $F | A \times \mathbf{I}$ is a homotopy of u. But there may be homotopies of u which do not arise in this way.

Fig. 7.1

For example, let $X = \mathbf{L} \cup [-1, 0]$ and let $A = [-1, 0]$. The inclusion $u : A \to X$ extends to $1 : X \to X$. But the homotopy $(a, t) \rightsquigarrow a - ta - t$

(which deforms u to the constant map at -1) does not extend to a homotopy of $1 : X \to X$.

Definition Let A be a subspace of X. We say (X, A) has the *homotopy extension property* (which we abbreviate to HEP) if for all spaces Y and all maps $f : X \to Y$ any homotopy of $u = f \mid A$ extends to a homotopy of f.

We shall show in the next section that this condition is satisfied for a useful class of pairs (X, A). In this section, we are concerned more with the interpretation of the HEP, and its application to the problem we started with.

The HEP is clearly satisfied for all homotopies on A if it is satisfied for homotopies of length 1, and it will usually be convenient to restrict ourselves to such homotopies.

Consider the diagram of inclusions

$$\begin{array}{ccc} A \times 0 & \longrightarrow & X \times 0 \\ \downarrow & & \downarrow \\ A \times \mathbf{I} & \longrightarrow & X \times \mathbf{I} \end{array} \qquad (7.2.1)$$

(where, as often in this chapter, we find it convenient to denote a space $\{x\}$ by x). A map $f : X \to Y$ is determined by the map $f' : X \times 0 \to Y$ such that $f' : (x, 0) \rightsquigarrow fx$; similarly, $u : A \to Y$ is determined by $u' : A \times 0 \to Y$. A homotopy U of $u : A \to Y$ is a map $U : A \times \mathbf{I} \to A$ whose restriction to $A \times 0$ is u'. Thus the HEP for (X, A) is equivalent to the condition that (7.2.1) is a *weak pushout* [cf. p. 111].

7.2.2 *Let A be a closed subspace of X. The following conditions are equivalent.*

(*a*) *(X, A) has the HEP.*

(*b*) *For any space Y, any map $X \times 0 \cup A \times \mathbf{I} \to Y$ extends over $X \times \mathbf{I}$.*

(*c*) *$X \times 0 \cup A \times \mathbf{I}$ is a retract of $X \times \mathbf{I}$.*

Proof (*a*) $\Leftrightarrow$ (*b*) Because A is closed in X the following square of inclusions is a pushout

$$\begin{array}{ccc} A \times 0 & \longrightarrow & X \times 0 \\ \downarrow & & \downarrow \\ A \times \mathbf{I} & \longrightarrow & X \times 0 \cup A \times \mathbf{I} \end{array} \qquad (*)$$

It follows easily that (7.2.1) is a weak pushout if and only if any map $X \times 0 \cup A \times \mathbf{I} \to Y$ extends over $X \times \mathbf{I}$.

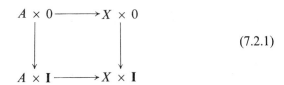

(*b*) ⇔ (*c*) This is a special case of the following proposition: *if* $i : C \to Z$ *is the inclusion map of the subspace C of Z then C is a retract of Z if and only if any map* $C \to Y$ *extends over Z*. The proof is as follows. If $r : Z \to C$ is a retraction, and $u : C \to Y$ is a map, then $ur : Z \to Y$ extends u. Conversely, if any map from C extends then, in particular, $1_C : C \to C$ extends; any extension of 1_C over Z is retraction $Z \to C$. □

When considering the HEP for a pair (X, A) we shall *always assume A is closed in X*. This assumption is no great loss [Exercise 2]. It ensures, because (*) of 7.2.2 is a pushout, that $X \times 0 \cup A \times \mathbf{I}$ is homeomorphic to the mapping cylinder $M(i)$ of the inclusion $i : A \to X$—a more general theory can be obtained by replacing $X \times 0 \cup A \times \mathbf{I}$ by this mapping cylinder. In any case the following picture for $X \times 0 \cup A \times \mathbf{I}$ should be kept in mind.

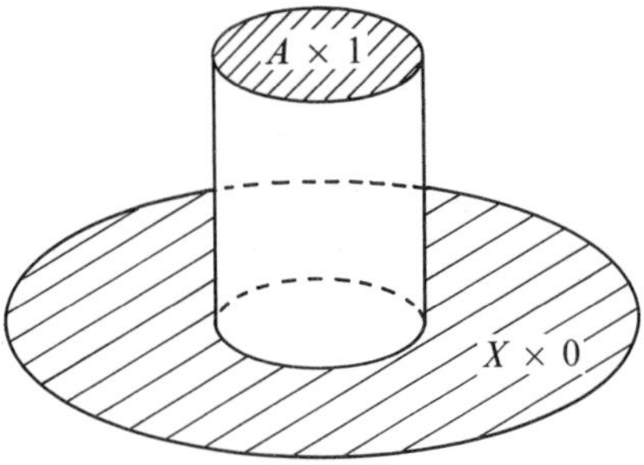

Fig. 7.2

Actually, if (X, A) has the HEP then a stronger condition than 7.2.2 (*c*) is satisfied.

If $i : C \to Z$ is an inclusion, we say C is a *deformation retract* of Z if there is a retraction $r : Z \to C$ such that $1_Z \simeq ir$ rel C. In this case, a homotopy $1_Z \simeq ir$ rel C is called a *retracting homotopy* of Z onto C and r is called a *deformation retraction*. Clearly, to prove C is a deformation retract of Z it is sufficient to exhibit such a retracting homotopy, that is, a homotopy $R_t : Z \to Z$ rel C such that

$$R_0 = 1_Z, \qquad R_1[Z] \subset C.$$

For example, 0 is a deformation retract of $\mathbf{I}$. More generally, if a is a point of the convex set X in a normed vector space, then a is a deformation retract of X since the homotopy $(x, t) \rightsquigarrow (1 - t)x + ta$ is a retracting homotopy of X onto a.

It is not true, in general, that C is a deformation retract of Z ⇔ the inclusion $C \to Z$ is a homotopy equivalence. However, the implication ⇒ trivially holds, and we shall prove later that the converse implication holds if (Z, C) has the HEP.

7.2.3 *If (X, A) has the HEP, then $X \times 0 \cup A \times \mathbf{I}$ is a deformation retract of $X \times \mathbf{I}$.*

Proof Let $\rho : X \times \mathbf{I} \to X \times 0 \cup A \times \mathbf{I}$ be a retraction and let ρ_1, ρ_2 be the two components of ρ (i.e., $\rho z = (\rho_1 z, \rho_2 z)$ where $\rho_1 z \in X$, $\rho_2 z \in \mathbf{I}$ for each z in $X \times \mathbf{I}$). We define

$$R_t : X \times \mathbf{I} \to X \times \mathbf{I}$$

$$(x, s) \rightsquigarrow (\rho_1(x, st), s(1 - t) + t\rho_2(x, s)).$$

Since ρ is a retraction, we have for each $x \in X$, $a \in A$, $s \in \mathbf{I}$

$$R_0(x, s) = (x, s), \qquad R_1(x, s) = \rho(x, s)$$

$$R_t(x, 0) = (\rho_1(x, 0), t\rho_2(x, 0)) = (x, 0)$$

$$R_t(a, s) = (a, s(1 - t) + ts) = (a, s).$$

Therefore R_t is the required retracting homotopy. □

We now come to the main result of this section.

7.2.4 *Let (X, A) have the HEP and let Y be a topological space. The function $u \rightsquigarrow X/\!/u$ is the object function of a functor*

$$X/\!/ : \pi Y^A \to \mathcal{S}\text{et}.$$

Proof Let u, v be maps $A \to Y$ and let $\gamma \in \pi Y^A(u, v)$. If $X/\!/u$ is empty, we define $X/\!/\gamma : X/\!/u \to X/\!/v$ to be the unique function. Suppose then that $X/\!/u$ is non-empty.

Let $f \in\in X/\!/u$ (by which we mean f represents an element of $X/\!/u$) so that f is a map $X \to Y$ whose restriction to A is u. Let U be a homotopy $u \simeq v$ of length 1 representing an element γ of πY^A. The maps f and U determine (as in the proof of 7.2.2) a map

$$L : X \times 0 \cup A \times \mathbf{I} \to Y.$$

By the HEP, L extends to a homotopy $F : X \times \mathbf{I} \to Y$ of f; let g be the final map of F. We define the element of $X/\!/u$

$$U_{*}f = \text{cls}\, g.$$

We now show that $U_{*}f$ is independent of the choices of f and U in their respective homotopy classes.

Suppose $f \simeq f_1$ rel A and $U \simeq U_1$ rel $A \times \dot{\mathbf{I}}$. Since A is closed in X these homotopies (supposed both of length 1) define a homotopy $L \simeq L_1$ rel $A \times \dot{\mathbf{I}}$ where L_1 is defined by f_1 and U_1.

Let $i : X \times 0 \cup A \times \mathbf{I} \to X \times \mathbf{I}$ be the inclusion and let $\rho : X \times \mathbf{I} \to X \times 0 \cup A \times \mathbf{I}$ be a retraction. By 7.2.3 $1 \simeq i\rho$ rel $X \times 0 \cup A \times \mathbf{I}$.

Therefore, if F, $F_1 : X \times \mathbf{I} \to Y$ are extensions of L, L_1 respectively, then

$$\begin{aligned} F &\simeq Fi\rho \quad \text{rel } X \times 0 \cup A \times \mathbf{I} \\ &= L\rho \\ &\simeq L_1\rho \quad \text{rel } A \times \dot{\mathbf{I}} \\ &= F_1 i\rho \\ &\simeq F_1 \quad \text{rel } X \times 0 \cup A \times \mathbf{I}. \end{aligned}$$

Hence $F \simeq F_1$ rel $A \times \dot{\mathbf{I}}$. It follows that $g \simeq g_1$ rel A (where g, g_1 are the final maps of F, F_1 respectively). Hence

$$\begin{aligned} U_* : X/\!/u &\to X/\!/v \\ \text{cls } f &\rightsquigarrow U_* f \end{aligned}$$

is well-defined and depends only on the homotopy class rel $A \times \dot{\mathbf{I}}$ of U. Similar considerations to the above apply if U is of length q for any $q \geqslant 0$. Notice that if U is a constant homotopy of length q then we can always extend U over $X \times [0, q]$ by the constant homotopy of f and so in such case U_* is the identity. Again, if $U : u \simeq v$ is of length q and $V : v \simeq w$ is of length r then U extends to a homotopy $F : f \simeq g$, V extends to a homotopy $G : g \simeq h$ and $V + U : u \simeq w$ extends to $G + F : f \simeq g$. This proves that

$$(V + U)_* = V_* U_* .$$

It follows easily that U_* depends only on the equivalence class of U—that is, we can define

$$X/\!/\gamma = U_*$$

for any U representing γ. The preceding remarks show immediately that $X/\!/0$ is the identity function, and that

$$X/\!/(\delta + \gamma) = (X/\!/\delta)(X/\!/\gamma)$$

whenever $\delta + \gamma$ is defined. Thus $X/\!/$ is a functor. □

As a very simple application of this result we have:

7.2.4 (*Corollary 1*) *Let $\gamma \in \pi Y^A(u, v)$. Then*

$$X/\!/\gamma : X/\!/u \to X/\!/v$$

is a bijection.

Proof This follows since γ is invertible in πY^A. □

7.2.4 (*Corollary 2*) *Let $u, v : A \to Y$ be homotopic maps. Then u extends* *over X if and only if v extends over X. In particular, any inessential map extends over X.*

Proof By 7.2.4 (Corollary 1) there is a bijection $X/\!/u \to X/\!/v$; hence $X/\!/u$ is non-empty if and only if $X/\!/v$ is non-empty, and this proves the first statement. The second statement follows since any constant map $A \to Y$ extends over X (for example, to the constant map $X \to Y$). □

Suppose now that $f : Y \to Z$ is a map. Then it is easy to prove that f induces by composition a morphism of groupoids

$$\begin{aligned} f_* : \pi Y^A &\to \pi Z^A \\ \text{cls } U &\rightsquigarrow \text{cls } fU \end{aligned}$$

and also for each $u : A \to Y$ a function, which we also write f_*,

$$\begin{aligned} f_* : X/\!/u &\to X/\!/fu \\ \text{cls } g &\rightsquigarrow \text{cls } fg. \end{aligned}$$

7.2.5 *Let (X, A) have the HEP and let $f, f' : Y \to Z$ be homotopic maps. Then for each map $u : A \to Y$ there is a track θu in $\pi Z^A(fu, f'u)$ such that the following diagram commutes*

$$\begin{array}{ccc} X/\!/u & \xrightarrow{f_*} & X/\!/fu \\ & \underset{f'_*}{\searrow} & \downarrow X/\!/\theta u \\ & & X/\!/f'u \end{array} \qquad (7.2.6)$$

Proof Let F_t be a homotopy $f \simeq f'$ and for each $u : A \to Y$ let θu in $\pi Z^A(fu, f'u)$ be the class of the homotopy $F_t u$. For each $g \in\in X/\!/u$ the homotopy $F_t g$ is a homotopy $fg \simeq f'g$ which extends the homotopy $F_t u$. It follows that

$$X/\!/\theta u\,(\text{cls } fg) = \text{cls } f'g. \quad \square$$

7.2.5 (*Corollary 1*) *If $f \simeq 1 : Y \to Y$, then for each $u : A \to Y$*

$$f_* : X/\!/u \to X/\!/fu$$

is a bijection.

Proof In diagram (7.2.6) we take $Y = Z$ and $f' = 1$; the result follows since $f'_* = 1$ and $X/\!/\theta u$ is a bijection. □

7.2.5 (*Corollary 2*) *If $f : Y \to Z$ is a homotopy equivalence, then for each $u : A \to Y$ the function*

$$f_* : X/\!/u \to X/\!/fu$$

is a bijection.

Proof This follows from 7.2.5 (Corollary 1) in a similar manner to the proof of 6.5.10. □

We now show that these results are not only obviously analogous to results on the fundamental groupoid of a space Y but, in fact, generalize these results exactly.

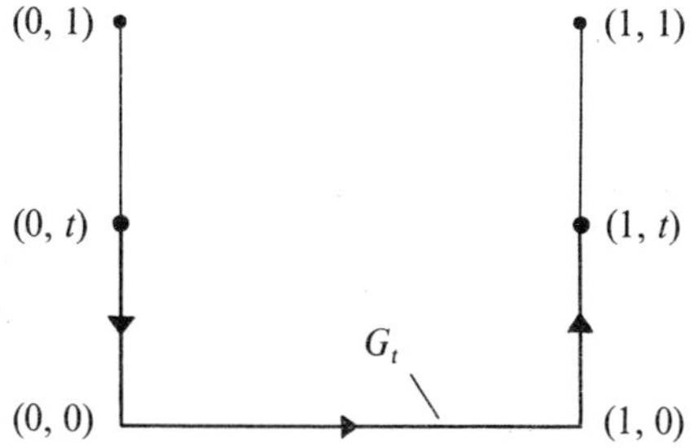

Fig. 7.3

In the groupoid πY, let

$$\beta \in \pi Y(x, x'), \qquad \gamma \in \pi Y(y, y').$$

In terms of β, γ we defined in 6.3.1 the function

$$\begin{aligned} \varphi : \pi Y(x, y) &\to \pi Y(x', y') \\ \alpha &\rightsquigarrow \gamma + \alpha - \beta. \end{aligned}$$

Our aim is to express φ as a function $X/\!/\theta : X/\!/u \to X/\!/u'$ for certain X, u, u', θ.

We take $X = \mathbf{I}, A = \dot{\mathbf{I}}$, and $u, u' : \dot{\mathbf{I}} \to Y$ the maps which send $0, 1 \rightsquigarrow x, y$ and $0, 1 \rightsquigarrow x', y'$ respectively. We can identify $\pi Y(x, y)$ with $X/\!/u$, and $\pi Y(x', y')$ with $X/\!/u'$. Let b, c be paths of length 1 respresenting β, γ respectively. Then b, c define a homotopy

$$\begin{aligned} F : \dot{\mathbf{I}} \times \mathbf{I} &\to Y \\ (0, t) &\rightsquigarrow bt \\ (1, t) &\rightsquigarrow ct \end{aligned}$$

whose class is a track θ in $\pi Y^{\dot{\mathbf{I}}}(u, u')$. We assert that, with the above identifications, $X/\!/\theta = \varphi$.

For the proof, let a be a path of length 1 representing an element α of $\pi Y(x, y)$ (which we identify with $X/\!/u$). Then F is a homotopy of $a \mid \dot{\mathbf{I}}$. We extend F to a homotopy G of a by letting G_t be the path shown above in Fig. 7.3—explicit formulae for G_t are easy to write down, and with these it is easy to prove that G is continuous. It is clear that the class of G_1 in $X/\!/u' = \pi Y(x', y')$ is $\varphi\alpha = \gamma + \alpha - \beta$. This completes the proof.

On one occasion in 7.4 we will need a stronger application of the HEP than that given solely by the operation $X//\gamma : X//u \to X//v$ for $\gamma \in \pi Y^A(u, v)$. In fact, suppose U represents γ, $f \in\in X//u$ and that $g \in\in X//v$ satisfies

$$X//\gamma\,(\text{cls}\, f) = \text{cls}\, g.$$

Then we can conclude that U extends to a homotopy F of f such that the final map g' of F is homotopic rel A to g, but we cannot conclude that U extends to a homotopy $f \simeq g$. However, if s denotes a constant homotopy $v \simeq v$ of length greater than 0, then g' is homotopic to g by a homotopy extending s—that is, *$s + U$ extends to a homotopy $f \simeq g$.*

Another way of stating this is as follows. A homotopy U is called *semi-constant* if U is a constant homotopy on some interval $[r, q]$ with $0 \leqslant r < q$, where q is the length of U. So, by replacing $s + U$ by U in the above we obtain:

7.2.7 *Let (X, A) have the HEP, let $\gamma \in \pi Y^A(u, v)$ and let $f \in\in X//u$, $g \in\in X//v$.* If $X//\gamma$ (cls f) $=$ cls g, *then any semi-constant representative of γ extends to a homotopy $f \simeq g$.*

Notice also that the restriction to semi-constant homotopies is no grave loss, since any homotopy is homotopic rel end points to a semi-constant homotopy.

EXERCISES

1. (i) Prove that if A is a deformation retract of B, and B is a deformation retract of C, then A is a deformation retract of C. (ii) Prove that if A is a deformation retract of X, and B is a deformation retract of Y, then $A \times B$ is a deformation retract of $X \times Y$.
2. Prove that if $X \times 0 \cup A \times \mathbf{I}$ is a retract of $X \times \mathbf{I}$, and X is Hausdorff, then A is closed in X.
3. Prove that there is a homeomorphism $\varphi : \mathbf{I} \times \mathbf{I} \to \mathbf{I} \times \mathbf{I}$ which maps $0 \times \mathbf{I} \cup \mathbf{I} \times \dot{\mathbf{I}}$ homeomorphically onto $0 \times \mathbf{I}$. Prove that $1 \times \varphi$ is a homeomorphism $X \times \mathbf{I} \times \mathbf{I} \to X \times \mathbf{I} \times \mathbf{I}$ which for any subspace A of X maps the space $L = (X \times 0 \cup A \times \mathbf{I}) \times \mathbf{I} \cup X \times \mathbf{I} \times \dot{\mathbf{I}}$ homeomorphically onto $(X \times 0 \cup A \times \mathbf{I}) \times \mathbf{I}$. Hence show that if $X \times 0 \cup A \times \mathbf{I}$ is a retract of $X \times \mathbf{I}$, then L is a retract of $X \times \mathbf{I} \times \mathbf{I}$. Use this fact to give another proof of 7.2.3.
4. A map $i : Y \to X$ is called a *cofibration* if for any map $f : X \to Z$ and any homotopy U_t of $u = fi$, there is a homotopy F_t of f such that $F_t i = U_t$. Let $e : M(i) \to X \times \mathbf{I}$ be the map defined by the maps.

$$\begin{array}{ll} X \to X \times \mathbf{I} & Y \times \mathbf{I} \to X \times \mathbf{I} \\ x \leadsto (x, 0) & (y, t) \leadsto (iy, t). \end{array}$$

Prove that i is a cofibration if and only if e has a left-inverse, and show that in such case i is a homeomorphism into whose image, in the case X is Hausdorff, is closed in X.

5. Let $D \subset C \subset Z$ and let C be a deformation retract of Z. Let $f, g : C \to Y$ be maps which are homotopic rel D. Prove that f, g extend over Z, and that if f', g' are extensions of f, g respectively then $f' \simeq g'$ rel D.
6. Let (X, A) have the HEP and let $u : A \to Y$ be a map. Prove that if $f: Y \to Z$ is a homotopy equivalence, then u extends over X if and only if fu extends over X.
7. Let A be a subspace of both X and X'. Suppose given maps $f: X \to X', g: X' \to X$ such that $fa = ga = a$, all a in A. If $gf \simeq 1$ rel A then we say X' *dominates* X rel A; if further $fg \simeq 1$ rel A then we say f is a *homotopy equivalence* rel A. Prove that if $f: X \to X'$ is a homotopy equivalence rel A, then f induces a bijection

$$\begin{aligned} X'//u &\to X//u \\ \text{cls } g &\rightsquigarrow \text{cls } gf \end{aligned}$$

for all maps $u : A \to Y$. Use such a homotopy equivalence to define an 'operation' of πY^A on the sets $X'//u$ when (X, A) is given to have the HEP.
8. A pair (X', A) is said to have the *weak* HEP (written WHEP) if for any map $f: X' \to Y$ and any homotopy F of $u = f \mid A$, there is a real number $r \geqslant 0$ such that the homotopy $F + r$ of u extends to a homotopy of f. Prove that if X' dominates X rel A and (X', A) has the WHEP, then (X, A) has the WHEP.
9. Let (X, A) have the HEP and consider the following diagram.

$$\pi Y^X\{f\} \xrightarrow{k^*} \pi Y^A\{u\} \xrightarrow{\partial} X//u \xrightarrow{j^*} [X, Y] \xrightarrow{i^*} [A, Y]$$

in which

(i) $u : A \to Y$ is a map which extends to a map $f : X \to Y$.
(ii) $[Z, Y]$ denotes the set of homotopy classes of maps $Z \to Y$.
(iii) i^* is defined by cls $g \rightsquigarrow$ cls gi; k^* is defined by cls $F_t \rightsquigarrow$ cls $F_t i$.
(iv) j^* sends the equivalence class of g in $X//u$ to its equivalence class in $[X, Y]$.
(v) ∂ is defined by $\partial\gamma = \gamma_*$ (cls f) where $\gamma_* = X//\gamma$.

Prove that (a) $(i^*)^{-1}[\text{cls } u] = \text{Im } j^*$, (b) $j^*x = j^*y$ if and only if there is a γ in $\pi Y^A\{u\}$ such that $\gamma_* x = y$, (c) $\partial\gamma = \partial\delta$ if and only if there is a β in $\pi Y^X\{f\}$ such that $k^*\beta = -\delta + \gamma$.
10. Let X, Y be pointed spaces and let $[X, Y], [X, Y]_\cdot$ denote respectively the (free) homotopy classes of all maps $X \to Y$, and the homotopy classes rel base point of all pointed maps $X \to Y$. Let $(X, \cdot)$ have the HEP. Prove that the function

$$i : [X, Y]_\cdot \to [X, Y]$$

which sends a based class to a free class is surjective if Y is path connected, and injective if Y is simply connected.
11. Prove that $\pi(Y \times Z)^A$ is isomorphic to $\pi Y^A \times \pi Z^A$. Hence prove that if $f \simeq f' : Y \to W$, then $f_* \simeq f'_* : \pi Y^A \to \pi W^A$.
12. Let (X, A) have the HEP. The *extension class groupoid* $X, Y// A$ is defined as follows. The objects are the sets $X//u$ for all maps $u : A \to Y$. If $X//u$ is empty, then there is only one morphism from $X//u$, namely, the identity at $X//u$. If $X//u$ is non-empty, then the morphisms $X//u \to X//v$ are the pairs $(\gamma, X//\gamma)$ for all $\gamma \in \pi Y^A(u, v)$. These morphisms are composed by $(\delta, X//\delta)(\gamma, X//\gamma) = (\delta + \gamma, X//(\delta + \gamma))$. Prove that a map $f : Y \to Z$ induces a morphism $f_* : X, Y // A \to X, Z// A$ which on objects sends $X//u \rightsquigarrow X//fu$; and that if $f \simeq f'$, then $f_* \simeq f'_*$.

7.3 Examples

We shall derive our main examples of pairs with the HEP from the following basic example, together with the trivial examples that for any X, the pairs (X, X) and $(X, \varnothing)$ have the HEP.

7.3.1 *The pair* $(\mathbf{I}, \dot{\mathbf{I}})$ *has the HEP.*

Proof

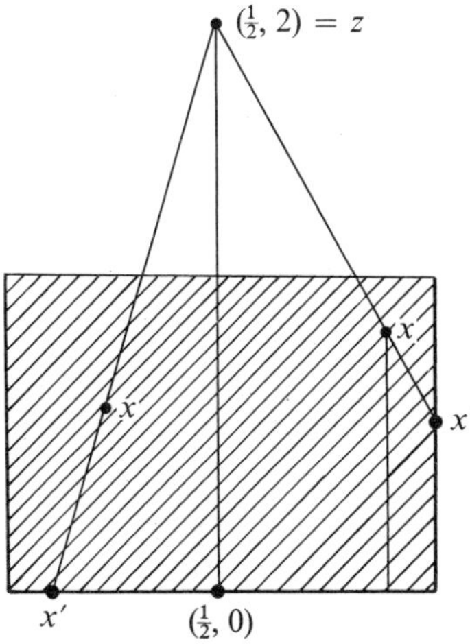

Fig. 7.4

We have to construct a retraction $\mathbf{I} \times \mathbf{I} \to \mathbf{I} \times 0 \cup \dot{\mathbf{I}} \times \mathbf{I}$. The latter space is shown as the thick line in Fig. 7.4. Let z be the point $(\frac{1}{2}, 2)$ in $\mathbf{R}^2$, and regard $\mathbf{I} \times \mathbf{I}$ as a subset of $\mathbf{R}^2$. For each x in $\mathbf{I} \times \mathbf{I}$ let x' be the unique point of $\mathbf{I} \times 0 \cup \dot{\mathbf{I}} \times \mathbf{I}$ such that z, x, x' are collinear. Then $\rho : x \rightsquigarrow x'$ is the required retraction. We leave the reader to work out a formula for x' (e.g., by writing $x = (\frac{1}{2} + s, 2 - t)$ where $|s| \leqslant \frac{1}{2}$ and $1 \leqslant t \leqslant 2$) and so to prove that ρ is continuous. $\square$

An alternative method of constructing the retraction required in 7.3.1 has been given in essence at the end of the last section.

The following propositions allow us to generate other examples from this basic example.

7.3.2 *Let (X, A), (Y, B) have the HEP. Then $(X \times Y, A \times B)$ has the HEP.*

7.3.3 *Let (X, B) and (B, A) have the HEP. Then (X, A) has the HEP.*

7.3.4 *Let (X, D) have the HEP. Let A be a closed subset of X such that*

$A \subset D$, and let $f : A \to B$ be a map. Then

$$(B\ {}_f\sqcup X,\ B\ {}_f\sqcup D)$$

has the HEP.

7.3.5 *For any X, A the pair $(X \sqcup A, A)$ has the HEP.*

Proof of 7.3.2 We may suppose given retractions

$$\rho : X \times \mathbf{I} \to X \times 0 \cup A \times \mathbf{I}, \qquad \sigma : Y \times \mathbf{I} \to Y \times 0 \cup B \times \mathbf{I}.$$

Let $\rho = (\rho_1, \rho_2)$, $\sigma = (\sigma_1, \sigma_2)$. Then it is easily verified that the following map

$$\begin{aligned} X \times Y \times \mathbf{I} &\to X \times Y \times 0 \cup A \times B \times \mathbf{I} \\ (x, y, t) &\rightsquigarrow (\rho_1(x, t), \sigma_1(y, t), \tfrac{1}{2}\{\sigma_2(x, t) + \rho_2(x, t)\}) \end{aligned}$$

is a retraction. □

Proof of 7.3.3 Let $f_0 : X \to Y$ be a map and let $g_t : A \to Y$ be a homotopy of $f_0 \mid A$. Then g_t extends to a homotopy h_t of $f_0 \mid B$ (since (B, A) has the HEP) and h_t extends to a homotopy f_t of f_0 (since (X, B) has the HEP). □

Remark 7.3.6 In order to prove 7.3.4 we need a method of constructing homotopies from adjunction spaces. Suppose there is given a pushout

$$\begin{array}{ccc} A & \xrightarrow{\ f\ } & B \\ {\scriptstyle i}\downarrow & & \downarrow{\scriptstyle \bar{i}} \\ X & \xrightarrow[\ \bar{f}\ ]{} & B\ {}_f\sqcup X \end{array}$$

We consider here homotopies of length 1 only, so that a homotopy $Z \times \mathbf{I} \to Y$ can be written as $H_t : Z \to Y$, it being understood that $t \in \mathbf{I}$. Since $\mathbf{I}$ is locally compact we can apply 4.6.6 which, in the above notation, becomes: *if we are given a commutative square*

$$\begin{array}{ccc} A & \xrightarrow{\ f\ } & B \\ {\scriptstyle i}\downarrow & & \downarrow{\scriptstyle G_t} \\ X & \xrightarrow[\ F_t\ ]{} & Y \end{array}$$

that is, if $F_t i = G_t f$ for each $t \in \mathbf{I}$, then there is a unique homotopy $H_t : B\ {}_f\sqcup X \to Y$ such that

$$H_t \bar{f} = F_t, \qquad H_t \bar{i} = G_t.$$

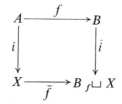

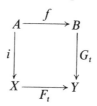

Proof of 7.3.4 Let $\varphi_0 : B\,{}_f\!\sqcup X \to Y$ be a map and let $\psi_t : B\,{}_f\!\sqcup D \to Y$ be a homotopy of $\varphi_0 \mid B\,{}_f\!\sqcup D$ ($B\,{}_f\!\sqcup D$ is a subspace of $B\,{}_f\!\sqcup X$ by 4.6.1). We consider the diagram

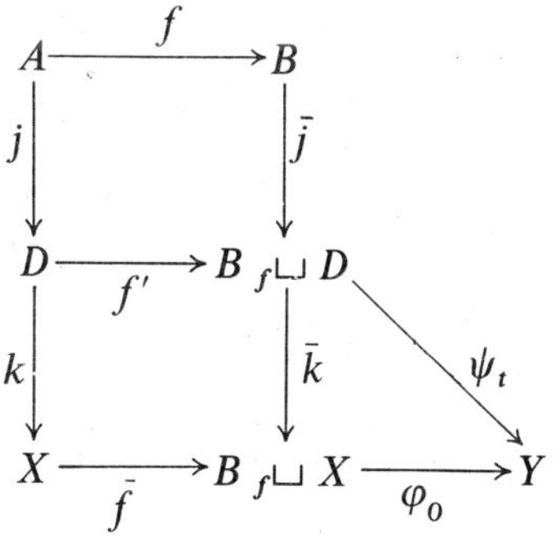

in which j, k are inclusions. The HEP for (X, D) shows that $\psi_t f'$ extends to a homotopy $\bar{\varphi}_t$ of $\varphi_0 \bar{f}$. Then, if $i : A \to X$ is the inclusion

$$\bar{\varphi}_t i = \bar{\varphi}_t kj = \psi_t f'j = \psi_t \bar{j} f.$$

If follows from Remark 7.3.6 that $\bar{\varphi}_t$ and $\psi_t \bar{j}$ define a homotopy $\varphi_t : B_f\!\sqcup X \to Y$ which, as is easily checked, is a homotopy of φ_0 extending ψ_t. □

Proof of 7.3.5 This follows from 7.3.4 on taking $D = A = \varnothing$. A direct proof also is easy. □

EXAMPLES 1. For any A, the pair $(A \times \mathbf{I}, A \times 0)$ has the HEP.

Proof We know that $(\mathbf{I}, \dot{\mathbf{I}})$ has the HEP. By 7.3.5, $(\dot{\mathbf{I}}, 0)$ has the HEP and so, by 7.3.3, $(\mathbf{I}, 0)$ has the HEP. Since (A, A) has the HEP, the example follows from 7.3.2. □

2. Let $f : A \to B$ be a map. We recall [Example 5 of 4.6] that the mapping cylinder $M(f)$ is $B\,{}_{f'}\!\sqcup (A \times \mathbf{I})$ where $f' : A \times 0 \to B$ is $(a, 0) \leadsto f(a)$. We shall identify A with the subset $A \times 1$ of $M(f)$. So we can state: the pairs $(M(f), A \cup B)$, $(M(f), A)$, $(M(f), B)$ all have the HEP.

Proof Certainly $(A \times \mathbf{I}, A \times \dot{\mathbf{I}})$ has the HEP, by 7.3.2. It follows from 7.3.4 with $D = A \times 0 \cup A \times 1$ $(= A \times 0 \cup A)$ that $(M(f), A \cup B)$ has the HEP. But A, B are disjoint closed sets of $M(f)$, so that $A \cup B = A \sqcup B$. So the remaining results follow from 7.3.5 and 7.3.3. □

3. For any A, the pair (CA, A) has the HEP (here we have identified A with the subset $A \times 1$ of CA).

Proof Let v be the vertex of CA. Since $(A \times \mathbf{I}, A \times \dot{\mathbf{I}})$ has the HEP, we obtain by 7.3.4 that $(CA, A \cup \{v\})$ has the HEP. Here again A and $\{v\}$ are disjoint closed subsets of CA, so that (CA, A) has the HEP. □

4. Two pairs (X, A), (X', A') are *homeomorphic* if there is a homeomorphism $f : X \to X'$ such that $f[A] = A'$. It is clear that, in such case, (X, A) has the HEP if and only if (X', A') has the HEP. Now $(\mathbf{E}^n, \mathbf{S}^{n-1})$ is homeomorphic to $(C\mathbf{S}^{n-1}, \mathbf{S}^{n-1})$. By the previous example, $(\mathbf{E}^n, \mathbf{S}^{n-1})$ has the HEP. This, with 7.3.4, shows that for any B and any map $f : \mathbf{S}^{n-1} \to B$, the pair $(B\ {}_f\!\sqcup \mathbf{E}^n, B)$ has the HEP.
5. Let X be a cell complex and let A be a subcomplex of X. Then (X, A) has the HEP.

Proof Let $f_0 : X \to Y$ be a map and let $g_t : A \to Y$ be a homotopy of $f_0 \mid A$. Let $X_n = X^n \cup A$ where (we recall) X^n is the n-skeleton of X. Then we can write X_n as an adjunction space

$$X_n = X_{n-1} \cup e_1^n \cup \cdots \cup e_{r_n}^n.$$

For $n = 0$, $X_0 = A \cup e_1^0 \cup \cdots \cup e_{r_0}^0$, and the homotopy g_t is extended arbitrarily over each e_i^0 to a homotopy of $f_0 \mid X_0$.

Suppose inductively that g_t has been extended to a homotopy g_t^{n-1} of $f_0 \mid X_{n-1}$. By the HEP applied in turn to each pair $(X_{n-1} \cup e_i^n, X_{n-1})$, g_t^{n-1} can be extended to a homotopy g_t^n of $f_0 \mid X_n$. Since $X_n = X$ for some n, the result follows. □

6. However, not all pairs have the HEP as we showed by an example in 7.2.

We now derive some interesting conditions for (X, A) to have the HEP; these will imply an important product rule for the HEP. However, this rule will not itself be used until the last part of 7.5, and the reader may therefore turn directly to applications of the HEP in the next section.

Definition The subspace A of X is said to be a *weak neighbourhood deformation retract* (WNDR) of X if there is a homotopy $\psi_t : 1_X \simeq \psi_1$ rel A and a map $v : X \to \mathbf{I}$ such that

$$v(x) < 1 \Rightarrow \psi_1(x) \in A$$

$$x \in A \Rightarrow v(x) = 0.$$

Remark This property is sometimes defined by the following conditions: *there is a neighbourhood V of A in X, a map $u : X \to \mathbf{I}$ such that $u[X \setminus V] = \{1\}$, $u[A] = \{0\}$, and a homotopy $\varphi_t : V \to X$* rel A *such that φ_0 is the inclusion $V \to X$ and $\varphi_1[V] \subset A$.*

However these two definitions are equivalent.

Proof We can assume V is closed in X, for otherwise we can replace V by $u^{-1}[0, \frac{1}{2}]$ and u by $x \rightsquigarrow \min(2u(x), 1)$.

Let $u'(x) = \min(2 - 2u(x), 1)$. Then we define

$$\psi : X \times \mathbf{I} \to X$$

$$(x, t) \rightsquigarrow \begin{cases} \varphi(x, tu'(x)), & x \in V \\ x, & u(x) = 1 \end{cases}$$

$$v : X \to \mathbf{I}$$

$$x \rightsquigarrow \min\{2u(x), 1\}.$$

Then ψ is well-defined and, since V is closed in X, is continuous. It is easily verified that ψ, v satisfy the conditions given in our first definition of WNDR.

On the other hand, given ψ, v as in this first definition, we can define $V = v^{-1}[0, 1[$, $\varphi = \psi \mid V \times \mathbf{I}$, and so satisfy the conditions of our second definition. □

7.3.6 *Let A be a closed subspace of X. The following conditions are equivalent.*
(a) (X, A) has the HEP.
(b) A is a WNDR of X and there is a map $w : X \to \mathbf{I}$ *such that* $A = w^{-1}[0]$.
(c) There is a map $u : X \to \mathbf{R}^{\geqslant 0}$ *and a map* $\varphi : X_u \to X$, where

$$X_u = \{(x, t) \in X \times \mathbf{I} : 0 \leqslant t \leqslant \min\{u(x), 1\}\}$$

such that (i) $u[A] = \{0\}$, *(ii)* $\varphi(x, 0) = x$ *for all x in X, (iii)* $\varphi(x, u(x)) \in A$ *for* $u(x) \leqslant 1$.

Proof (*a*) ⇒ (*b*) By assumption, there is a retraction

$$\rho : X \times \mathbf{I} \to X \times 0 \cup A \times \mathbf{I}$$

and we write $\rho = (\rho_1, \rho_2)$, where ρ_1, ρ_2 are the components of ρ. We define maps

$$\psi : X \times \mathbf{I} \to X \qquad\qquad v : X \to \mathbf{I}$$

$$(x, t) \rightsquigarrow \rho_1(x, t) \qquad\qquad x \rightsquigarrow 1 - \rho_2(x, 1).$$

It is easily checked that the conditions for the definition of WNDR are satisfied.

The function w of (*b*) is defined by

$$w : X \to \mathbf{I}$$

$$x \rightsquigarrow \sup_{t \in I} (t - \rho_2(x, t)).$$

We postpone the proof that w is continuous to the end of this section. That $w(x) \geqslant 0$ follows from $\rho_2(x, 0) = 0$. Clearly, $w[A] = \{0\}$. Suppose $w(x) = 0$; then $\rho_2(x, t) \geqslant t$ for all t, whence $\rho(x, t) \in A \times \mathbf{I}$ for $t > 0$.

Since $A \times \mathbf{I}$ is closed this implies that $\rho(x, t) \in A \times \mathbf{I}$ for $t = 0$; but $\rho(x, 0) = (x, 0)$, so $x \in A$.
(*b*) $\Rightarrow$ (*c*) We suppose given the ψ, v of the definition of WNDR, and also the map w of (*b*). Let

$$u(x) = 2 \max \{v(x), w(x)\}$$

and define

$$\begin{aligned} q : X \times \mathbf{I} &\to \{(x, t) \in X \times \mathbf{R}^{\geqslant 0} : 0 \leqslant t \leqslant u(x)\} \\ (x, s) &\rightsquigarrow (x, su(x)). \end{aligned}$$

Then q is an identification map—we postpone the proof to the end of this section. Now $q(x, s) = q(x', s')$ is equivalent to $x = x'$, and $s = s'$ or $u(x) = 0$; so if two points have the same image under q, then they also have the same image under ψ. Hence, there is a map φ' such that $\psi = \varphi' q$. The restriction $\varphi = \varphi' \mid X_u$ and map u have the properties required by (*c*).
(c) $\Rightarrow$ (a) We define a retraction

$$\rho : X \times \mathbf{I} \to X \times 0 \cup A \times \mathbf{I}$$

$$(x, t) \rightsquigarrow \begin{cases} (\varphi(x, t), 0), & t \leqslant ux \\ (\varphi(x, ux), t - ux), & t \geqslant ux. \end{cases} \qquad \square$$

The following result is a simple corollary of condition (*c*) of 7.3.6.

7.3.7 *Let* (X, A), (Y, B) *have the HEP. Then*

$$(X \times Y, X \times B \cup A \times Y)$$

has the HEP.

Proof Let u, φ be given for (X, A) as in 7.3.6 (*c*), and let v, ψ be the corresponding maps for (Y, B). We define

$$\begin{aligned} w : X \times Y &\to \mathbf{R}^{\geqslant 0} \\ (x, y) &\rightsquigarrow \min \{ux, vy\} \\ \chi : (X \times Y)_w &\to X \times Y \\ (x, y, t) &\rightsquigarrow (\varphi(x, t), \psi(y, t)). \end{aligned}$$

The conditions (i), (ii), (iii) of 7.3.6 (*c*) are easily verified. $\square$

Remark Let $w : X \to \mathbf{I}$ be a map and let $A = w^{-1}[0]$. Then A is the intersection of the sets $w^{-1}[0, n^{-1}[$. Thus A is a G_δ-set (the intersection of a countable family of open sets). It may be proved [cf. Exercise 16 of 3.6] that if A is a closed, G_δ-set in a normal space X, then there is a map $w : X \to \mathbf{I}$ such that $A = w^{-1}[0]$.

In order to complete the proof of 7.3.6 (and so of 7.3.7) we have two continuity lemmas to prove. The first of these follows immediately from

7.3.8 *Let $\varphi : X \times C \to \mathbf{R}$ be a map, let C be compact and define*

$$\begin{aligned} w : X &\to \mathbf{R} \\ x &\rightsquigarrow \sup_{c \in C} \varphi(x, c). \end{aligned}$$

Then w is well-defined and continuous.

Proof For each x in X, $\varphi[x \times C]$ is a compact and hence bounded subset of $\mathbf{R}$. Therefore w is well-defined.

Suppose that $wx = r$ and that $N = [r - \varepsilon, r + \varepsilon]$ is a neighbourhood of r. By definition of r, $c \in C \Rightarrow \varphi(x, c) \leqslant r$, and so

$$x \times C \subset \varphi^{-1}]\leftarrow, r + \varepsilon[$$

By 3.5.6 (Corollary 2) there is an open neighbourhood U_1 of x such that

$$U_1 \times C \subset \varphi^{-1}]\leftarrow, r + \varepsilon[.$$

and this implies that

$$w[U_1] \subset]\leftarrow, r + \varepsilon]. \qquad (*)$$

However, there is a c in C such that $\varphi(x, c) \in \operatorname{Int} N$ and so there is a neighbourhood U_2 of x such that

$$\varphi[U_2 \times c] \subset N. \qquad (**)$$

So, if $y \in U_1 \cap U_2$, then (*) implies $w(y) \leqslant r + \varepsilon$ while (**) implies $w(y) \geqslant r - \varepsilon$; hence $w[U_1 \cap U_2] \subset N$. □

The second lemma is the following.

7.3.9 *Let $u : X \to \mathbf{R}^{\geqslant 0}$ be a map, let*

$$Y = \{(x, t) \in X \times \mathbf{R}^{\geqslant 0} : 0 \leqslant t \leqslant u(x)\}$$

and let

$$\begin{aligned} q : X \times \mathbf{I} &\to Y \\ (x, s) &\rightsquigarrow (x, su(x)). \end{aligned}$$

Then q is an identification map.

Proof Let $V \subset Y$ be such that $q^{-1}[V]$ is open in $X \times \mathbf{I}$. We must prove that V is open in Y.

Let $(x, t) \in Y$ and suppose first that $ux > 0$. Then there is a neighbourhood W of x such that $w \in W \Rightarrow uw > 0$. For this W, $q \mid W \times \mathbf{I}$ is a homeomorphism onto an open subset of Y and so V is a neighbourhood of (x, t).

Suppose now that $ux = 0$. Then $t = 0$ and

$$q^{-1}(x, 0) = x \times \mathbf{I} \subset q^{-1}[V].$$

Since **I** is compact there is a neighbourhood M of x such that $M \times \mathbf{I} \subset q^{-1}[V]$ [3.5.6, Corollary 2]. But

$$q[M \times \mathbf{I}] = (M \times \mathbf{I}) \cap Y.$$

Therefore V is a neighbourhood of $(x, 0)$. □

EXERCISES

1. Give an example of a pair (X, A) and maps $u, v : A \to Y$ such that $u \simeq v$ and v extends over X, but u does not extend over X.
2. Let A, B be disjoint closed subspaces of X such that there are maps $\lambda, \mu : X \to [0, 1]$ with $A = \lambda^{-1}[0]$, $B = \mu^{-1}[0]$. Prove that there is a map $\nu : X \to [0, 1]$ such that $A = \nu^{-1}[0]$, $B = \nu^{-1}[1]$.
3. Prove that if A, B are disjoint closed subspaces of X such that (X, A), (X, B) have the HEP, then $(X, A \cup B)$ has the HEP. [It seems likely that if A, B are closed subspaces of X, then $(X, A \cup B)$ has the HEP whenever (X, A), (X, B), $(A, A \cap B)$, $(B, A \cap B)$ do so—but I know of no proof.]
4. Suppose there is given the pushout square of 7.3.6, and that (X, A) has the HEP. Let $u : B \to Y$ be a map. Prove that the function

$$\begin{aligned}(B\,{}_f\!\sqcup X)/\!/u &\to X/\!/uf,\\ \text{cls}\, g &\rightsquigarrow \text{cls}\, g\bar{f}\end{aligned}$$

is a bijection.
5. Let M be the topological product of **I** with itself uncountably many times. Let $\mathbf{0}$ be the element of M with all components 0. Prove that (i) $\{\mathbf{0}\}$ is a WNDR of M, (ii) $M \times 0 \cup \mathbf{0} \times \mathbf{I}$ is not a neighbourhood retract (and hence not a WNDR) of $M \times \mathbf{I}$, (iii) $(M, \mathbf{0})$ has the WHEP, but $\{\mathbf{0}\}$ is not a G_δ-set in M and so $(M, \mathbf{0})$ does not have the HEP.
*6. Let (X, A) have the HEP, let $B \subset V \subset X$ and let $\tau : X \to \mathbf{I}$ be a map such that $B \subset \tau^{-1}[1]$, $X \setminus V \subset \tau^{-1}[0]$. Let $f : X \to Y$ be a map and let U_t be a homotopy of $f \mid A$. Prove that if G_t is any homotopy of $f \mid V$ such that $G_t \mid A \cap V = U_t \mid A \cap V$, then there is a homotopy F_t of f such that $F_t \mid A = U_t$ and $F_t \mid B = G_t \mid B$.
7. A *pair of pointed spaces* consists of a pointed space X, and a subspace A of X containing the base point of X; this base point is also taken as base point of A. We say (X, A) has the *pointed* HEP, written $\text{HEP}_{\bullet}$, if it has the HEP for pointed maps and pointed homotopies (i.e., homotopies rel base point). Give an example of a pair of pointed spaces which satisfies the pointed, but not the usual, HEP. Prove also that if (P, Q) is any pair, and (Y, B) has the HEP, then the pair (X, A) has the pointed HEP, where

$$\begin{aligned}X &= P \times Y / Q \times Y,\\ A &= (P \times B \cup Q \times Y) / Q \times Y = P \times B / Q \times B.\end{aligned}$$

7.4 Applications of the HEP

In 2.7.3 we proved a rather trivial glueing theorem for homeomorphisms: the main objective of this section is a glueing theorem for homotopy equivalences. The statement of this theorem is clarified by introducing the category of M-ads of spaces where M is an indexing set.

An *M-ad* $\mathbf{X} = (X; X_\mu)_{\mu \in M}$ consists of a topological space X and a family $(X_\mu)_{\mu \in M}$ of subspaces of X. If $M = \varnothing$ then we identify $\mathbf{X}$ and X. A *map* $\mathbf{f} : \mathbf{X} \to \mathbf{Y}$ of M-ads consists of the M-ads $\mathbf{X}$, $\mathbf{Y}$ and a map $f : X \to Y$ such that

$$f[X_\mu] \subset Y_\mu, \quad \mu \in M.$$

Thus $\mathbf{f}$ determines not only f but also maps

$$f^\mu = f \mid X_\mu, Y_\mu$$

for each $\mu \in M$.

It is easily verified that these objects and maps form a category in which $\mathbf{f} : \mathbf{X} \to \mathbf{Y}$ is an isomorphism if and only if $f : X \to Y$ is a homeomorphism and $f[X_\mu] = Y_\mu$ for each μ in M.

A homotopy $\mathbf{F} : \mathbf{f} \simeq \mathbf{g}$ (of length 1) of maps $\mathbf{f}, \mathbf{g} : \mathbf{X} \to \mathbf{Y}$ of M-ads is a homotopy $F : f \simeq g$ such that

$$F[X_\mu \times \mathbf{I}] \subset Y_\mu, \mu \in M;$$

such a homotopy is often written $\mathbf{F}_t$. Notice that a homotopy $\mathbf{F} : \mathbf{f} \simeq \mathbf{g}$ determines homotopies $F^\mu : f^\mu \simeq g^\mu$ for each μ in M.

As usual, when homotopy has been defined, we have the notions of *homotopic maps*, *domination*, *homotopy equivalence* of M-ads. We leave the reader to give the definitions and basic properties of these notions.

Let m be a positive integer. An *m-ad* is an M-ad in which M is the set $\{1, \ldots, m-1\}$. Such an m-ad is often written $(X; X_1, \ldots, X_{m-1})$; in particular, a 2-ad is called a *pair* and is usually written (X, X_1); a 3-ad $(X; X_1, X_2)$ is called a *triad*. Of course, a pair (X, X_1) in which X_1 consists of a single point of X is simply a pointed space.

The notion of homotopy equivalence for pairs is definitely more restrictive than that for spaces. For example, suppose $\mathbf{f} : (X, X_1) \to (Y, Y_1)$ is a map. If $\mathbf{f}$ is a homotopy equivalence, then so also are $f : X \to Y$ and $f^1 : X_1 \to Y_1$. The converse is not necessarily true; if we are given that f and f^1 are homotopy equivalences then we know there are homotopy inverses g of f, g^1 of f^1, but we do not know that g^1 is the restriction of g, indeed we do not even know that $g[Y_1] \subset X_1$.

However, the converse is true if (X, X_1) and (Y, Y_1) have the HEP. This fact is both a special case, and a crucial step, in the proof of our glueing theorem, 7.4.1 below.

A triad $(X; X_1, X_2)$ is called a *Mayer–Vietoris triad*, in brief an *MV-triad*, if the following conditions hold:
(*a*) $X = X_1 \cup X_2$,
(*b*) X_1, X_2 are closed in X,
(*c*) if $X_0 = X_1 \cap X_2$, then (X_1, X_0), (X_2, X_0) have the HEP.
(The name Mayer–Vietoris is explained by the fact that these conditions are sufficient to ensure the existence of the Mayer–Vietoris sequence in homology [Eilenberg–Steenrod [1]].) Notice that if $\mathbf{f} : (X; X_1, X_2) \to (Y; Y_1, Y_2)$ is a map of triads then $\mathbf{f}$ also restricts to a map $f^0 : X_0 \to Y_0$ where $X_0 = X_1 \cap X_2$, $Y_0 = Y_1 \cap Y_2$. Also, results on triads can be applied to pairs by considering, instead of the pair (X, X_1), the triad $(X; X, X_1)$.

7.4.1 *Let* $\mathbf{f} : \mathbf{X} \to \mathbf{Y}$ *be a map of MV-triads such that*

$$f^\mu : X_\mu \to Y_\mu, \quad \mu = 0, 1, 2$$

is a homotopy equivalence. Then $\mathbf{f}$ *is a homotopy equivalence, and in particular*

$$f : X \to Y$$

is a homotopy equivalence.

Actually our proof will give more information which is used in 7.4.1 (Corollary 1).

7.4.1 (*Addendum*) *Let* $g^0 : Y_0 \to X_0$ *be any homotopy inverse of* f^0 *and let* $H_t^0 : f^0 g^0 \simeq 1$, $K_t^0 : g^0 f^0 \simeq 1$ *be semi-constant homotopies. Then* g^0 *extends to a homotopy inverse* $\mathbf{g}$ *of* $\mathbf{f}$ *such that the homotopy* $\mathbf{fg} \simeq 1$ *extends* H_t^0 *while the homotopy* $\mathbf{gf} \simeq 1$ *extends the sum*

$$K^0 + g^0 H^0 f^0 - g^0 f^0 K^0$$

of the homotopies

$$g^0 f^0 = g^0 f^0 1_{X_0} \simeq g^0 f^0 g^0 f^0 \simeq g^0 1_{Y_0} f^0 \simeq 1_{X_0}$$

determined by H_t^0 *and* K_t^0.

Proof of 7.4.1 *Step 1.* We first prove the result for the case

$$X = X_1, \quad X_2 = X_0; \qquad Y = Y_1, \quad Y_2 = Y_0$$

so that we are in effect dealing with the case of pairs (X, X_0), (Y, Y_0) which have the HEP. Consider the diagram

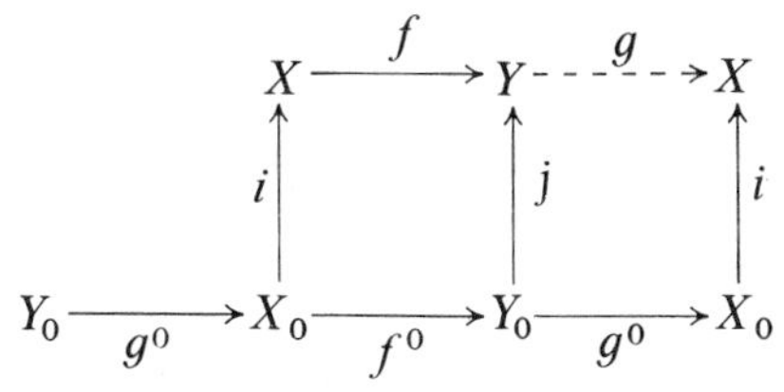

in which i, j are the inclusions. Since **f** is a map, we have

$$jf^0 = fi.$$

Consider the diagram

$$Y/\!/ig^0 \overset{f_*}{\rightarrow} Y/\!/jf^0g^0 \overset{Y/\!/\alpha}{\rightarrow} Y/\!/j$$

where α is the class in πY^{Y_0} of the homotopy $jH_t^0 : jf^0g^0 \simeq j$. Since $f : X \rightarrow Y$ is a homotopy equivalence, f_* is a bijection [7.2.5 (Corollary 2)]. Also $Y/\!/\alpha$ is a bijection. So there is an element cls g in $Y/\!/ig^0$ such that

$$(Y/\!/\alpha)f_* (\text{cls } g) = \text{cls } 1_Y. \tag{*}$$

Then $g : Y \rightarrow X$ extends ig^0 and the above relation with 7.2.7 shows that there is a homotopy $fg \simeq 1$ which agrees on Y_0 with the given homotopy $H_t^0 : f^0g^0 \simeq 1$. Notice also that g is a homotopy inverse of f since if $\bar{g}$ is any homotopy inverse of f then

$$gf = 1_X(gf) \simeq \bar{g}f(gf) = \bar{g}(fg)f \simeq \bar{g}1_Yf \simeq 1$$

(this argument is really the same as that in 6.1.1).

Step 2. We now apply Step 1 with X replaced by X_μ, Y replaced by Y_μ and f replaced by $f^\mu : X_\mu \rightarrow Y_\mu$ for $\mu = 1, 2$. We deduce that each f^μ ($\mu = 1, 2$) has a homotopy inverse g^μ which agrees with g^0 on Y^0; and also there is a homotopy $H_t^\mu : f^\mu g^\mu \simeq 1$ which agrees on Y^0 with the given homotopy $H_t^0 : f^0g^0 \simeq 1$. Since X_1, X_2 are closed in X, the maps g^μ and the homotopies H_t^μ together define a map $g : Y \rightarrow X$ and homotopy $H_t : fg \simeq 1$. Clearly, g and H_t define $\mathbf{g} : \mathbf{Y} \rightarrow \mathbf{X}$ and $\mathbf{H}_t : \mathbf{fg} \simeq 1$.

Step 3. We now apply the same process to **g** instead of **f**, and deduce that there is a map $\bar{\mathbf{f}} : \mathbf{X} \rightarrow \mathbf{Y}$ which agrees with f^0 on X, and there is a homotopy $\mathbf{K}_t : \mathbf{g}\bar{\mathbf{f}} \simeq 1$ which agrees on X_0 with the given homotopy $K_t^0 : g^0f^0 \simeq 1$. It follows that

$$\mathbf{gf} = \mathbf{gf}1_X \simeq \mathbf{gfg}\bar{\mathbf{f}} \simeq \mathbf{g}\bar{\mathbf{f}} \simeq 1.$$

This proves that **g** is a homotopy inverse of **f** and confirms the statement about the homotopies in 7.4.1 (addendum). □

7.4.1 (*Corollary 1*) *Suppose further that there is a subset A of both X_0 and Y_0 such that* (*i*) (X_0, A), (Y_0, A) *have the HEP,* (*ii*) $f \mid A, A$ *is the identity. Then* **f** *is a homotopy equivalence* rel A, *that is, there is a map* $\mathbf{g} : \mathbf{Y} \to \mathbf{X}$ *such that*

$$\mathbf{fg} \simeq 1 \text{ rel } A, \qquad \mathbf{gf} \simeq 1 \text{ rel } A.$$

Proof Notice first of all that if in 7.4.1 (addendum) the given homotopies are rel A, then so also are the homotopies constructed in 7.4.1 (addendum). It follows, by applying 7.4.1, 7.4.1 (addendum) to the triads $(X_0; X_0, A)$, $(Y_0; Y_0, A)$ that $f^0 : X_0 \to Y_0$ has a homotopy inverse $g^0 : Y_0 \to X_0$ such that $f^0g^0 \simeq 1$ rel A, $g^0f^0 \simeq 1$ rel A. The result follows easily from this. □

7.4.1 (*Corollary 2*) *Let (X, A) have the HEP. Then the inclusion $i : A \to X$ is a homotopy equivalence if and only if A is a deformation retract of X.*

Proof The forward implication follows by applying the previous corollary to the inclusion map $(A; A, A) \to (X; X, A)$. The other implication is obvious. □

This corollary is not true for all pairs (X, A). For example, if $X = \mathbf{C}L$, $A = \{0\}$, then the inclusion $A \to X$ is a homotopy equivalence since both A and X are contractible. But A is not a deformation retract of X (the proof of this is left as an exercise).

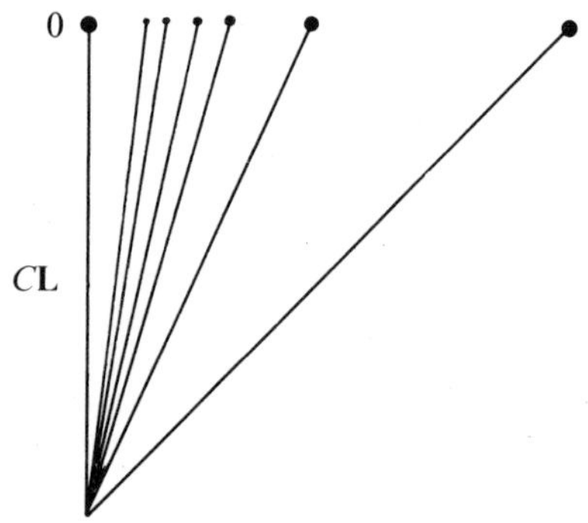

Fig. 7.5

This corollary has a more direct proof [Exercise 5]. Other applications of 7.4.1 will be given in the next section.

Our next results will be used in 7.6. It is convenient to write now α_* for $X//\alpha$ where α is a track.

Let $\mathbf{f} : \mathbf{X} \to \mathbf{Y}$ be a map of m-ads, and let $B \subset Y$. We say $\mathbf{f}$ is *deformable into* B if there is a homotopy $\mathbf{f} \simeq \mathbf{g}$ such that $g[X] \subset B$. If, further, $A \subset X$ and the homotopy $\mathbf{f} \simeq \mathbf{g}$ is rel A, then we say $\mathbf{f}$ is *deformable into* B rel A.

7.4.2 *Let (X, A) have the HEP and let $\mathbf{f} : (X, A) \to (Y, B)$ be a map of pairs such that $\mathbf{f}$ is deformable into B. Then $\mathbf{f}$ is deformable into B* rel A.

Proof Let $j : B \to Y$ be the inclusion and let $f^1 : A \to B$ be the map determined by $\mathbf{f}$. The homotopy $\mathbf{f} \simeq \mathbf{g}$ which deforms $\mathbf{f}$ into B determines a homotopy $f^1 \simeq g^1$ whose class in πB^A we denote by α. Let $\beta = j_*\alpha$ in πY^A. We consider the diagram

$$\begin{array}{ccccc} X/\!/jf^1 & \xrightarrow{\beta_*} & X/\!/jg^1 & \xrightarrow{(-\beta)_*} & X/\!/jf^1 \\ & & \uparrow{\scriptstyle j_*} & & \uparrow{\scriptstyle j_*} \\ & & X/\!/g^1 & \xrightarrow[(-\alpha)_*]{} & X/\!/f^1 \end{array}$$

Let θ in $X/\!/jf^1$ be the class of $f : X \to Y$ so that

$$\beta_*\theta = \text{cls } g.$$

But $g[X] \subset B$, and so $g = jg'$, where $g' = g \mid X, B$. Let $\varphi = \text{cls } g'$. Then

$$\beta_*\theta = j_*\varphi$$

and so, since the above diagram is obviously commutative,

$$\begin{aligned} \theta = (-\beta)_*\beta_*\theta &= (-\beta)_*j_*\varphi \\ &= j_*(-\alpha)_*\varphi. \end{aligned}$$

So if $h : X \to B$ represents $(-\alpha)_*\varphi$ then jh represents $\theta = \text{cls } f$. Hence f is homotopic rel A to jh. □

It should be confessed that a simpler proof of this result is possible [Exercise 3]. The above proof is given here because it illustrates the use of the operations. A similar use occurs in the proof of the next result.

7.4.3 *Let (X, x_0) have the HEP and let $f : X \to Y$ be inessential. Then f is inessential* rel x_0, *that is, there is a homotopy $f \simeq f'$* rel x_0 *such that f' is constant.*

Proof Let $F_t : f \simeq g$ be a homotopy such that g is constant. Let

$$u = f \mid \{x_0\}, v = g \mid \{x_0\}$$

and let α be the class in $\pi Y^{\{x_0\}}(u, v)$ of the homotopy $U_t = F_t \mid \{x_0\}$. Now the homotopy $-U_t = U_{1-t}$ has an obvious extension to a homotopy of g, namely, $(x, t) \rightsquigarrow U_{1-t}x_0$. It follows that the class $(-\alpha)_* \operatorname{cls} g$ contains a constant map f'. But

$$\begin{aligned} \operatorname{cls} f &= (-\alpha)_* \alpha_* \operatorname{cls} f \\ &= (-\alpha)_* \operatorname{cls} g \\ &= \operatorname{cls} f'. \end{aligned}$$

Therefore $f \simeq f'$ rel x_0. □

EXERCISES

1. Generalize 7.4.1 to the case of m-ads $(X; X_1, \ldots, X_{m-1})$ such that, if $X_0 = X_1 \cap \cdots \cap X_{m-1}$, then
   (i) (X_λ, X_0) has the HEP for $\lambda = 1, \ldots, m-1$,
   (ii) $X_\lambda \cap X_\mu = X_0$ for $\lambda \neq \mu$,
   (iii) $X = X_1 \cup \cdots \cup X_{m-1}$,
   (iv) X_λ is closed in X.
2. Let $\mathbf{f} : (X, A) \to (Y, B)$ be a map such that B is a deformation retract of Y. Prove that $\mathbf{f}$ is deformable into B rel A.
3. Let $R_t : X \times \mathbf{I} \to X \times \mathbf{I}$ be a homotopy retracting $X \times \mathbf{I}$ onto $X \times 1 \cup A \times \mathbf{I}$. Let $F_t : (X, A) \to (Y, B)$ be a homotopy deforming F_0 into B. Prove that the homotopy $(x, t) \rightsquigarrow FR_t(x, 0)$ deforms F_0 into B rel A.
4. Let (X, A) have the HEP, let $B \subset Y$ and let $F_t : X \to Y$ be a homotopy such that (i) $F_1[X] \subset B$, (ii) $F_0 \mid A = F_1 \mid A$, (iii) $F_0 \mid A, B$ is a homeomorphism. Prove that F_0 is homotopic to F_1 rel A.
5. Prove 7.4.1 (corollary 2) and 7.4.3 by using the previous exercise.
6. Let (X, A) have the HEP. Prove that the function

$$X \mathbin{/\!/} (1 : A \to A) \to [X, A]$$

which sends a homotopy class rel A to a (free) homotopy class, is injective.
7. Let (X, X_0), (Y, Y_0) have the HEP. Prove that a map $\mathbf{f} : (X, X_0) \to (Y, Y_0)$ has a right homotopy inverse if (i) $f : X \to Y$ is a homotopy equivalence and $f^0 : X_0 \to Y_0$ has a right homotopy inverse, or (ii) there is a map $\mathbf{g} : (Y, Y_0) \to (X, X_0)$ such that $fg \simeq 1_Y$ and $f^0 g^0 \simeq 1_{Y_0}$.
8. Prove that $\{0\}$ is not a deformation retract of $\mathbf{CL}$.
9. Let $\mathbf{Y}$ be a triad and $\mathbf{X}$ an MV-triad. Let $A \subset X_0$ and for each map $u : A \to Y_0$ (where $Y_0 = Y_1 \cap Y_2$) let $\mathbf{X}, \mathbf{Y} \mathbin{/\!/} u$ denote the triad homotopy classes rel A of triad maps $\mathbf{X} \to \mathbf{Y}$ which agree on A with u. Prove that there is a functor $\pi(Y_0)^A \to \mathcal{S}et$ which on objects sends $u \rightsquigarrow \mathbf{X}, \mathbf{Y} \mathbin{/\!/} u$. Prove also, that if $\mathbf{f} : \mathbf{Y} \to \mathbf{Z}$ is a homotopy equivalence of triads, then $\mathbf{f}$ induces for each map $u : A \to Y_0$ a bijection $f_* : \mathbf{X}, \mathbf{Y} \mathbin{/\!/} u \to \mathbf{X}, \mathbf{Z} \mathbin{/\!/} f_0 u$ which sends $\operatorname{cls} g \rightsquigarrow \operatorname{cls} fg$.
10. Prove that a map $f: A \to Y$ is inessential if and only if f extends over CA (where A is identified with the subspace $A \times 1$ of CA). Hence show that if $\mathbf{X}$ is an MV-triad such that X_1 and X_2 are contractible, then X is of the homotopy type of SX_0.

7.5 The homotopy type of adjunction spaces

Suppose given an adjunction space $B\ {}_f\sqcup X$ so that we have a pushout

$$\begin{array}{ccc} A & \xrightarrow{f} & B \\ {\scriptstyle i}\downarrow & & \downarrow{\scriptstyle \bar{i}} \\ X & \xrightarrow[\bar{f}]{} & B\ {}_f\sqcup X \end{array} \qquad (7.5.1)$$

In this section, we shall apply 7.4.1 to show that the homotopy type of $B\ {}_f\sqcup X$ depends only on the homotopy class of f and the homotopy types of B and (X, A).

We first need some simple lemmas on deformation retracts.

7.5.2 *Let $A \subset D \subset X$ and let D be a deformation retract of X. Then $B\ {}_f\sqcup D$ is a deformation retract of $B\ {}_f\sqcup X$.*

Proof Let $R_t : X \to X$ be a retracting homotopy of X onto D. Let

$$F_t = \bar{f}R_t : X \to B\ {}_f\sqcup X,\ G_t = \bar{i} : B \to B\ {}_f\sqcup X.$$

Then $F_t i = G_t f$ (since D contains A); by Remark 7.3.6, F_t and G_t define a homotopy $H_t : B\ {}_f\sqcup X \to B\ {}_f\sqcup X$. It is easily checked that H_t is a retracting homotopy of $B\ {}_f\sqcup X$ onto $B\ {}_f\sqcup D$. □

Intuitively, 7.5.2 is 'obvious', since the homotopy which retracts X down onto D also retracts $B\ {}_f\sqcup X$ onto $B\ {}_f\sqcup D$. But this last statement, although it contains the essential idea, is not accurate and also does not indicate why the resulting homotopy is continuous.

7.5.2 (*Corollary 1*) *If A is a deformation retract of X, then B is a deformation retract of $B\ {}_f\sqcup X$.*

Proof Take $A = D$ in 7.5.1. □

We recall that $\{0\}$ is a deformation retract of $\mathbf{I}$. It follows easily that, for any A, $A \times 0$ is a deformation retract of $A \times \mathbf{I}$; so the following result is immediate.

7.5.2 (*Corollary 2*) *For any A, the vertex of the cone CA is a deformation retract of CA.*

A similar argument shows that if $f : A \to B$ is a map then B is a deformation retract of the mapping cylinder [Example 2 of 7.3]

$$M(f) = B\ {}_{f'}\sqcup (A \times \mathbf{I})$$

where $f' : A \times 0 \to B$ is $(a, 0) \rightsquigarrow fa$. In fact, if $r : A \times \mathbf{I} \to A \times 0$ is the map $(a, t) \rightsquigarrow (a, 0)$, then r is a deformation retraction, and therefore so also is the map $q : M(f) \to B$ which is defined by the maps 1_B and $f'r$. Let $j : A \to M(f)$ be the map $a \rightsquigarrow (a, 1)$ by which we identify A as a subspace of $M(f)$. Then we have a commutative diagram,

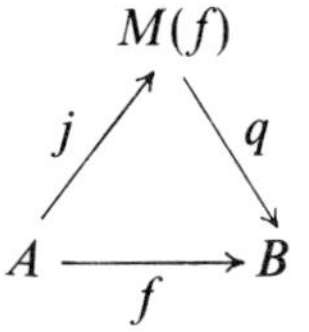

in which j is an inclusion and q is a deformation retraction. This 'factorization' of f is very useful.

7.5.3 *The map $f : A \to B$ is a homotopy equivalence if and only if A is a deformation retract of $M(f)$.*

Proof We know that q is a homotopy equivalence. Therefore, j is a homotopy equivalence if and only if f is a homotopy equivalence. But $(M(f), A)$ has the HEP [7.3]. By 7.4.1 (Corollary 2) j is a homotopy equivalence if and only if A is a deformation retract of $M(f)$. □

This result shows that two spaces are of the same homotopy type if and only if there is a third space containing each as a deformation retract.

Consider again the adjunction space $B\, {}_f\!\sqcup X$ of diagram (7.5.1). In order to study the homotopy type of this space it is convenient to use some auxiliary spaces, namely, $M(f)$, $M(f) \cup X$ and $M(\bar{f})$; these spaces are represented in the following figure.

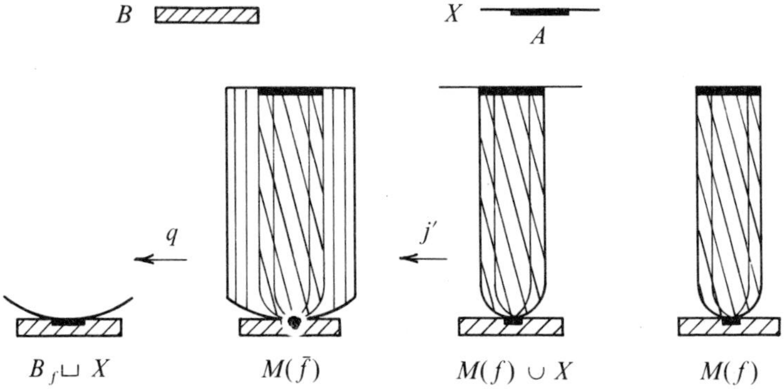

Fig. 7.6

By 4.6.1 $M(f) \cup X$ is a subspace of $M(\bar{f})$—we write j' for the inclusion map. Let $q : M(\bar{f}) \to B \,{}_f\!\sqcup X$ be the standard deformation retraction. It turns out that if (X, A) has the HEP then qj' is a homotopy equivalence, and therefore, in this case, $B \,{}_f\!\sqcup X$ can be replaced by $M(f) \cup X$. Also, the triad $(M(f) \cup X; M(f), X)$ is then an MV-triad; and this triad is a convenient one on which to apply the glueing theorem 7.4.1.

7.5.4 *If (X, A) has the HEP, then*

$$p = qj' : M(f) \cup X \to B \,{}_f\!\sqcup X$$

is a homotopy equivalence rel B.

Proof We know q is a homotopy equivalence rel B. It follows from 7.2.3 that $X \times 1 \cup A \times \mathbf{I}$ is a deformation retract of $X \times \mathbf{I}$. By 7.5.2 with A, D, X replaced by $A \times 1, X \times 1 \cup A \times \mathbf{I}, X \times \mathbf{I}$ respectively (and with $X \times 1$ identified with X), $M(f) \cup X$ is a deformation retract of $M(\bar{f})$. Thus j' also is a homotopy equivalence rel B. □

As a simple example of the previous result, let $X = \mathbf{S}^2$, let A consist of the North and South Poles of $\mathbf{S}^2$, and let B consist of a single point. Then the adjunction space $B \,{}_f\!\sqcup X$ is simply $\mathbf{S}^2$ with North and South Poles identified. On the other hand, $M(f) \cup X$ is homeomorphic to $\mathbf{S}^2$ with an arc C joining the North to the South Pole. 7.5.4 shows that the map $M(f) \cup X \to B \,{}_f\!\sqcup X$ which shrinks C to a point is a homotopy equivalence.

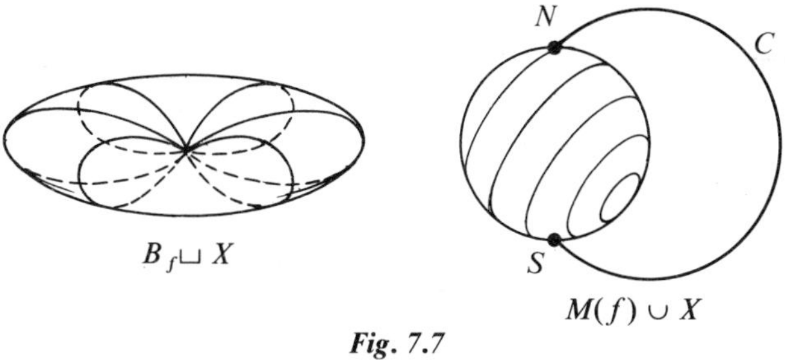

Fig. 7.7

The preceding result is false without some assumptions on the pair (X, A) [cf. Example 1 on p. 251].

We now show that the homotopy type of $M(f) \cup X$ depends only on the homotopy class of f.

Suppose that $f_t : f_0 \simeq f_1$ is a homotopy of maps $A \to B$. Let

$$F : M(f_0) \cup X \to M(f_1) \cup X$$

be the identity on B and on X, and on the part $A \times \,]0, 1]$ of the mapping cylinder be given by

$$F(a, t) = \begin{cases} f_{2t}a, & 0 < t \leqslant \frac{1}{2} \\ (a, 2t - 1), & \frac{1}{2} \leqslant t \leqslant 1 \end{cases}$$

(where $(a, 1)$ is identified with a). The map F is illustrated for the case $X = A = \{a\}$ in Fig. 7.8 which amalgamates the pictures of $M(f_0)$, $M(f_1)$. The proof of the continuity of F is left as an exercise.

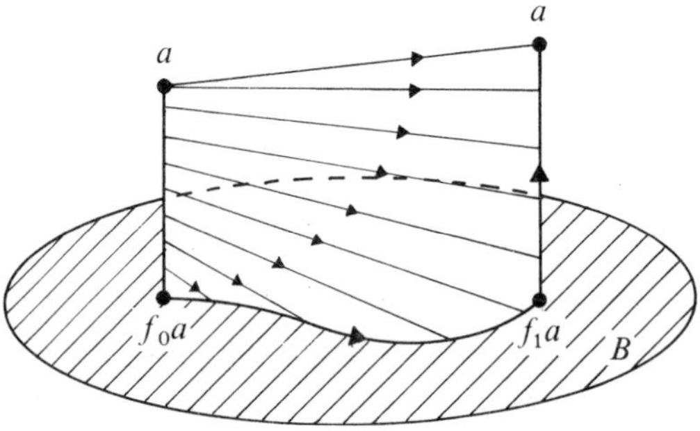

Fig. 7.8

7.5.5 *The above map F is a homotopy equivalence* rel $(B \cup X)$.

Proof Let $G = F \mid M(f_0), M(f_1)$, and let $i_\varepsilon : B \to M(f_\varepsilon)$ $(\varepsilon = 0, 1)$ be the inclusion. Since G is the identity on B, $Gi_0 = i_1$. Since i_0, i_1 are homotopy equivalences, so also is G. Since $G \mid B \cup A, B \cup A$ is the identity, and $(M(f_\varepsilon), B \cup A)$ has the HEP, G is actually a homotopy equivalence rel $(B \cup A)$ [7.4.1 Corollary 1]. Since F is the identity on X, it follows that F is a homotopy equivalence rel $(B \cup X)$. □

7.5.5 (*Corollary 1*) *If (X, A) has the HEP and $f_0 \simeq f_1 : A \to B$, then there is a homotopy equivalence*

$$B \,{}_{f_0}\!\sqcup X \to B \,{}_{f_1}\!\sqcup X \text{ rel } B.$$

Proof This follows from 7.5.5 and 7.5.4. □

As a simple application of this fact, note that if B is path-connected then $B \,{}_f\!\sqcup \mathbf{E}^1$ (where $f : \mathbf{S}^0 \to B$) is always of the homotopy type of $B \vee \mathbf{S}^1$.

We now show the dependence of the homotopy type of $B \,{}_f\!\sqcup X$ on the homotopy types of B and (X, A). Suppose given a commutative diagram of maps

$$\begin{array}{ccccc} X & \xleftarrow{\;i\;} & A & \xrightarrow{\;f\;} & B \\ \downarrow{\scriptstyle\varphi} & & \downarrow{\scriptstyle\varphi^1} & & \downarrow{\scriptstyle\varphi^2} \\ Y & \xleftarrow[\;j\;]{} & C & \xrightarrow[\;g\;]{} & D \end{array} \tag{7.5.6}$$

where j, like i, is the inclusion map of a closed subspace. The map φ determines a map of pairs.

$$\boldsymbol{\varphi} : (X, A) \to (Y, C)$$

while $\boldsymbol{\varphi}$ and φ^2 determine a map

$$\Phi : B\,{}_f\!\sqcup X \to D\,{}_g\!\sqcup Y.$$

7.5.7 *Let (X, A), (Y, C) have the HEP and suppose φ, φ^1, and φ^2 are homotopy equivalences. Then Φ is a homotopy equivalence.*

Proof The diagram (7.5.6) determines a map

$$\Psi : M(f) \cup X \to M(g) \cup Y$$

which agrees with φ^2 on B, with φ on X and with $\varphi^1 \times 1$ on $A \times]0, 1[$. There is a commutative diagram

$$\begin{array}{ccc} M(f) & \xrightarrow{\;\Psi^1\;} & M(g) \\ \downarrow{\scriptstyle q} & & \downarrow{\scriptstyle q} \\ B & \xrightarrow[\;\varphi^2\;]{} & D \end{array}$$

in which Ψ^1 is the restriction of Ψ and the vertical maps are the standard deformation retractions. Since φ^2 is a homotopy equivalence it follows that Ψ^1 is a homotopy equivalence.

Now $\Psi \mid A, C = \varphi^1$ and $\Psi \mid X, Y = \varphi$; both of these maps are homotopy equivalences. Further, $(M(f) \cup X; M(f), X)$ and $(M(g) \cup Y; M(g), Y)$ are MV-triads. It follows from 7.4.1 that Ψ is a homotopy equivalence.

However the following diagram is commutative

$$\begin{array}{ccc} M(f) \cup X & \xrightarrow{\;\Psi\;} & M(g) \cup Y \\ \downarrow{\scriptstyle p} & & \downarrow{\scriptstyle p} \\ B\,{}_f\!\sqcup X & \xrightarrow[\;\Phi\;]{} & D\,{}_g\!\sqcup Y \end{array}$$

where the vertical maps are the homotopy equivalences of 7.5.4. It follows that Φ is a homotopy equivalence. □

In the following corollaries we assume that the pairs occurring have the HEP.

7.5.7 (*Corollary 1*) *Let* $\boldsymbol{\varphi} : (X, A) \to (Y, C)$ *be a homotopy equivalence of pairs and let* $g : C \to D$ *be a map. Then the map*

$$\Phi : D\ {}_{g\varphi^1}\!\sqcup X \to D\ {}_g\!\sqcup Y$$

defined by $\boldsymbol{\varphi}$ *is a homotopy equivalence.*

Proof This is the case $B = D$, $\varphi^2 = 1$ of 7.5.7. □

7.5.7 (*Corollary 2*) *If* $f : A \to B$ *is a homotopy equivalence, then*

$$\bar{f} : X \to B\ {}_f\!\sqcup X$$

is a homotopy equivalence.

Proof In (7.5.6) we replace the first row by $X \xleftarrow{i} A \xrightarrow{1} A$, the second row by $X \xleftarrow{i} A \xrightarrow{f} B$ and take $\boldsymbol{\varphi} = 1$, $\varphi^2 = f$. □

7.5.7 (*Corollary 3*) *If* A *is contractible then the identification map* $p : X \to X/A$ *is a homotopy equivalence.*

Proof This is the case of 7.5.7 (Corollary 2) when B consists of a single point. □

This last corollary has the following simple proof. Let $F_t : A \to A$ be a homotopy such that $F_0 = 1$ and F_1 is constant. Let $i : A \to X$ be the inclusion. Then iF_t extends to a homotopy $G_t : X \to X$ such that $G_0 = 1$ and $G_1[A]$ is a single point of A. Therefore, G_1 defines a map $g : X/A \to X$ such that $gp = G_1$—whence $gp \simeq 1_X$. Also, since $G_t[A] \subset A$, G_t defines a homotopy $H_t : X/A \to X/A$ such that $H_t p = pG_t$. Here $H_0 = 1_{X/A}$ and $H_1 p = pG_1 = pgp$, whence $H_1 = pg$. This proves that g is a homotopy inverse of p.

Similar proofs can be given for some other cases of 7.5.7 (Corollary 2) (this remark is useful in the solution of Exercise 8).

7.5.7 (Corollary 2) is often useful when combined with 4.5.8 as follows. Suppose given a Hausdorff space Q, a closed subspace B of Q and a closed subspace A of the compact space X. Let $h : X \to Q$ be a map such that $h[A] \subset B$ and $h \mid X \setminus A, Q \setminus B$ is defined and is a bijection. Let $f = h \mid A, B$. Finally, let (X, A) have the HEP.

7.5.8 *Under these conditions, if* $f : A \to B$ *is a homotopy equivalence, then so also is* $h : X \to Q$.

Proof By 4.5.8 there is a homeomorphism $g : B\ {}_f\!\sqcup X \to Q$ such that

$g\bar{f} = h$. By 7.5.7 (Corollary 2) $\bar{f}$ is a homotopy equivalence. Therefore h is a homotopy equivalence. □

The previous results are false without some conditions on the pair (X, A).

EXAMPLE Consider the subspaces of $\mathbf{R}$

$$Y = \{0\} \cup \bigcup_{n \geqslant 1} \left[\frac{1}{2n}, \frac{1}{2n-1}\right], \qquad B = \{0, 1\} \cup \bigcup_{n \geqslant 1} \left[\frac{1}{2n+1}, \frac{1}{2n}\right].$$

Let $i : \mathbf{L} \to B$ be the inclusion, let $f = q : M(i) \to B$. $A = M(i)$, $X = A \cup Y$. Then $B\ {}_f\!\sqcup X \approx B\ {}_i\!\sqcup Y \approx \mathbf{I}$. But X is not path-connected (draw the picture and compare with Example 2, p. 55). So $\bar{f}$ is not a homotopy equivalence even though f is.

We now give some applications of the preceding results, first of all using only 7.5.5 (Corollary 1).

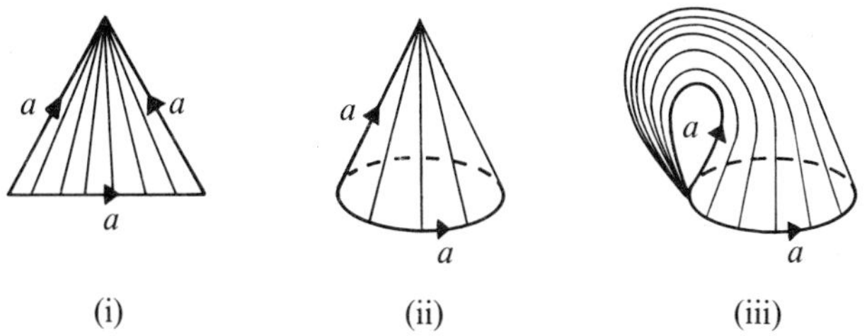

Fig. 7.9

The *dunce's hat* is obtained from a triangle by identifying the sides according to the pattern shown in the above figure. Two steps in the identification are shown; the final step is the identification of the two thickly drawn circles in (iii). The following, somewhat intuitive, discussion shows that the dunce's hat is contractible.

We can think of the dunce's hat as an adjunction space $\mathbf{S}^1\ {}_f\!\sqcup \Delta^2$ where $f : \dot{\Delta}^2 \to \mathbf{S}^1$ is determined by the identification shown in (i). In fact, regarding each side a of Δ^2 as a map of a line segment onto $\mathbf{S}^1$, we can write

$$f = a + a - a.$$

It is clear that f is homotopic to the obvious homeomorphism $g : \dot{\Delta}^2 \to \mathbf{S}^1$. But $\mathbf{S}^1\ {}_g\!\sqcup \Delta^2$ is homeomorphic to $\mathbf{E}^2$. By 7.5.5 (Corollary 1), the dunce's hat is of the homotopy type of $\mathbf{E}^2$ and so is contractible.

Our next applications are to joins and smashed products. For the remainder of this section *all spaces will be assumed pointed, compact, and Hausdorff*. The base points of X, Y are to be x_0, y_0 respectively. The base point of $X * Y$ will then be $e = \frac{1}{2}x_0 + \frac{1}{2}y_0$; the base point of SX will be the top vertex v_1 (i.e., the set $X \times 1$). We will also identify $X \vee Y$ with the space obtained from $X \sqcup Y$ by identifying the base point of X with that of Y. In CZ we identify Z with the subspace $Z \times 1$ by the map $z \rightsquigarrow (z, 1)$.

7.5.9 *There is a homotopy equivalence*

$$S(X \times Y) \to (X * Y) \vee SX \vee SY.$$

Proof

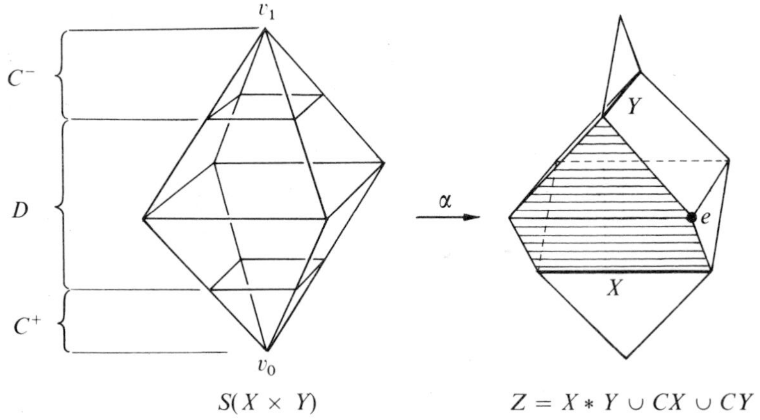

Fig. 7.10

The points of $S(X \times Y)$ we write as (z, t) for z in $X \times Y$ and t in $\mathbf{I}$, so that $(z, 0) = v_0$, $(z, 1) = v_1$, for any z in $X \times Y$. Let C^-, C^+, D be respectively the set of points (z, t) such that $t \geqslant \frac{3}{4}$, $t \leqslant \frac{1}{4}$, $\frac{1}{4} \leqslant t \leqslant \frac{3}{4}$. Obviously, we have a homeomorphism from $S(X \times Y)$ to

$$W = (X \times Y \times \mathbf{I}) \cup C(X \times Y \times 0) \cup C(X \times Y \times 1)$$

in which, C^-, C^+, D are mapped to $C(X \times Y \times 1)$, $C(X \times Y \times 0)$ and $X \times Y \times \mathbf{I}$ respectively. So we now replace $S(X \times Y)$ by W.

Consider the maps

$$p : C(X \times Y \times 0) \to CX \qquad p' : C(X \times Y \times 1) \to CY$$
$$(x, y, 0, t) \rightsquigarrow (x, t) \qquad (x, y, 1, t) \rightsquigarrow (y, t)$$

$$q : X \times Y \times \mathbf{I} \to X * Y$$
$$(x, y, s) \rightsquigarrow (1 - s)x + sy.$$

Clearly p, p' agree with q on $X \times Y \times \dot{\mathbf{I}}$; therefore, these maps define a map

$$\alpha : W \to Z = X * Y \cup CX \cup CY.$$

Now p, p' are homotopy equivalences, since they each are maps from a contractible space to a contractible space. By 7.5.8 α is a homotopy equivalence (in 7.5.8 take $X = W$, $A = C(X \times Y \times 0) \cup C(X \times Y \times 1)$; the fact that (X, A) has the HEP is an easy consequence of results of 7.3). We complete the proof by showing that Z is of the homotopy type of $(X * Y) \vee SX \vee SY$.

The inclusion $t : X \to X * Y$ is homotopic to the constant map c with value y_0, since a homotopy $i \simeq c$ is given by

$$(x, t) \rightsquigarrow (1 - t)x + ty_0$$

(in Fig. 7.10, the shaded face of $X * Y$ is homeomorphic to CX, and this homotopy is simply sliding X up the cone). Also, c itself is homotopic to the constant map with value e (such a homotopy is given by $(x, t) \rightsquigarrow (1 - t)y_0 + te$). Since i is the attaching map of the cone CX and CX/X is homeorphic to SX, it follows from 7.5.5 (Corollary 1) that Z is of the homotopy type of

$$((X * Y) \vee SX) \cup CY.$$

A similar argument shows that this space is of the homotopy type of $(X * Y) \vee SX \vee SY$. □

A pointed space X is called *well-pointed* if $(X, \{x_0\})$ has the HEP. We recall that all spaces are assumed compact and Hausdorff.

7.5.10 *There is an identification map*

$$X * Y \to \Sigma(X \circledast Y)$$

which is a homotopy equivalence if X and Y are well-pointed.

Proof We proved in chapter 5 that $\Sigma(X \circledast Y)$ is homeomorphic to an identification space $(X * Y)/R$. Thus, to complete the proof we have only to show that if X, Y are well-pointed then R is contractible and $(X * Y, R)$ has the HEP.

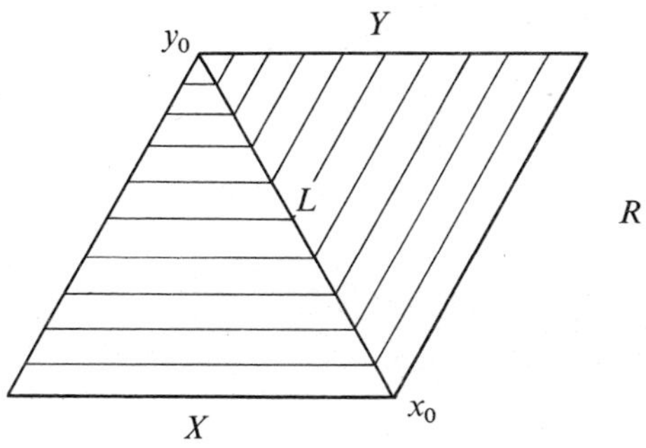

Fig. 7.11

Let L be the subspace of R of points $rx_0 + sy_0$. Then R is the union of two cones with a common 'generator' L. The inclusion $L \to CX$ is a homotopy equivalence, since both L and CX are contractible. Also $(X \times \mathbf{I}, X \times 0 \cup x_0 \times \mathbf{I})$ has the HEP by 7.3.7 and hence so also does $(CX, Cx_0) = (CX, L)$. By 7.4.1 (Corollary 2) L is a deformation retract of CX. Similarly, L is a deformation retract of CY and it follows that L is a deformation retract of R. Therefore, R is contractible, since L is contractible.

Let $J : X \times Y \times \Delta^1 \to X * Y$ be the identification map of 5.7.2 (Corollary 1). Then

$$J^{-1}[R] = x_0 \times Y \times \Delta^1 \cup X \times y_0 \times \Delta^1 \cup X \times Y \times \dot{\Delta}^1.$$

Now 7.3.7 applied to the product of the three pairs (X, x_0), (Y, y_0) and $(\Delta^1, \dot{\Delta}^1)$ shows that $(X \times Y \times \Delta^1, J^{-1}[R])$ has the HEP. Since J is an identification map, it follows that $(X * Y, R)$ has the HEP, and the proof is complete. □

7.5.11 *If X is well-pointed then the identification map*

$$SX \to \Sigma X = SX/Sx_0$$

is a homotopy equivalence.

Proof By an argument similar to that in 7.5.10 the pair (SX, Sx_0) has the HEP. Since Sx_0 is contractible, the result follows. □

7.5.12 *If X and Y are well-pointed then there is a homotopy equivalence*

$$\Sigma(X \times Y) \to \Sigma(X \divideontimes Y) \vee \Sigma X \vee \Sigma Y.$$

Proof As base point of $X \times Y$ we take (x_0, y_0) where x_0, y_0 are the base points of X, Y respectively. By 7.3.2, $X \times Y$ is well-pointed and so $\Sigma(X \times Y)$ is of the same homotopy type as $S(X \times Y)$. Similarly, we have a map

$$(X * Y) \vee SX \vee SY \to \Sigma(X \divideontimes Y) \vee \Sigma X \vee \Sigma Y$$

which is a homotopy equivalence on each 'summand' of the wedge. By two applications of the glueing rule 7.4.1, this map is a homotopy equivalence (we leave the reader to verify the necessary HEP's). The result now follows from 7.5.9. □

7.5.12 (*Corollary 1*) *There is a homotopy equivalence*

$$\Sigma(\mathbf{S}^m \times \mathbf{S}^n) \to \mathbf{S}^{m+n+1} \vee \mathbf{S}^{m+1} \vee \mathbf{S}^{n+1}.$$

Proof This is the case $X = \mathbf{S}^m$, $Y = \mathbf{S}^n$ of 7.5.12. □

EXERCISES

1. Let X, Y, Z be spaces and $h : X \to Z$ a map. (i) Let $f : X \to Y$ be a map; prove that there is a map $g : Y \to Z$ such that $gf \simeq h$ if and only if h extends over $M(f)$. (ii) Let $g : Y \to Z$; prove that there is a map $f : X \to Y$ such that $gf \simeq h$ if and only if $jh : X \to M(g)$ is deformable into Y (where $j : Z \to M(g)$ is the inclusion). Interpret these statements in the case $X = Z$, $h = 1$.

2. Let $\mathbf{L}$ have base point 0, and let $C\mathbf{L}$ have the same base point as $\mathbf{L}$. Prove that $C\mathbf{L} \vee C\mathbf{L}$ is not contractible. [Let $Z = C\mathbf{L} \vee C\mathbf{L}$ and let v_1, v_{-1} be the vertices of the two cones of Z. Let F_t be a homotopy $Z \to Z$ such that $F_0 = 1$, F_1 is constant. Without loss of generality it may be assumed that F_t0 takes the values v_1, v_{-1}. Let $\alpha_\varepsilon = \inf \{t : F_t(0) = v_\varepsilon\}$ for $\varepsilon = 1, -1$. Show that either assumption $\alpha_1 < \alpha_{-1}, \alpha_{-1} < \alpha_1$ leads to a contradiction.]

3. (i) Let $f : \mathbf{L} \to \mathbf{L} \times C\mathbf{L}$ be the map $x \rightsquigarrow (x, 0)$. Prove that $\bar{f} : \mathbf{I} \to (\mathbf{L} \times C\mathbf{L})_f \sqcup \mathbf{I}$ is not a homotopy equivalence. (ii) Prove that if $e \in \mathbf{S}^n$, then $\mathbf{S}^n/(\mathbf{S}^n \setminus \{e\})$ is contractible.

4. Prove that if X, Y are compact and Hausdorff then there is a homotopy equivalence

$$SX \times Y/v_1 \times Y \to (X * Y) \vee SX.$$

Deduce that if, further, X, Y are well-pointed, then there is a homotopy equivalence

$$\Sigma X \times Y/. \times Y \to \Sigma(X \circledast Y) \vee \Sigma X.$$

5. Let $X_1, \ldots, X_n$ be compact, Hausdorff, well-pointed spaces. For any non-empty subset N of $\{1, \ldots, n\}$ let Z_N denote the smashed product of the spaces X_i for i in N. Prove that there is a homotopy equivalence from $\Sigma(X_1 \times \cdots \times X_n)$ to the wedge of the spaces ΣZ_N for all non-empty subsets N of $\{1, \ldots, n\}$.

6. Let A be a set of m points in $\mathbf{S}^n$ ($n \geqslant 1$). Prove that $\mathbf{S}^n/A$ has the homotopy type of the wedge of $\mathbf{S}^n$ with $(m - 1)$ circles.

7. Let A be a subspace of both X and X'. Prove that if A is a WNDR of X' and X' dominates X rel A, then A is a WNDR of X.

8. Let $i : A \to X$ be the inclusion of the closed subspace A of X and let A be identified as usual with the subspace $A \times 1$ of $M(i)$. Let $p : M(i) \to X$ be the projection. Prove that the following conditions are equivalent: (i) A is a WNDR of X, (ii) there is a homotopy $h_t : M(i) \to M(i)$ of the identity rel A such that

$$h_1[A \times \mathbf{I}] \subset A, \qquad h_t[a \times \mathbf{I}] \subset a \times \mathbf{I}, \qquad a \in A,$$

(iii) $p : M(i) \to X$ is a homotopy equivalence rel A, (iv) there is a homotopy equivalence $M(i) \to X$ rel A, (v) (X, A) has the WHEP.

9. Consider the following properties of a pair (X, A), which we label as the *rather*, *very*, and *completely* WHEP. (RWHEP) If $f : X \to Y$ is a map and $U_t : A \to Y$ is a homotopy of $f \mid A$, then U_t is homotopic rel end maps to a homotopy which extends to a homotopy of f. (VWHEP) If $f : X \to Y$ is a map, $u = f \mid A$ and $u \simeq v$, then some homotopy $u \simeq v$ extends to a homotopy of f. (CWHEP) If $u, v : A \to Y$ are maps and $u \simeq v$, then u extends over X if and only if v extends over X. Prove that WHEP $\Rightarrow$ RWHEP $\Rightarrow$ VWHEP $\Rightarrow$ CWHEP. Prove also that CWHEP is not a property invariant under homotopy equivalence of pairs.

10. Show that 7.5.7 (Corollary 3) is true if (X, A) satisfies only the VWHEP.
11. Let $\mathbf{f} : (X, A) \to (Y, B)$ have homotopy inverse $\mathbf{g}$ such that $\mathbf{fg} \simeq 1$ rel B. Prove that if (Y, B) has the VWHEP, then so also does (X, A).
12. Let (X, A), (Y, B) have the WHEP, and let $Z = (X \times Y)/(X \times B \cup A \times Y)$, with the usual base point. Prove that $(Z, \cdot)$ has the WHEP.
13. If $f : X \to Y$ is a map, then there is an inclusion map $f' : Y \to C(f)$ where $C(f)$ is the mapping cone [Example 5 of 4.6]. The following sequence of maps

$$X \xrightarrow{f} Y \xrightarrow{f'} C(f) \xrightarrow{f''} C(f') \xrightarrow{f'''} C(f'') \to \cdots \xrightarrow{f^{(n)}} C(f^{(n-1)})$$

is called the (*unpointed*, or *free*) *Puppe sequence*; here $f^{(n)}$ is defined inductively by $f^{(n)} = (f^{(n-1)})'$. Prove that there is a diagram, commutative up to homotopy

$$\begin{array}{ccc} C(f') & \xrightarrow{f'''} & C(f'') \\ \downarrow g & & \downarrow g' \\ SX & \xrightarrow[Sf]{} & SY \end{array}$$

in which g, g' are homotopy equivalences.
14. Prove a result similar to that of the previous exercise for the pointed case, in which C is replaced by Γ and S by Σ.
15. A sequence $A_1 \xleftarrow{a_1} A_2 \xleftarrow{a_2} A_3 \leftarrow \cdots$ of pointed sets and pointed functions is called *exact* if for each $i \geqslant 2$, $\operatorname{Im} a_i = a_{i-1}^{-1}[\cdot]$. A sequence $X_1 \xrightarrow{f_1} X_2 \xrightarrow{f_2} X_3 \to \cdots$ of pointed maps of pointed spaces is called *exact* if the induced sequence of sets

$$[X_1, Z]_\bullet \xleftarrow{f_1^*} [X_2, Z]_\bullet \xleftarrow{f_2^*} [X_3, Z]_\bullet \leftarrow \cdots$$

is exact for any pointed space—here the base point of $[X_n, Z]_\bullet$ is the class of the constant map and f_i^* is cls $g \rightsquigarrow$ cls gf_i. Prove that the (pointed) Puppe sequence is exact. Deduce that for any pointed map $f : X \to Y$ there is an exact sequence

$$X \xrightarrow{f} Y \xrightarrow{f'} \Gamma(f) \to \Sigma X \xrightarrow{\Sigma f} \Sigma Y \xrightarrow{\Sigma f'} \Sigma\Gamma(f) \to \cdots.$$

Prove further that if X is a closed subspace of Y, $f : X \to Y$ is the inclusion and (Y, X) has the HEP., then there is an exact sequence

$$X \xrightarrow{f} Y \xrightarrow{p} Y/X \to \Sigma X \xrightarrow{\Sigma f} \Sigma Y \to \Sigma(Y/X) \to \cdots.$$

16. Compare the last sequence with the sequence of sets in Exercise 8 of 7.2.
17. Let $i : X \to Y$ be the inclusion of the closed subspace X of Y. Let (Y, X) be a pointed pair with the HEP.. We say X is *retractile* in Y if $\Sigma i : \Sigma X \to \Sigma Y$ has a left-homotopy inverse. Prove that the following conditions are equivalent (i) X is retractile in Y, (ii) ΣX is a retract of ΣY, (iii) for each Z the following sequence is exact

$$0 \to [\Sigma(Y/X), Z]_\bullet \xrightarrow{p^*} [\Sigma Y, Z]_\bullet \xrightarrow{i^*} [\Sigma X, Z]_\bullet \to 0$$

where 0 denotes a trivial group.

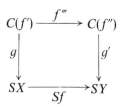

18. Prove that if X, Y are well pointed, then $X \vee Y$ is retractile in $X \times Y$.
19. Let $f : X \to Y$ be a pointed map. Prove that Y is retractile in $\Gamma(f)$ if and only if Σf is inessential.
20. Let X, Y be well pointed. The homotopy equivalences $X * Y \to \Sigma(X \circledast Y)$, $SX \to \Sigma X$, $SY \to \Sigma Y$ and the Whitehead product map $w : X * Y \to SX \vee SY$ determine a Whitehead product map $w' : \Sigma(X \circledast Y) \to \Sigma X \vee \Sigma Y$ uniquely up to homotopy. Prove that $\Sigma w'$ is inessential.
21. Let X, Y be well-pointed, and let $Z = \Sigma X \vee \Sigma Y$. Let $\rho_1 : \Sigma(X \times Y) \to Z$ be $i_1(\Sigma p_1)$ where $p_1 : X \times Y \to X$ is the projection and $i_1 : \Sigma X \to Z$ is the inclusion. Let $\rho_2 : \Sigma(X \times Y) \to Z$ be defined similarly as $i_2(\Sigma p_2)$. Prove that the following sequence

$$0 \to [\Sigma(X \circledast Y), Z]_{\bullet} \xrightarrow{p^*} [\Sigma(X \times Y), Z]_{\bullet} \xrightarrow{i^*} [Z, Z]_{\bullet} \to 0$$

is exact, and that if w' is defined as in the previous exercise then

$$p^*(\text{cls } w') = -\sigma_2 - \sigma_1 + \sigma_2 + \sigma_1 \text{ where } \sigma_i = \text{cls } \rho_i.$$

7.6 The cellular approximation theorem

As motivation for the work of this section we consider the following question: suppose X, B are cell complexes, A is a subcomplex of X and $f : A \to B$ is a map; is then $B \,_f\!\sqcup X$ of the homotopy type of a cell complex? We know that $B \,_f\!\sqcup X$ *is* a cell complex if f is a cellular map. Also we know that the homotopy type of $B \,_f\!\sqcup X$ depends only on the homotopy class of f (by 7.5.5 (Corollary 1) and since (X, A) has the HEP). Our question is thus answered by the *cellular approximation theorem*, a special case of which asserts that any map $f : A \to B$ is homotopic to a cellular map. The word approximation here is used in a rough sense only—we will not be concerned with questions of metrics nor with a real number $\varepsilon > 0$.

The main technical work is in the following result; the elegant formulation of the proof is due to J. F. Adams.

7.6.1 *The following statements are true for each $n \geqslant 1$.*
$\alpha(n)$ *Any map $\mathbf{S}^r \to \mathbf{S}^n$ with $r < n$ is inessential.*
$\beta(n)$ *Any map $\mathbf{S}^r \to \mathbf{S}^n$ with $r < n$ extends over $\mathbf{E}^{r+1}$.*
$\gamma(n)$ *Let B be path-connected and let Q be formed by attaching a finite number of n-cells to B. Then any map*

$$(\mathbf{E}^r, \mathbf{S}^{r-1}) \to (Q, B)$$

with $r < n$ is deformable into B.

The proof is by induction by means of the implications

$$\gamma(n) \Rightarrow \alpha(n) \Leftrightarrow \beta(n) \Rightarrow \gamma(n+1)$$

the only difficult step being the proof of $\beta(n) \Rightarrow \gamma(n+1)$. The start of the induction—the proof of $\gamma(1)$—is easy; in fact, since $\mathbf{E}^0$ consists of a single point and $\mathbf{S}^{-1}$ is the empty set, $\gamma(1)$ is equivalent to the statement that Q is path-connected, and this is a special case of 4.6.3.

Proof of $\gamma(n) \Rightarrow \alpha(n)$

Let $f : \mathbf{S}^r \to \mathbf{S}^n$ be a map such that $r < n$. Let $p : \mathbf{E}^r \to \mathbf{S}^r$ be an identification map which shrinks the boundary $\mathbf{S}^{r-1}$ of $\mathbf{E}^r$ to a point x of $\mathbf{S}^r$, and let $e^0 = \{fx\}$. Then $\mathbf{S}^n$ has a cell structure

$$\mathbf{S}^n = e^0 \cup e^n$$

and so $fp : \mathbf{E}^r \to \mathbf{S}^n$ defines a map

$$\mathbf{g} : (\mathbf{E}^r, \mathbf{S}^{r-1}) \to (e^0 \cup e^n, e^0).$$

The hypothesis $\gamma(n)$ implies that $\mathbf{g}$ is deformable into e^0. By 7.4.2 $\mathbf{g}$ is deformable into e^0 rel $\mathbf{S}^{r-1}$. Throughout this homotopy $\mathbf{S}^{r-1}$ is mapped to the point fx, and so, by Remark 7.3.6, this homotopy defines a homotopy $f \simeq f' : \mathbf{S}^r \to \mathbf{S}^n$ rel $\{x\}$ such that f' is constant. Thus f is inessential. □

Proof of $\alpha(n) \Leftrightarrow \beta(n)$

The proof depends on the fact that there is a homeomorphism $\mathbf{E}^{r+1} \to C\mathbf{S}^r$ which is the identity on $\mathbf{S}^r$ (where $\mathbf{S}^r$ is identified with the subset $\mathbf{S}^r \times 1$ of $C\mathbf{S}^r$). Let $p : \mathbf{S}^r \times \mathbf{I} \to C\mathbf{S}^r$ be the identification map and let $g : \mathbf{S}^r \to X$ denote a constant map.

A homotopy $g \simeq f : \mathbf{S}^r \to X$ is a map $\mathbf{S}^r \times \mathbf{I} \to X$ which is $(x, 0) \rightsquigarrow gx$ on $\mathbf{S}^r \times 0$ and $(x, 1) \rightsquigarrow fx$ on $\mathbf{S}^r \times 1$. Since g is constant, such a homotopy defines an extension $C\mathbf{S}^r \to X$ of f. Conversely, if $f : \mathbf{S}^r \to X$ is a map and $F : C\mathbf{S}^r \to X$ is an extension of f, then the composite Fp is a homotopy $g \simeq f$ such that g is constant. □

Proof of $\beta(n) \Rightarrow \gamma(n + 1)$

Let $\dot{\mathbf{I}}^r$ denote the boundary of the r-cube $\mathbf{I}^r$, that is, the set of points x of $\mathbf{I}^r$ which have at least one coordinate with a value 0 or 1. The pairs $(\mathbf{E}^r, \mathbf{S}^{r-1})$ and $(\mathbf{I}^r, \dot{\mathbf{I}}^r)$ are homeomorphic and so we can assume $\beta(n)$ in the form that *any map* $\dot{\mathbf{I}}^{r+1} \to \mathbf{S}^n$ *with* $r < n$ *extends over* $\mathbf{I}^{r+1}$. Further, since $(\mathbf{I}^{r+1}, \dot{\mathbf{I}}^{r+1})$ has the HEP, we can by 7.2.6 (Corollary 2) assume that this is true not only for $\mathbf{S}^n$ but also for any space of the homotopy type of $\mathbf{S}^n$.

We assume that

$$Q = B \,{}_{k_1}\!\sqcup \mathbf{E}^{n+1} \,{}_{k_2}\!\sqcup \mathbf{E}^{n+1} \cdots {}_{k_m}\!\sqcup \mathbf{E}^{n+1}.$$

Let $\bar{k}_i : \mathbf{E}^{n+1} \to Q$ be the usual extension of $k_i : \mathbf{S}^n \to B$. We use these maps to define an open cover of Q.

Let U_i be the image under $\bar{k}_i$ of the set $\{x \in \mathbf{E}^{n+1} : |x| < \frac{2}{3}\}$, and let U_i' be the image under $\bar{k}_i$ of the set $\{x \in \mathbf{E}^{n+1} : |x| > \frac{1}{3}\}$. Let $U = B \cup U_1' \cdots \cup U_m'$. Then U_i and U are open in Q and we set

$$\mathscr{U} = \{U, U_1, \ldots, U_m\}.$$

Notice that $U \cap U_i$ is homeomorphic to the space of points x in $\mathbf{E}^{n+1}$ such that $\frac{1}{3} < |x| < \frac{2}{3}$, and this space is homeomorphic to $\mathbf{S}^n \times]\frac{1}{3}, \frac{2}{3}[$. Hence $U \cap U_i$ is of the homotopy type of $\mathbf{S}^n$, and so we have
(*) : *$\beta(n)$ can be applied for maps into $U \cap U_i$ rather than $\mathbf{S}^n$.*

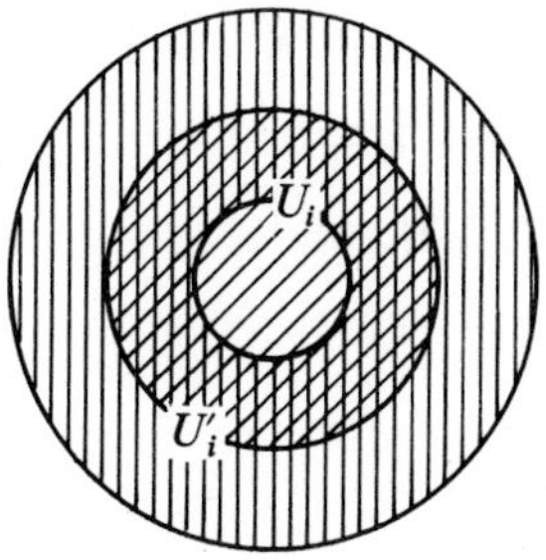

Fig. 7.12

Suppose now we are given a map

$$\mathbf{f} : (\mathbf{I}^r, \dot{\mathbf{I}}^r) \to (Q, B)$$

where $r < n + 1$. We must prove that $\mathbf{f}$ is deformable into B.

By means of hyperplanes in $\mathbf{R}^r$ with equations of the form

$$x_j = s/N, \qquad s = 1, \dots, N-1, \quad j = 1, \dots, r$$

we may subdivide $\mathbf{I}^r$ into cubes of diameter $\leqslant \sqrt{r}/N$. By the Lebesgue covering lemma, N may be chosen so large that each such cube is mapped by f into some set of $\mathscr{U}$.

Let A be the union of all cubes J of all dimensions in this subdivision such that $f[J] \subset U$. Let K^q be the union of all cubes J of dimension $\leqslant q$ (so that $K^{-1} = \varnothing$, K^0 consists of isolated points, and $K^r = \mathbf{I}^r$), and let $K_q = K^q \cup A$.

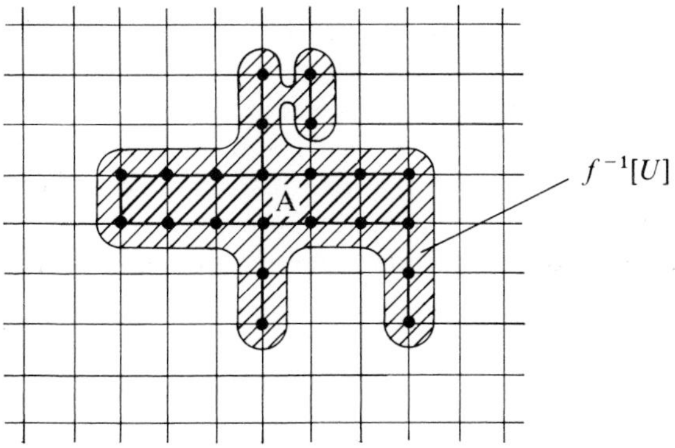

Fig. 7.13

We construct maps $g_q : K_q \to Q$ by induction on q so that the following conditions are fulfilled:

1_q) g_q agrees with f on A, and $g_q \mid K_{q-1} = g_{q-1}$,
2_q) if $x \in K_q$ and $fx \in U_i$ then $g_q x \in U \cap U_i$.

To start the induction we define g_{-1} to be $f \mid A$—clearly conditions 1_{-1}) and 2_{-1}) are satisfied. Suppose g_q has been defined and satisfies 1_q) and 2_q); we extend g_q over K_{q+1}.

Let J^{q+1} be a $(q + 1)$-cube of K_{q+1} which is not contained in A. Then for some unique i, $f[J^{q+1}] \subset U_i$ and it follows from 2_q) that

$$g_q[\dot{J}^{q+1}] \subset U \cap U_i.$$

But $q + 1 \leqslant r < n + 1$, whence $q < n$. By (*) above, g_q extends to a map $J^{q+1} \to U \cap U_i$, and we define $g_{q+1} : K_{q+1} \to Q$ to agree on J^{q+1} with this map. Clearly, conditions 1_{q+1}) and 2_{q+1}) are satisfied, so the induction is complete.

Let $g = g_{r+1} : \mathbf{I}^{r+1} \to Q$. We prove that $f \simeq g$.

The map $\bar{k}_i : \mathbf{E}^{n+1} \to Q$ maps the set of points x with $|x| < \frac{2}{3}$ bijectively onto U_i: we suppose there is given a linear structure on U_i by means of this bijection. We then define a homotopy $h_t : \mathbf{I}^{r+1} \to Q$ by

$$h_t x = \begin{cases} fx, & fx \in \text{no } U_i \\ (1 - t)fx + tgx, & fx \in \text{some } U_i. \end{cases}$$

Each $(r + 1)$-cube J^{r+1} in the given subdivision of $\mathbf{I}^{r+1}$ is mapped by f into U or into some U_i; thus the formula for h_t shows that h_t is continuous on J^{r+1}. Since these cubes form a cover of $\mathbf{I}^{r+1}$ by closed subsets, this implies the continuity of h_t, *qua* function $\mathbf{I}^{r+1} \times \mathbf{I} \to Q$.

Clearly $h_0 = f$. We prove that $h_1 = g$.

Suppose J is an $(r + 1)$-cube of the subdivision and $x \in J$. If fx belongs to some U_i, then the formula for h_1 shows that $h_1 x = gx$. Suppose fx belongs to no U_i (in which case $h_1 x = fx$). Then $f[J]$ is contained in no U_i and so $f[J] \subset U$. Hence $x \in A$ and so $h_1 x = fx = gx$.

Since $\dot{\mathbf{I}}^{r+1} \subset A$ the homotopy h_t is rel $\dot{\mathbf{I}}^{r+1}$; hence h_t defines a homotopy $\mathbf{f} \simeq \mathbf{g}$ where $\mathbf{g} : (\mathbf{I}^{r+1}, \dot{\mathbf{I}}^{r+1}) \to (Q, B)$ is the map defined by g.

Finally, Im $g \subset U$ and B is a deformation retract of U. Therefore, g is deformable into B [cf. Exercise 2 of 7.4]. □

The cellular approximation theorem itself is a consequence of the following deformation theorem. (We recall now that the r-skeleton of a cell complex K is written K^r.)

7.6.2 *Let L be a space and $(L_r)_{r \geqslant 0}$ a sequence of subspaces of L such that for all $r \geqslant 0$*

(*a*) $L_r \subset L_{r+1}$,

(*b*) *any map* $(\mathbf{E}^{r+1}, \mathbf{S}^r) \to (L, L_r)$ *is deformable into* L_{r+1}.
Let K be a cell complex, A a subcomplex of K and $f : K \to L$ *a map such that*

$$f[A^r] \subset L_r, \quad r = 0, 1, \ldots.$$

Then f is homotopic rel *A to a map* $g : K \to L$ *such that*

$$g[K^r] \subset L_r, \quad r = 0, 1, \ldots.$$

Proof Let $K_r = A \cup K^r$. We construct a sequence of maps $f^r : K \to L$ and homotopies $f^{r-1} \simeq f^r$ such that
$1_r)$, $f^r[K_s] \subset L_s$, $0 \leqslant s \leqslant r$,
$2_r)$, f^r agrees with f^{r-1} on K^{r-1},
$3_r)$, $f^{r-1} \simeq f^r$ rel K^{r-1}.
The induction is started with $f^{-1} = f$ when the above conditions are vacuously satisfied.

By condition (*b*) any map $(\mathbf{E}^{r+1}, \mathbf{S}^r) \to (L, L_r)$ is deformable into L_{r+1} rel $\mathbf{S}^r$. It follows easily, using Remark 7.3.6, that if e^{r+1} is any $(r+1)$-cell of $K \setminus A$ then $f^r \mid K^r \cup e^{r+1}$ is deformable into L_{r+1} rel K_r. By applying this to each $(r+1)$-cell of $K \setminus A$ in turn we obtain a homotopy $f^r \mid K_{r+1} \simeq f'$ rel K_r such that $f'[K_{r+1}] \subset L_{r+1}$. By the HEP for (K, K_{r+1}) this homotopy extends to a homotopy $f^r \simeq f^{r+1}$ rel K_r. Clearly, f^{r+1} satisfies $1_{r+1})$, $2_{r+1})$, $3_{r+1})$.

Since $K = K^N$ for some N, the map $g = f^N$ is the required map. □

7.6.2 (*Corollary* 1) (*the cellular approximation theorem*). *Let K, L be complexes and A a subcomplex of K. If* $f : K \to L$ *is a map such that* $f \mid A$ *is cellular, then f is homotopic* rel *A to a cellular map.*

Proof This follows from 7.6.1 $\gamma(n)$ and 7.6.2 with $L_r = L^r$. □

EXERCISES

1. Let L_0 be a subcomplex of L such that for all $r \geqslant 0$ any map $(\mathbf{E}^{r+1}, \mathbf{S}^r) \to (L, L_0)$ is deformable into L_0. Prove that L_0 is a deformation retract of L.
2. Let $f_0, f_1 : K \to L$ be cellular maps and A a subcomplex of K. Suppose that $F : f_0 \simeq f_1$ is a homotopy such that for all $t \in \mathbf{I}$, $F_t[A^r] \subset L^{r+1}$, $r = 0, 1, \ldots$. Prove that F is homotopic rel $A \times \mathbf{I} \cup K \times \dot{\mathbf{I}}$ to a homotopy $G : f_0 \simeq f_1$ such that for all $t \in \mathbf{I}$, $G_t[K^r] \subset L^{r+1}$, $r = 0, 1, \ldots$.
3. Let $i : K^2 \to K$ be the inclusion of the 2-skeleton of the cell complex K. Prove that for any subset A of K^2, i induces an isomorphism $\pi K^2 A \to \pi K A$.
4. Prove that $\mathbf{S}^n$ is simply connected for $n > 1$. Prove also that for any point x of $\mathbf{S}^n$ the ith homotopy group $\pi_i(\mathbf{S}^n, x)$ [cf. 7.1] is 0 for $i < n$.
5. Prove the cellular approximation theorem for (infinite) CW-complexes.
6. Let K be a connected subcomplex of the complex L. Let $j : K \to L$ be the

inclusion and let $x \in K^0$, $e \in \mathbf{S}^r$. Prove that any map $(\mathbf{E}^{r+1}, \mathbf{S}^r) \to (L, K)$ is homotopic to a map h such that $he = x$.

7. Continuing the notation of the previous exercise, suppose that $j_* : \pi_r(K, x) \to \pi_r(L, x)$ is injective. Let $\mathbf{h} : (\mathbf{E}^{r+1}, \mathbf{S}^r) \to (L, K)$ be a map such that $he = x$. Prove that $h^1 : \mathbf{S}^r \to K$ is inessential rel e, and hence show that $\mathbf{h}$ is homotopic rel e to a map $\mathbf{k}$ such that $k[\mathbf{S}^r] = \{x\}$.

8. Suppose further to the previous exercise, that $j_* : \pi_{r+1}(K, x) \to \pi_{r+1}(L, x)$ is surjective. Prove that $\mathbf{h}$ is deformable into K.

9. Let $j : K \to L$ be the inclusion map of the subcomplex K of L. Let K, L be connected, and suppose $j_* : \pi_r(K, x) \to \pi_r(L, x)$ is an isomorphism for all $r \geqslant 1$ (where $x \in K^0$). Prove that K is a deformation retract of L.

10. Let $f : K \to K'$ be a map of connected complexes such that $f_* : \pi_r(K, x) \to \pi_r(K', fx)$ is an isomorphism for all $r \geqslant 1$ (where $x \in K^0$). Prove that f is a homotopy equivalence.

NOTES

In this book we are only dealing with half of the story. The HEP for pairs generalizes slightly to cofibrations; the 'dual' notion of fibration is more important. An introductory account of this duality is given in Cockcroft–Jarvis [1]—the duality itself is due to Eckmann–Hilton [cf. Eckmann [1], Hilton [2]].

Track groups (i.e. the object groups of πY^X) were first studied extensively by M. G. Barratt 'Track groups D, II' Proc. London Math. Soc. 5 (1955) 71–106, 285–329. The exact sequences he used were generalised to the relative case by Spanier–Whitehead [1]; the elegant formulation of these sequences given in Exercise 15 of 7.5 is due to D. Puppe [1].

I owe 7.2.3 to D. Puppe. Parts of 7.3.6 have appeared at various times; I have taken this version, and also 7.3.7 from Puppe [2] [cf. also Steenrod [2]]. The glueing theorem is proved in a very general form, but for CW-complexes, in Spanier–Whitehead [1]. The present version, and its applications, seems to be new.

I have introduced the notation $X/\!/u$ mainly for its brevity and appearance—a 'dual' version of this concept is written $[X, Y; u]$ by James and Thomas [1].

The excellent notes of Dold [2] form a natural continuation of some of the topics studied here.

8. Computation of the fundamental groupoid

In order to describe the various groupoids which arise as πXA (for 'nice' spaces X) we consider two ways of obtaining new groupoids from old. The formation of quotient groupoids is rather similar to the formation of quotient groups. The extra power which the use of groupoids as against groups gives us is in the construction of universal groupoids.

In 8.4 we prove the van Kampen theorem, which allows us to compute the fundamental groupoid of adjunction spaces, and hence of cell complexes.

8.1 Universal morphisms

We have seen that groupoids model well the product and sum of topological spaces. In order to model processes of identification, we construct for groupoids two kinds of identifications which may be thought of as identifications in dimensions 0 and 1—in dimension 0 we have *universal groupoids* and in dimension 1 *quotient groupoids*. Here we are concerned with the former process.

Let G be a groupoid, X a set and

$$\sigma : \mathrm{Ob}(G) \to X$$

a function. We shall construct from G and σ a groupoid U and morphism $\sigma^* : G \to U$ such that (i) $\mathrm{Ob}(U) = X$ and (ii) σ^* is exactly σ on objects.

The idea of this construction is as follows (we use multiplicative notation for groupoids throughout most of this chapter). Let $a_1 \in G(x_1, y_1)$, $a_2 \in G(x_2, y_2)$. Then $a_2 a_1$ is defined if and only if $y_1 = x_2$; in such a case, a morphism $\sigma^* : G \to U$ will satisfy

$$\sigma^* a_2 \, \sigma^* a_1 = \sigma^*(a_2 a_1).$$

Suppose, however, that $y_1 \neq x_2$ but $\sigma y_1 = \sigma x_2$. Then $\sigma^* a_2 \, \sigma^* a_1$ will be defined in U but will not necessarily be $\sigma^* b$ for any b. So, in order to construct U from G and σ, we must have a method of constructing elements such as $\sigma^*a_2\, \sigma^*a_1$. This is done by means of 'words' in the elements of G.

Let $x, x' \in X$. A *word of length n* ($n \geqslant 1$) from x to x' is a sequence

$$a = (a_n, \ldots, a_1)$$

of elements of G such that, if $a_i \in G(x_i, x_i')$, $i = 1, \ldots, n$, then
(*a*) $x_i' \neq x_{i+1}$, $i = 1, \ldots, n-1$,
(*b*) $\sigma x_i' = \sigma x_{i+1}$, $i = 1, \ldots, n-1$,
(*c*) $\sigma x_1 = x$, $\sigma x_n' = x'$,
(*d*) no a_i is an identity.
These conditions ensure that no product $a_{i+1}a_i$ is defined in G, but that $\sigma^*a_{i+1}\, \sigma^*a_i$ will be defined in U.

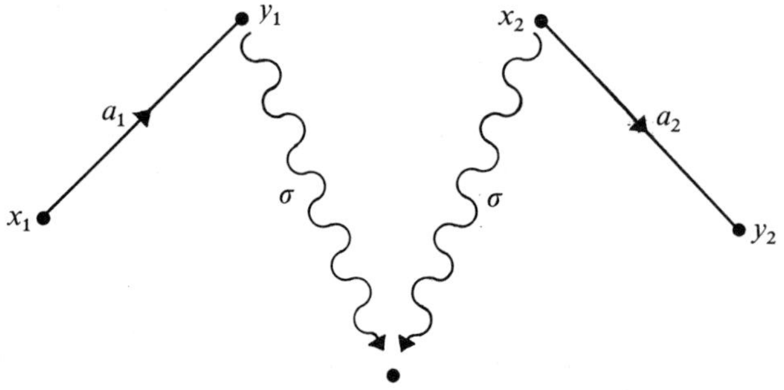

Fig. 8.1

The set of all words from x to x' of length $\geqslant 1$ is written $U(x, x')$; except that, if $x = x'$, then we also include in $U(x, x)$ the empty word of length 0 which we write $(\ )_x$ (and thus suppose that $(\ )_x \neq (\ )_y$ when $x \neq y$).

We now define by induction on length a multiplication

$$U(x', x'') \times U(x, x') \to U(x, x'').$$

The empty word in $U(x, x)$ is to act as identity; that is, if $a \in U(x, x')$ then

$$(\ )_{x'}a = a,\ a(\ )_x = a.$$

Suppose that a is as above and $b = (b_m, \ldots, b_1) \in U(x', x'')$, where $b_j \in G(y_j, y_j')$. We define ba to be, in the various cases,

$(b_m, \ldots, b_1, a_n, \ldots, a_1)$ if $y_1 \neq x_n'$,

$(b_m, \ldots, b_1a_n, \ldots, a_1)$ if $y_1 = x_n'$ but $b_1a_n \neq 1$,

$(b_m, \ldots, b_2)(a_{n-1}, \ldots, a_1)$ by induction if $y_1 = x_n'$ and $b_1a_n = 1$.

This can be expressed as: multiply the two words by putting them end to end, computing in G and cancelling identities where possible.

The definition of multiplication shows that

$$(a_n, \ldots, a_1)(a_1^{-1}, \ldots, a_n^{-1}) = (\,)_{x'},$$
$$(a_1^{-1}, \ldots, a_n^{-1})(a_n, \ldots, a_1) = (\,)_x.$$

Thus each element has a left and right inverse.

We now show that multiplication is associative. Let

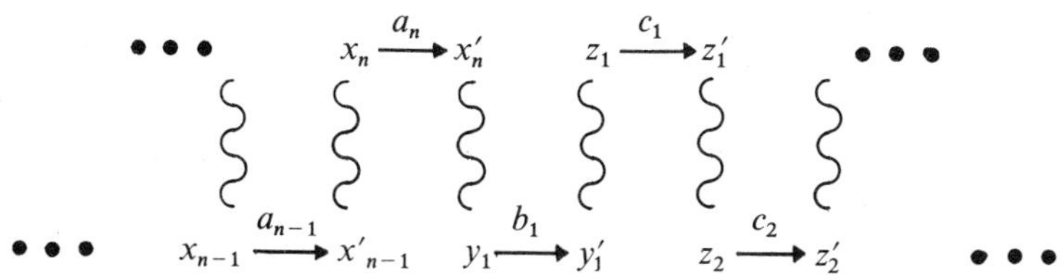

Fig. 8.2

$c = (c_r, \ldots, c_1) \in U(x'', x''')$ where $c_k \in G(z_k, z_k')$. If b is of length 0 then certainly $c(ba) = ca = (cb)a$; this is true, similarly, if a or c is of length 0. Suppose that $m = 1$; we check the value of $c(ba)$ and $(cb)a$ in each case that can arise. These values are

$(c_r, \ldots, c_1, b_1, a_n, \ldots, a_1)$	if $x_n' \neq y_1, y_1' \neq z_1$
$(c_r, \ldots, c_1, b_1a_n, a_{n-1}, \ldots, a_1)$	if $y_1' \neq z_1, x_n' = y_1, b_1a_n \neq 1$
$(c_r, \ldots, c_1b_1, a_n, \ldots, a_1)$	if $y_1' = z_1, c_1b_1 \neq 1, x_n' \neq y_1$
$(c_r, \ldots, c_1)(a_{n-1}, \ldots, a_1)$	if $y_1' \neq z_1, b_1a_n = 1$
$(c_r, \ldots, c_2)(a_n, \ldots, a_1)$	if $x_n' \neq y_1, c_1b_1 = 1$
$(c_r, \ldots, c_1b_1a_n, a_{n-1}, \ldots, a_1)$	if $y_1' = z_1, x_n' = y_1, c_1b_1a_n \neq 1$
$(c_r, \ldots, c_2)(a_{n-1}, \ldots, a_1)$	if $y_1' = z_1, x_n' = y_1, c_1b_1a_n = 1.$

If $m > 1$ we proceed by induction. Write $b = b''b'$ with b', b'' shorter than b. Then

$$\begin{aligned} c(ba) &= c((b''b')a) = c(b''(b'a)) = (cb'')(b'a) \\ &= ((cb'')b')a = (cb)a \end{aligned}$$

as was to be shown.

So if we take X as the set of objects of U we have proved that U is a groupoid.

Notice that if x, or x', does not belong to Im σ, then $U(x, x')$ is either empty (if $x \neq x'$), or contains the identity alone (if $x = x'$). Thus $U = U' \sqcup U''$ where U' is the full subgroupoid on Im σ and U'' is the discrete groupoid on $X \setminus \text{Im } \sigma$.

The groupoid U depends on G and on σ, and we therefore write $U_\sigma(G)$ for U.

We define a morphism $\sigma^* : G \to U_\sigma(G)$ to be σ on objects and to be defined for a in $G(x_1, x_1')$ by

$$\sigma^* a = \begin{cases} (a), & a \neq 1 \\ (\,)_{\sigma x_1}, & x_1 = x_1' \text{ and } a = 1. \end{cases}$$

Thus σ^* is injective on the set of non-identities of G and also maps no non-identity to an identity. It is immediate from the definition of the multiplication of $U_\sigma(G)$ that σ^* is a morphism.

Although the above construction of $U_\sigma(G)$ gives a good idea of its structure, the important property of $U_\sigma(G)$ from a general point of view is that σ^* satisfies a universal property.

Let us, temporarily, identify the set of objects $\text{Ob}(G)$ of a groupoid G with the wide subgroupoid of G whose elements are all identities; thus we regard $\text{Ob}(G)$ as a subgroupoid of G. (A reader who dislikes this convention may instead call this wide subgroupoid $\text{Id}(G)$, and replace suitably Ob by Id in what follows.)

A morphism $f : G \to H$ is called *universal* if the following square,

$$\begin{array}{ccc} \text{Ob}(G) & \xrightarrow{\text{Ob}(f)} & \text{Ob}(H) \\ {\scriptstyle i}\downarrow & & \downarrow{\scriptstyle j} \\ G & \xrightarrow[f]{} & H \end{array}$$

in which the vertical morphisms are inclusions, is a pushout (in the category of groupoids). A useful restatement of this definition is

8.1.1 *A morphism $f : G \to H$ is universal if and only if for any morphism $g : G \to K$ and any function $\tau : \text{Ob}(H) \to \text{Ob}(K)$ such that*

$$\text{Ob}(g) = \tau \, \text{Ob}(f)$$

there is a unique morphism $g^* : H \to K$ such that $\text{Ob}(g^*) = \tau$ and $g^* f = g$.

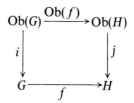

Proof Let k be the inclusion $\text{Ob}(K) \to K$. Clearly the function $\tau : \text{Ob}(H) \to \text{Ob}(K)$ and the morphism $\tau' : \text{Ob}(H) \to K$ determine each other by the rule $\tau' = k\tau$. The condition $\text{Ob}(g) = \tau \, \text{Ob}(f)$ is equivalent to $gi = \tau' \operatorname{Ob}(f)$, and the condition $\operatorname{Ob}(g^*) = \tau$ is equivalent to $g^*j = \tau'$. So the result follows from the definition of pushouts. □

A universal morphism $f : G \to H$ is clearly determined (up to isomorphism on H) by $\operatorname{Ob}(f)$ and G. On the other hand, given $\operatorname{Ob}(f)$ and G there is a universal morphism $f : G \to H$ for some H. This is a consequence of

8.1.2 *Let G be a groupoid and* $\sigma : \operatorname{Ob}(G) \to X$ *a function. The morphism* $\sigma^* : G \to U_\sigma(G)$ *is universal.*

Proof We use 8.1.1. Let $g : G \to K$ be a morphism and $\tau : X \to \operatorname{Ob}(K)$ a function such that $\operatorname{Ob}(g) = \tau\sigma$. Let $U = U_\sigma(G)$.

We wish a morphism $g^* : U \to K$ such that $\operatorname{Ob}(g^*) = \tau$. Thus g is determined on the identities of U by τ. An element of U which is not an identity is a word

$$a = (a_n, \ldots, a_1), \quad a_i \in G(x_i, x'_i)$$

such that $\sigma x'_i = \sigma x_{i+1}$, $i = 1, \ldots, n - 1$. If $n = 1$, then we set

$$g^*(a_1) = ga_1;$$

this is the only definition consistent with $g^*\sigma^* = g$. If $n > 1$, then

$$g x'_i = \tau\sigma x'_i = \tau\sigma x_{i+1} = g x_{i+1};$$

therefore we can set

$$\begin{aligned} g^*a &= g a_n \ldots g a_1 \\ &= g^*(a_n) \ldots g^*(a_1) \end{aligned}$$

the product on the right being defined in K. Clearly, this is the only possible definition of a morphism g^* consistent with $g^*\sigma^* = g$, $\operatorname{Ob}(g^*) = \tau$. To complete this proof we verify that this g^* is a morphism.

Suppose a is as above, b is a word $(b_m, \ldots, b_1)$ where $b_j \in G(y_j, y'_j)$ and ba is defined. The equation $g^*(ba) = g^*b\, g^*a$ is clearly true if m or n is 0. Suppose it is true for $m = p - 1$, $n = q - 1$. Then for $m = p$, $n = q$, referring to the definition of ba, we find that $g^*(ba)$ is

$$\begin{array}{ll} g\, b_m \ldots g\, b_1\, g\, a_n \ldots g\, a_1 & \text{if } y_1 \neq x'_n \\ g\, b_m \ldots g\, b_2\, g(b_1 a_n)\, g\, a_{n-1} \ldots g\, a_1 & \text{if } y_1 = x'_n,\ b_1 a_n \neq 1 \\ g^*((b_m, \ldots, b_2)(a_{n-1}, \ldots, a_1)) & \text{if } y_1 = x'_n,\ b_1 a_n = 1. \end{array}$$

This is the same as $g^*b\, g^*a$, by the inductive hypothesis and the fact that g is a morphism. □

From now on, we identify each non-identity a_1 of G with its image (a_1) in $U_\sigma(G)$.

Let $f : G \to H$ be a morphism. We say f is *strictly universal* if for any morphism $g : G \to K$ such that $\mathrm{Ob}(g)$ factors through $\mathrm{Ob}(f)$, there is a unique morphism $g^* : H \to K$ such that $g^*f = g$ (this is the same as the condition in 8.1.1 except that we drop the requirement $\mathrm{Ob}(g^*) = \tau$).

8.1.3 *A morphism $f : G \to H$ is strictly universal $\Leftrightarrow f$ is universal and $\mathrm{Ob}(f)$ is surjective*

Proof $\Leftarrow$ Suppose given $g : G \to K$ and $\tau : \mathrm{Ob}(H) \to \mathrm{Ob}(K)$ such that $\mathrm{Ob}(g) = \tau\,\mathrm{Ob}(f)$. Since f is universal there is a unique morphism $g^* : H \to K$ such that $g^*f = g$ and $\mathrm{Ob}(g^*) = \tau$. But the last condition is redundant, since $g^*f = g$ implies $\mathrm{Ob}(g^*)\ \mathrm{Ob}(f) = \mathrm{Ob}(g) = \tau\,\mathrm{Ob}(f)$ and so, since $\mathrm{Ob}(f)$ is surjective, that $\mathrm{Ob}(g^*) = \tau$.

$\Rightarrow$ Suppose $f : G \to H$ is strictly universal. Let $\sigma = \mathrm{Ob}(f)$. The given universal properties imply that there is an isomorphism $f^* : H \to U_\sigma(G)$ such that $f^*f = \sigma^*$, $\mathrm{Ob}(f^*) = 1$ the identity on $\mathrm{Ob}(H)$. It follows easily that $\sigma^* : G \dashrightarrow U_\sigma(G)$ is strictly universal. This implies that $\sigma = \mathrm{Ob}(f)$ is surjective, for otherwise the condition $g^*\sigma^* = g$ does not always determine g^* on the identities of $U_\sigma(G)$ which are not images by σ^* of identities of G. □

EXAMPLES 1. Let G be a groupoid and σ the unique function from $\mathrm{Ob}(G)$ to a single point set $\{x\}$. Then $U_\sigma(G)$ is a groupoid with only one object, that is, $U_\sigma(G)$ is a group. This group is called the *universal group* of G and is written UG. The morphism $\sigma^* : G \to UG$ is universal for morphisms from G to groups, i.e., if $g : G \to K$ is a morphism to a group K, then there is a unique morphism $g^* : UG \to K$ such that $g^*\sigma^* = g$. By the construction of UG, the elements of this group are the identity and also all words $a_n \dots a_1$ such that $a_i \in G$, no a_i is the identity and no $a_{i+1}a_i$ is defined in G.

2. In particular, consider the groupoid I with two objects 0, 1 and two non-identities ι, ι^{-1} from 0 to 1, and 1 to 0 respectively. A word of length n in UI is

$$\iota\iota\dots\iota \quad \text{or} \quad \iota^{-1}\iota^{-1}\dots\iota^{-1}$$

and these we can write ι^n and ι^{-n} respectively. Clearly, there is an isomorphism $UI \to \mathbf{Z}$ which sends $\iota^{\pm n} \rightsquigarrow \pm n$.

Notice that the computation of $\pi(\mathbf{S}^1, 1)$ in 6.7.5 is now easy—it is immediate from 6.7.4 that there is a universal morphism $I \to \pi(\mathbf{S}^1, 1)$.

3. Let A be a set. A *free group on* A is a group FA and a function $\lambda : A \to FA$ which is universal for functions from A into groups, that is, if $\mu : A \to G$ is any function to a group G then there is a unique morphism $\mu^* : FA \to G$ such that $\mu^*\lambda = \mu$. We prove that *a free group on A always exists.*

Proof Let A be regarded as a discrete groupoid, let FA be the universal group of $A \times I$. In the following diagram

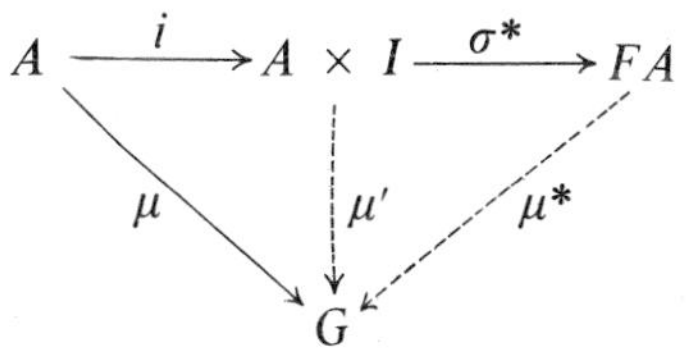

i is the function $a \rightsquigarrow (a, \iota)$, σ^* is the universal morphism and μ is a given function to a group G. The elements μa of G define uniquely a morphism $\mu' : A \times I \to G$ which sends $(a, \iota) \rightsquigarrow \mu a$ (so that $\mu' i = \mu$). Since μ' is a morphism to a group, μ' defines uniquely $\mu^* : FA \to G$ such that $\mu^*\sigma^* = \mu'$. Hence $\mu^*\sigma^* i = \mu' i = \mu$.

Suppose $\bar{\mu} : FA \to G$ is any morphism such that $\bar{\mu}\sigma^* i = \mu$. Then $\bar{\mu}\sigma^* = \mu'$ whence $\bar{\mu} = \mu^*$. This shows that FA with $\sigma^* i : A \to FA$ is a free group on A. □

In the above proof, note that each (a, ι) is a non-identity of $A \times I$; so σ^* is injective on Im i, whence $\lambda = \sigma^* i$ is injective. Let us identify each a in A with λa, so that $(\lambda a)^{-1}$ is written a^{-1} (note that $a^{-1} = \sigma^*(a, \iota^{-1})$). Then the non-identity elements of FA are uniquely written as products

$$a_n^{\varepsilon_n} \ldots a_1^{\varepsilon_1}, \qquad a_i \in A, \quad \varepsilon_i = \pm 1$$

such that for no i is it true that both $a_i = a_{i+1}$ and $\varepsilon_{i+1} = -\varepsilon_i$.

4. Let G, H be groupoids, and let

$$G \xrightarrow{j_1} K \xleftarrow{j_2} H$$

be morphisms. We say these morphisms present K as the *free product* of G and H if the following property is satisfied: if $g : G \to L$, $h : H \to L$ are any morphisms which agree on $\mathrm{Ob}(G) \cap \mathrm{Ob}(H)$, then there is a unique morphism $k : K \to L$ such that $kj_1 = g$, $kj_2 = h$. We prove that *such a free product always exists*.

Proof Let $X = \mathrm{Ob}(G) \cup \mathrm{Ob}(H)$ and let

$$\sigma : \mathrm{Ob}(G) \sqcup \mathrm{Ob}(H) \to X$$

be the function defined by the two inclusions into X. Note that σ is always a surjection, and σ is a bijection if and only if $\mathrm{Ob}(G)$, $\mathrm{Ob}(H)$ are disjoint.

In the following diagram

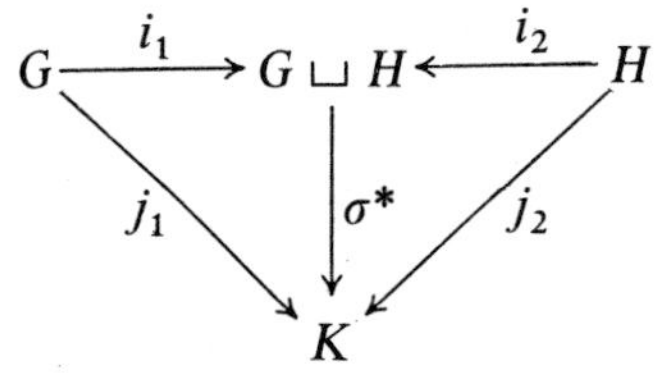

i_1, i_2 are the injections of the coproduct, $K = U_\sigma(G \sqcup H)$ and $j_1 = \sigma^* i_1$, $j_2 = \sigma^* i_2$. The universal property for j_1, j_2 is trivial to verify. □

Notice that the universal property of the free product can also be expressed by saying that the following square is a pushout

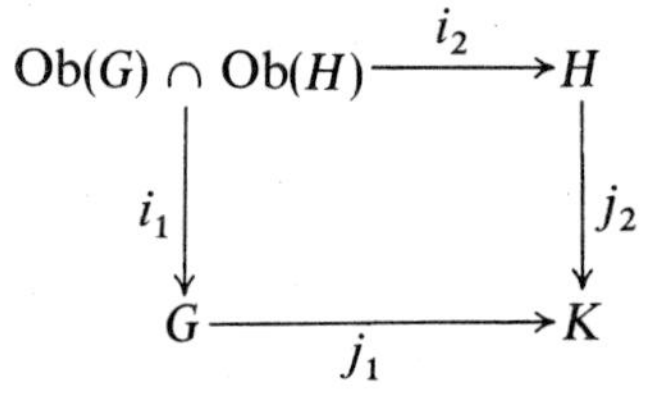

where i_1, i_2 are the inclusions.

The free product of G and H is usually written $G * H$. In particular, let G, H be groups (supposed to have the same object). Then it is clear that the *injections into the free product $G * H$ form a coproduct of groups.* If G, H have no common elements, then the elements of $G * H$ are the identity and all products

$$k_n \ldots k_2 k_1$$

where (i) each k_i belongs to one or other of G, H, (ii) no k_i is an identity, (iii) for no i do k_i, k_{i+1} belong to the same group. (When we write $G * H$ for groups G, H we will always assume that this is the coproduct of groups in the above sense.)

The aim of the rest of this section is to determine the universal group of any connected groupoid. This is useful for the topological applications.

8.1.4 *If $f : G \to H$, $g : H \to K$ are universal morphisms, then so also is gf.*

Proof This is immediate from the definition and 6.6.5. □

8.1.4 (*Corollary 1*) *Let G be a groupoid and $\sigma : \mathrm{Ob}\,(G) \to X$, $\tau : X \to Y$ functions. Then $U_\tau U_\sigma(G)$ is isomorphic to $U_{\tau\sigma}(G)$.*

Proof Both $\tau^*\sigma^* : G \to U_\tau U_\sigma(G)$ and $(\tau\sigma)^* : G \to U_{\tau\sigma}(G)$ are universal. □

8.1.4 (*Corollary 2*) *Let G, H be groupoids. Then the groups*

$$U(G \sqcup H),\ U(G * H),\ UG * UH$$

are all isomorphic.

Proof Let X be a set with one object. Consider the commutative diagram

$$\begin{array}{ccccc} \{x, y\} & \xleftarrow{\sigma} & \mathrm{Ob}(G) \sqcup \mathrm{Ob}(H) & \xrightarrow{\tau} & \mathrm{Ob}(G) \cup \mathrm{Ob}(H) \\ & \searrow & \downarrow & \swarrow & \\ & & X & & \end{array}$$

in which the downward functions are constant, τ is the inclusion on $\mathrm{Ob}(G)$ and on $\mathrm{Ob}(H)$, and σ maps $\mathrm{Ob}(G)$ to x and $\mathrm{Ob}(H)$ to y (where $x \neq y$). Then we obtain a diagram of strictly universal morphisms.

$$\begin{array}{ccccc} UG \sqcup UH & \longleftarrow & G \sqcup H & \longrightarrow & G * H \\ \downarrow & & \downarrow & & \downarrow \\ UG * UH & & U(G \sqcup H) & & U(G * H) \end{array}$$

By 8.1.4 (Corollary 1), all of the bottom groups in this diagram are isomorphic. □

We now show how to determine UG for any connected groupoid G. First we prove:

8.1.5 *Let G be a groupoid and T a wide, tree subgroupoid of G. Then for any object x_0 of G, there is an isomorphism*

$$G \cong G\{x_0\} * T.$$

Proof Let $j_1 : G\{x_0\} \to G, j_2 : T \to G$ be the two inclusions. Each element a of $G(x, y)$ can be written uniquely as

$$\tau_y a' \tau_x^{-1}$$

for $a' \in G\{x_0\}$ and $\tau_y, \tau_x \in T$. Therefore, morphisms $f_1 : G\{x_0\} \to K, f_2 : T \to K$ which agree on x_0 define a morphism $f : G \to K$ by

$$fa = f_2(\tau_y)f_1(a')f_2(\tau_x^{-1})$$

and f is the only morphism such that $fj_1 = f_1, fj_2 = f_2$. □

The proof of 8.1.5 shows that the isomorphism $G \to G\{x_0\} * T$ is given by $a \rightsquigarrow \tau_y a' \tau_x^{-1}$ $(a \in G(x, y))$.

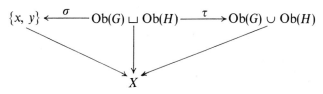

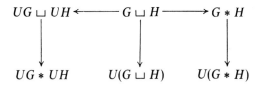

8.1.6 *If T is a tree groupoid, then UT is a free group. If further T has n objects, then UT is a free group on $(n-1)$ elements.*

Proof Let x_0 be an object of T and for each object x of T let τ_x be the unique element of $T(x_0, x)$. Let A be the set of these τ_x for all $x \neq x_0$, and in the groupoid $A \times I$ let ι_x, ι_x^{-1} denote respectively (τ_x, ι) and its inverse (τ_x, ι^{-1}).

Let $f : A \times I \to T$ be the morphism which sends $\iota_x \rightsquigarrow \tau_x$, so that $\sigma = \text{Ob}(f)$ simply identifies all $(\tau_x, 0)$ to x_0. In $U_\sigma(A \times I)$ the only non-identity words are $\iota_x, \iota_x^{-1}, \iota_y\iota_x^{-1}$ $(x \neq y)$. So the morphism $\varphi : U_\sigma(A \times I) \to T$ which sends $\iota_x \rightsquigarrow \tau_x$, $\iota_x^{-1} \to \tau_x^{-1}$, $\iota_y\iota_x^{-1} \rightsquigarrow \tau_y\tau_x^{-1}$ is an isomorphism such that $\varphi\sigma^* = f$. Therefore f is universal. Hence the composite $A \times I \xrightarrow{f} T \to UT$ is universal and so UT is isomorphic to FA, the free group on A. Finally, if T has n objects, then A has $(n-1)$ elements. □

8.1.6 (*Corollary 1*) *If G is a connected groupoid and x_0 is an object of G, then UG is isomorphic to $G\{x_0\} * F$ where F is a free group.*

Proof By 8.1.4 (Corollary 2), 8.1.5, and 8.1.6 it is enough to find a wide tree subgroupoid T of G. This can be done by choosing for each object $x \neq x_0$ of G an element τ_x of $G(x_0, x)$ and defining T to have all elements τ_x, their inverses, and their products. The only element of $T(x, y)$ is then $\tau_y\tau_x^{-1}$ and so T is a wide, tree subgroupoid of G. □

EXERCISES

1. Define the coproduct $\sqcup_{\alpha \in A} G_\alpha$ for an arbitrary family in (i) the category of groupoids, (ii) the category of groups. Prove that if $(G_\alpha)_{\alpha \in A}$ is a family of groupoids, then the universal group of their coproduct is isomorphic to the coproduct (i.e., free product) of their universal groups. Hence show that, if G is a groupoid, then UG is isomorphic to the free product of the groups UG_α for all components G_α of G.
2. Let $\mathscr{C}$ be a category, let X be a set and $\sigma : \text{Ob}(\mathscr{C}) \to X$ a surjection. Prove that there is a category $\mathscr{U}$ and functor $\sigma^* : \mathscr{C} \to \mathscr{U}$ such that (i) $\text{Ob}(\mathscr{U}) = X$, $\text{Ob}(\sigma^*) = \sigma$, (ii) if $\tau : \mathscr{C} \to \mathscr{D}$ is any functor such that $\text{Ob}(\tau)$ factors through σ, then there is a unique functor $\tau : \mathscr{U} \to \mathscr{D}$ such that $\tau^*\sigma^* = \tau$.
3. Let $f : G \to H$ be a morphism of groupoids such that $\text{Ob}(G) = \text{Ob}(H) = X$ and $\text{Ob}(f) = 1_X$. Let $\sigma : X \to Y$ be a surjection. Prove that there is a unique morphism $f' : U_\sigma(G) \to U_\sigma(H)$ such that $f'\sigma^* = \sigma^* f$, and that f' is injective (on elements) if f is.
4. Prove that if $f : G \to H$ is a strictly universal morphism then f is epic in the category of groupoids.
5. Prove that if $f : G \to H$, $g : H \to K$ are morphisms of groupoids such that gf is strictly universal and f is epic, then g is strictly universal.
6. A groupoid G is called the *internal free product* of subgroupoids G_λ if (i) each identity of G lies in some G_λ, (ii) for each non-identity element x of G there is a

unique sequence $\lambda_1, \ldots, \lambda_n$ $(n \geqslant 1)$ with $\lambda_i \neq \lambda_{i+1}$, and unique non-identity elements x_i in G_{λ_i} such that $x = x_n x_{n-1} \ldots x_1$. Prove that G is the internal free product of the G_λ if and only if the canonical morphism $\sqcup_\lambda G_\lambda \to G$ is universal. Prove that in such case, any two distinct G_λ meet in a discrete groupoid.

7. Let (G_λ) be a family of subgroupoids of the groupoid G such that any element x of G is a product of elements of various G_λ. Prove that G is the internal free product of the G_λ if and only if the following condition is satisfied: if $x_i \in G_{\lambda_i}$ $(i = 1, \ldots, n;\ n \geqslant 1)$ with $\lambda_i \neq \lambda_{i+1}$ $(i = 1, \ldots, n-1)$ and if $x_n \ldots x_1$ is defined in G and is an identity element, then at least one of the x_i is an identity element.

8. Suppose there is given a square of groupoid morphisms which is a pushout in

$$\begin{array}{ccc} G_0 & \xrightarrow{i_1} & G_1 \\ {\scriptstyle i_2}\downarrow & & \downarrow{\scriptstyle u_1} \\ G_2 & \xrightarrow[u_2]{} & G \end{array}$$

the category of groupoids. Suppose also that $\sigma_1 = \text{Ob}(i_1)$ and $\sigma_2 = \text{Ob}(i_2)$ are surjective. Prove that the following square in which $j_1 = \sigma_1^* i_1, j_2 = \sigma_2^* i_2$, $v_1 = u_1^*$, $v_2 = u_2^*$, is also a pushout.

$$\begin{array}{ccc} G_0 & \xrightarrow{j_1} & U_{\sigma_1}(G_1) \\ {\scriptstyle j_2}\downarrow & & \downarrow{\scriptstyle v_1} \\ U_{\sigma_2}(G_2) & \xrightarrow[v_2]{} & G \end{array}$$

9. Suppose that the first square of the previous exercise is a pushout, that G_0, G_1 are discrete and that $\text{Ob}(i_2)$ is surjective. Prove that $u_2 : G_2 \to G$ is a universal morphism.

10. Let G' be a subgroupoid of G and let $f : G \to H$ be a universal morphism. Prove that f restricts to a universal morphism $G' \to H'$ where H' is a subgroupoid of H.

8.2 Free groupoids

The free groupoids generalize the free groups. As the reader will by now have expected, free groupoids are defined by means of a universal property —to express this we need a little of the language of graph theory.

A *graph* Γ consists of a set $\text{Ob}(\Gamma)$ of *objects* (or *vertices*) and for each x, y in $\text{Ob}(\Gamma)$ a set $\Gamma(x, y)$ (often called the set of *edges* from x to y). The sets $\Gamma(x, y)$ for various x, y in $\text{Ob}(\Gamma)$ are supposed disjoint. In particular,

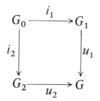

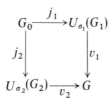

$\Gamma(x, y)$ does not meet $\Gamma(y, x)$, so that the graphs we are concerned with are often called *oriented* graphs. As we did for categories, we shall usually write Γ for the union of the sets $\Gamma(x, y)$ for all x, y in $\text{Ob}(\Gamma)$, so that $a \in \Gamma$, or a is an element of Γ, means $a \in \Gamma(x, y)$ for some x, y in $\text{Ob}(\Gamma)$.

Let Γ, Δ be graphs. A *graph morphism* $f : \Gamma \to \Delta$ assigns to each x in $\text{Ob}(\Gamma)$ an object fx of $\text{Ob}(\Delta)$ and to each a in $\Gamma(x, y)$ an element fa in $\Delta(fx, fy)$. It is easy to verify that the graphs and graph morphisms form a category.

If $\text{Ob}(\Gamma) \subset \text{Ob}(\Delta)$, $\Gamma \subset \Delta$ and the inclusion $\Gamma \to \Delta$ is a graph morphism, then we say Γ is a *subgraph* of Δ: further, Γ is *wide* in Δ if $\text{Ob}(\Gamma) = \text{Ob}(\Delta)$; and Γ is *full* in Δ if $\Gamma(x, y) = \Delta(x, y)$ for all x, y in $\text{Ob}(\Gamma)$.

Suppose Γ' is a subgraph of Γ and $f : \Gamma \to \Delta$ is a graph morphism. Then by $f[\Gamma']$ we mean the subgraph Δ' of Δ whose objects are fx for x in $\text{Ob}(\Gamma')$, and such that $\Delta'(w, z)$ is the union of the sets $f[\Gamma'(x, y)]$ for all x, y in $\text{Ob}(\Gamma')$ such that $fx = w$, $fy = z$. In particular, $\text{Im} f$ is defined to be the graph $f[\Gamma]$.

If $\mathscr{C}$ is a category then $\mathscr{C}$ defines a graph, also written $\mathscr{C}$, simply by forgetting about the composition of elements of $\mathscr{C}$. If $f : \mathscr{C} \to \mathscr{D}$ is a functor, then f defines also a graph morphism $\mathscr{C} \to \mathscr{D}$—the converse, of course, is false. Since groupoids are special kinds of categories, these remarks apply also to groupoids and morphisms of groupoids. In particular, we can talk about subgraphs of a groupoid. Notice that if $f : G \to H$ is a morphism of groupoids then $\text{Im} f$ is a subgraph of G but is not usually a subgroupoid. For example, if $f : I \to \mathbf{Z}$ is the universal morphism, then $\text{Im} f$ has only three elements $0, 1, -1$.

The graphs in a groupoid of the form $\text{Im} f$ are rather special. In fact, a subgraph Γ of a groupoid G is called *invertible* if (i) $x \in \text{Ob}(\Gamma) \Rightarrow 1_x \in \Gamma$, (ii) $a \in \Gamma \Rightarrow a^{-1} \in \Gamma$. If Γ is any subgraph of G there is clearly a smallest invertible graph containing Γ—this graph we write $\bar{\Gamma}$. Note that $\bar{\bar{\Gamma}} = \bar{\Gamma}$. If $f : G \to H$ is a morphism of groupoids then $\text{Im} f$ is an invertible subgraph of H.

Let Γ be as above a graph in a groupoid G. The subgroupoid of G *generated* by Γ is the intersection of all subgroupoids of G which contain Γ; it is thus the smallest subgroupoid of G containing Γ, and its elements are clearly all products

$$a_n \ldots a_1$$

which are well defined in G and for which $a_i \in \bar{\Gamma}$.

A sequence of elements $a_1, \ldots, a_n$ such that $a_i \in \Gamma(x_i, x_{i+1})$, $i = 1, \ldots, n$, is called a *path* in Γ joining x_1 to x_{n+1}; and Γ is said to be *connected* if any two objects of Γ can be joined by a path in Γ. If Γ is a graph in a groupoid G, then $\bar{\Gamma}$ is connected if and only if $\bar{\Gamma}$ generates a connected subgroupoid of G (since $a_1, \ldots, a_n$ is a path in $\bar{\Gamma}$ if and only if each $a_i \in \bar{\Gamma}$ and the product $a_n \ldots a_1$ is defined in G).

The result from graph theory that we need is the following:

8.2.1 *Let Γ be a graph in a groupoid G such that Γ is connected and invertible. Then Γ contains a tree groupoid which is wide in Γ.*

Proof We may suppose $\mathrm{Ob}(\Gamma)$ non-empty. Let $\mathscr{T}$ be the set of all tree groupoids contained in Γ; $\mathscr{T}$ is non-empty, since it contains for example the trivial group on one object of Γ.

Clearly, $\mathscr{T}$ is partially ordered by inclusion. We construct in $\mathscr{T}$ a maximal element T. This is easy when $\mathrm{Ob}(\Gamma)$ is finite, since any T in $\mathscr{T}$ with the maximum number of objects of all the elements of $\mathscr{T}$ must be maximal in $\mathscr{T}$. If $\mathrm{Ob}(\Gamma)$ is not finite, then we must apply Zorn's Lemma [cf. Glossary].

Let $\mathscr{C}$ be any ordered subject of $\mathscr{T}$ and let C be the union of the elements of $\mathscr{C}$. If x, y are objects of C, then x, y are objects of some element C' of $\mathscr{C}$. Since C' is a tree groupoid, $C'(x, y)$ has exactly one element. The same applies to any element of $\mathscr{C}$ which contains C', and therefore it applies also to C. Therefore, C is a tree groupoid. Clearly, C is contained in Γ, and so $C \in \mathscr{T}$. We have thus shown that any ordered subset of $\mathscr{T}$ has an upper bound in $\mathscr{T}$. By Zorn's Lemma, $\mathscr{T}$ has a maximal element, say T.

Suppose T is not wide in Γ. Since Γ is connected there is a path $a_1, \ldots, a_n$ such that $a_i \in \Gamma(x_i, x_{i+1})$, x_1 is an object of T but x_{n+1} is not an object of T. Let i_0 be the first i such that x_{i+1} is not an object of T. Then by adjoining to T the elements $a_{i_0}, a_{i_0}^{-1}$ and the identity at x_{i_0+1} we obtain an element of $\mathscr{T}$ which properly includes T. This contradicts the maximality of T, and it follows that T is wide in Γ. □

We come now to the free groupoids. Let Γ be a graph in a groupoid G. We say G is *free on* Γ if Γ is wide in G and for any groupoid H, any graph morphism $f : \Gamma \to H$ extends uniquely to a morphism $G \to H$; and if such Γ exists, we say G is a *free groupoid.*

8.2.2 *Let Γ be a graph in a groupoid G. The following conditions are equivalent.*

(a) G is free on Γ.

(b) If A is the set whose elements are those of Γ, then the morphism $\sigma^* : A \times I \to G$, such that $(a, \iota) \rightsquigarrow a$, is universal.

(c) Each non-identity of G can be written uniquely as a product

$$a_n^{\varepsilon_n} \ldots a_1^{\varepsilon_1}$$

such that $a_i \in \Gamma$, $\varepsilon_i = \pm 1$ and for no i is it true that $a_i = a_{i+1}$, $\varepsilon_i = -\varepsilon_{i+1}$.

Proof We prove $(a) \Leftrightarrow (b) \Leftrightarrow (c)$.
$(a) \Rightarrow (b)$ Let $\sigma = \text{Ob}(\sigma^*)$. Suppose $h : A \times I \to H$ is a morphism and $\tau : \text{Ob}(G) \to \text{Ob}(H)$ a function such that

$$\text{Ob}(h) = \tau\sigma.$$

We construct from h and τ a graph morphism $f : \Gamma \to H$.

On objects, f is to be τ; on elements

$$fa = h(a, \iota).$$

Clearly, f is a graph morphism. By the assumption that G is free on Γ, f extends uniquely to a morphism $h^* : G \to H$. If $a \in \Gamma$ then

$$h^*\sigma^*(a, \iota) = h^*a = fa = h(a, \iota);$$

it follows that $h^*\sigma^* = h$. Since Γ is wide in G we must also have $\text{Ob}(h^*) = \tau$.

If $\bar{h} : G \to H$ is any morphism such that $\bar{h}\sigma^* = h$, then $\bar{h}$ extends f and so $\bar{h} = h^*$. This proves that σ^* is universal.
$(b) \Rightarrow (a)$ Let $f : \Gamma \to H$ be a graph morphism and let $\tau = \text{Ob}(f)$. We define a morphism $h : A \times I \to H$ by $(a, \iota) \rightsquigarrow fa$. Then $\text{Ob}(h) = \tau\sigma$ and so there is a unique morphism $h^* : G \to H$ such that $\text{Ob}(h^*) = \tau$ and $h^*\sigma^* = h$. Thus $\text{Ob}\,(h^*) = \text{Ob}\,(f)$ and for each element a of Γ, $h^*a = h(a, \iota) = fa$. This shows that h^* extends f.

Let $\bar{h} : G \to H$ be any morphism extending f. Since Γ is wide in G, we have $\text{Ob}(\bar{h}) = \text{Ob}(f)$. Further, if $a \in \Gamma$, then

$$\bar{h}\sigma^*(a, \iota) = \bar{h}a = fa$$

whence $\bar{h}\sigma^* = h$. It follows that $\bar{h} = h^*$.
$(b) \Leftrightarrow (c)$ This follows from the explicit construction of $U_\sigma(A \times I)$. □

8.2.2 (*Corollary 1*) *Let G be a free groupoid on Γ. If $f : G \to H$ is strictly universal, then H is free on $f[\Gamma]$. In particular, UG is a free group.*

Proof This follows from 8.2.2 and 8.1.4. □

8.2.2 (*Corollary 2*) *Let G be a free groupoid on Γ and let Δ be a subgraph of Γ. If H is the subgroupoid of G generated by Δ, then H is free on Δ.*

Proof We use 8.2.2 (c). First of all, Δ is certainly wide in H. Next, if a is a non-identity of H, then, since Δ generates H, a is a product $a_n^{\varepsilon_n} \ldots a_1^{\varepsilon_1}$ such that $a_i \in \Delta$ and $\varepsilon_i = \pm 1$.

If for some i we have $a_i = a_{i+1}$, $\varepsilon_i = -\varepsilon_{i+1}$, then we can cancel $a_{i+1}^{\varepsilon_{i+1}} a_i^{\varepsilon_i}$. Further, we can repeat such cancellations until no relation of this form holds. However, since $\Delta \subset \Gamma$ and G is free on Γ, the resulting expression for a is unique (by 8.2.2 (c)). It follows again from 8.2.2 (c) that H is free on Δ. □

In fact, *any* subgroupoid of a free groupoid is free, but this is more difficult to prove [cf. Exercise 1 of 9.4].

Let G be a free groupoid on Γ. The cardinality of the set of elements of Γ is called the *rank* of G. Now it is easy to see from 8.2.2 (*b*) or (*c*) that no element of Γ is an identity of G; so, if $f : G \to H$ is universal, then f is injective on the elements of Γ and hence G and H have the same rank. In particular, G and its universal group UG have the same rank. But it may be proved that two free groups are isomorphic if and only if they have the same rank [Crowell and Fox [1] p. 48]. So the rank of G depends only on G and not on the particular choice of Γ freely generating G.

8.2.3 *If G is a connected, free groupoid then each object group $G\{x\}$ of G is free. If further G is of rank r_1 and has r_0 objects, then $G\{x\}$ is of rank $r_1 - r_0 + 1$.*

Proof Let $x \in \text{Ob}(G)$ and let Γ be a graph freely generating G. By 8.2.1, $\overline{\Gamma}$ contains a tree groupoid T. For each y in $\text{Ob}(G)$ there is a unique element τ_y of $T(x, y)$; these elements define by 6.5.11 retractions $r : G \to G\{x\}$, $s : T \to \{x\}$.

Let Δ be the wide subgraph of Γ whose elements are those a in Γ which do not lie in T. Let H be the (free) subgroupoid of G generated by Δ. We claim that G is the free product $T * H$.

Let us grant this claim for the moment. Consider the diagram

$$\begin{array}{ccccc} \text{Ob}(G) & \longrightarrow & T & \xrightarrow{\ s\ } & \{x\} \\ \downarrow & & \downarrow & & \downarrow \\ H & \xrightarrow[\ i\ ]{} & G & \xrightarrow[\ r\ ]{} & G\{x\} \end{array}$$

in which the left-hand square of inclusions is a pushout, since $G = T * H$, and the right-hand square is a pushout, by 6.7.3. Therefore the composite square is a pushout. It follows that $ri : H \to G\{x\}$ is universal, and so $G\{x\}$ is a free group on $r[\Delta]$.

If G is of rank r_1 then Γ has r_1 elements. If G has r_0 objects then $\Gamma \cap T$ has $r_0 - 1$ elements. Then Δ, and so also $r[\Delta]$, has $r_1 - (r_0 - 1)$ elements —hence $G\{x\}$ is of rank $r_1 - r_0 + 1$.

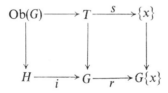

To complete the proof we must show that $G = T * H$. Let $t : T \to K$, $h : H \to K$ be morphisms which agree on $\text{Ob}(G)$. Let $\Gamma' = \Gamma \cap T$, so that T is free on Γ'. Then t, h restrict to graph morphisms $t' : \Gamma' \to K$, $h' : \Delta \to K$ which (since they agree on objects and $\Gamma' \cap \Delta$ has no elements) together define a graph morphism $f : \Gamma \to K$. Since G is free on Γ, f extends uniquely to a morphism $k : G \to K$. However $k \mid T$ extends t', $k \mid H$ extends h'. Therefore $k \mid T = t$, $k \mid H = h$ and this proves that t and h extend to a morphism $k : G \to H$.

If $\bar{k} : G \to H$ also extends t and h, then $\bar{k} \mid \Gamma = f$ and it follows that $\bar{k} = k$. □

EXERCISES

1. Prove that a groupoid G is free if and only if each component of G is free.
2. Prove that a simply-connected groupoid is free.
3. Prove that the coproduct, and free product of free groupoids is free.
4. Give another proof of 8.2.3 using 8.2.2(*b*), Exercise 3 of 6.7, and Exercise 9 of 8.1.
5. Give another proof of 8.2.2 (Corollary 2) using Exercise 10 of 8.1.

8.3 Quotient groupoids

Let G be a groupoid. A subgroupoid N of G is called *normal* if, for any objects x, y of G and a in $G(x, y)$,

$$aN\{x\}a^{-1} \subset N\{y\}$$

that is,

$$a_*[N\{x\}] \subset N\{y\}.$$

This last condition implies that $(a^{-1})_*[N\{y\}] \subset N\{x\}$ and hence, since $(a^{-1})_* = (a_*)^{-1}$, we will in fact have

$$a_*[N\{x\}] = N\{y\}.$$

EXAMPLE Let $f : G \to H$ be a morphism. Then $\operatorname{Ker} f$, the wide subgroupoid of G whose elements are all a in G such that fa is an identity of H, is a wide normal subgroupoid of G. In fact, it is obvious that $\operatorname{Ker} f$ is wide in G, and normality follows from

$$f(aba^{-1}) = fafbfa^{-1} = fafa^{-1} = 1, \qquad b \in N\{x\}, \quad a \in G(x, y).$$

We note also that if $\operatorname{Ob}(f)$ is injective then $\operatorname{Ker} f$ is totally disconnected.

A morphism $f : G \to H$ is said to *annihilate* a subgraph Γ of G if $f[\Gamma]$ is a discrete subgroupoid of H. Thus $\operatorname{Ker} f$ is the largest subgroupoid annihilated by f.

8.3.1 *Let N be a wide, totally disconnected, normal subgroupoid of G. Then there is a groupoid G/N and a morphism $p : G \to G/N$ such that p annihilates N and is universal for morphisms from G which annihilate N.*

Proof We define $\text{Ob}(G/N) = \text{Ob}(G)$. If $x, y \in \text{Ob}(G)$ we define $G/N(x, y)$ to consist of all cosets

$$aN\{x\}, \quad a \in G(x, y).$$

If $a \in G(x, y)$, $b \in G(y, z)$ then, by normality,

$$\begin{aligned} bN\{y\}aN\{x\} &= baN\{x\}N\{x\} \\ &= baN\{x\}. \end{aligned}$$

Therefore, multiplication of cosets again gives a coset. The associativity of multiplication is obvious. The identity element of $G/N(x, x)$ is the coset $N\{x\}$ and the inverse of $aN\{x\}$ is $a^{-1}N\{y\}(a \in G(x, y))$. So G/N is a groupoid.

The morphism $p : G \to G/N$ is the identity on objects, and on elements is defined by $a \rightsquigarrow aN\{x\}$—clearly p is a morphism and $\text{Ker}\, p = N$.

The universal property of p is that if $f : G \to H$ is any morphism which annihilates N, then there is a unique morphism $f^* : G/N \to H$ such that $f^*p = f$. Now the cosets of N are exactly the equivalence classes of the elements of G under the relation $a \sim b \Leftrightarrow ab^{-1}$ is defined and belongs to N; so $a \sim b \Leftrightarrow pa = pb$. The universal property follows easily from 4.6 of the Appendix. □

Remark 8.3.1 is true on the assumption only that N is normal [cf. Exercise 2]; but we shall have no need of this more complicated construction of G/N. We call G/N a *quotient groupoid* of G.

The usual homomorphism theorem for groups (that if $f : G \to H$ is a morphism then $\text{Im}\, f$ is isomorphic to $G/\text{Ker}\, f$) is false for groupoids, one reason being that $\text{Im}\, f$ need not be a subgroupoid of H; for example, the universal morphism $f : I \to \mathbf{Z}$ has $I/\text{Ker}\, f$ isomorphic to I. However we do have

8.3.2 *Let $f : G \to H$ be a morphism such that* $\text{Ob}(f)$ *is injective. Then* $\text{Im}\, f$ *is a subgroupoid of H and the canonical morphism*

$$G/\text{Ker}\, f \to \text{Im}\, f$$

is an isomorphism.

Proof To prove that $\text{Im}\, f$ is a subgroupoid of H, it is sufficient to prove that if $c, d \in \text{Im}\, f$ and $d^{-1}c$ is defined in H, then $d^{-1}c \in \text{Im}\, f$.

Suppose $c = fa$, $d = fb$ where $a \in G(x, y)$, $b \in G(z, w)$. Since $d^{-1}c$ is defined, $fy = fw$, which implies (since $\text{Ob}(f)$ is injective) that $y = w$. Hence, $b^{-1}a$ is defined and $d^{-1}c = f(b^{-1}a)$ which belongs to $\text{Im}\, f$.

Since f annihilates $\text{Ker}\, f$, which is a wide, totally disconnected and normal subgroupoid of G, there is a canonical morphism $f' : G/\text{Ker}\, f \to H$ such that $f'p = f$. Since f' is defined by $f'(a\, \text{Ker}\, f) = fa$, it is clear that

$\operatorname{Im} f' = \operatorname{Im} f$. Let $f'' : G/\operatorname{Ker} f \to \operatorname{Im} f$ be the restriction of f'. Then $\operatorname{Ob}(f'')$ is bijective, and for each $a, b \in G$

$$\begin{aligned} f''(a \operatorname{Ker} f) = f''(b \operatorname{Ker} f) &\Leftrightarrow fa = fb \\ &\Leftrightarrow a \operatorname{Ker} f = b \operatorname{Ker} f. \end{aligned}$$

It follows from 6.4.3 that f'' is an isomorphism. □

We now consider relations in a groupoid. Suppose given for each object x of the groupoid G a set $R\{x\}$ of elements of $G\{x\}$—thus R can be regarded as a wide, totally disconnected subgraph of G. The *normal closure* $N(R)$ of R is the smallest wide normal subgroupoid of G which contains R. This obviously exists since the intersection of any family of normal subgroupoids of G is again a normal subgroupoid of G. Further, $N(R)$ is totally disconnected since the family of object groups of any normal subgroupoid N of G is again a normal subgroupoid of G.

Alternatively, $N = N(R)$ can be constructed explicitly. Let x be an object of G. By a *consequence* of R at x is meant either the identity of G at x, or any product

$$\rho = a_n^{-1}\rho_n a_n \ldots a_1^{-1}\rho_1 a_1 \qquad (*)$$

for which $a_i \in G(x, x_i)$ and ρ_i, or ρ_i^{-1}, is an element of $R\{x_i\}$. Clearly, $N\{x\}$, the set of consequences of R at x, is a subgroup of $G\{x\}$ and the family N of these groups is a wide totally disconnected subgroupoid of G containing R. Also N is normal, since if $a \in G(y, x)$ then

$$a^{-1}\rho a = (a_n a)^{-1}\rho_n(a_n a)\ldots(a_1 a)^{-1}\rho_1 a_1 a$$

is an element of $N\{y\}$. On the other hand, any normal, wide subgroupoid of G which contains R must clearly contain all products such as (*) and so must contain N. Hence $N = N(R)$.

The projection $p : G \to G/N(R)$ clearly has the universal property: *if $f : G \to H$ is any morphism which annihilates R then there is a unique morphism $f' : G/N(R) \to H$ such that $f'p = f$.* We call $G/N(R)$ the groupoid G *with the relations* $\rho = 1, \rho \in R$.

In applications, we are often given G, R as above and wish to describe the object groups of $G/N(R)$. These are determined by the following result.

8.3.3 *Let G be connected, let $x \in \operatorname{Ob}(G)$ and let $r : G \to G\{x\}$ be a deformation retraction. Let $H = G/N(R)$. Then $H\{x\}$ is isomorphic to the group $G\{x\}$ with the relations*

$$r(\rho) = 1, \quad \rho \in R.$$

Proof The deformation retraction r and the morphism $p : G \to H$ determine as in 6.7.3 a deformation retraction $s : H \to H\{x\}$ such that the following square is a pushout.

$$\begin{array}{ccc} G & \xrightarrow{r} & G\{x\} \\ {\scriptstyle p}\downarrow & & \downarrow{\scriptstyle p'} \\ H & \xrightarrow[s]{} & H\{x\} \end{array} \qquad (*)$$

where p' is the restriction of p. We verify that p' satisfies the required universal property.

Let $f : G\{x\} \to K$ be a morphism such that f annihilates $r[R]$, i.e. fr annihilates R. Then there is a unique morphism $g : H \to K$ such that $gp = fr$. Since $(*)$ is a pushout, there is a unique morphism $f' : H\{x\} \to K$ such that $f'p' = f, f's = g$. To complete the proof we must show that the last condition is redundant as far as the uniqueness of f' is concerned.

Let $f'' : H\{x\} \to K$ be any morphism such that $f''p' = f$. Then $f''sp = f''p'r = fr = gp$. But p is surjective, so that $f''s = g$. The pushout property of $(*)$ now implies that $f'' = f'$. □

EXERCISES

1. Prove that there are morphisms $f : I \to \mathbf{Z}$, $g : I \to \mathbf{Z}_2$ such that $\operatorname{Ker} f = \operatorname{Ker} g$ and $\operatorname{Im} f$ is not (graph) isomorphic to $\operatorname{Im} g$.
2. Let N be any normal subgroupoid of the groupoid G. Prove that there is a groupoid G/N and morphism $p : G \to G/N$ such that p annihilates N and is universal for this property. [The elements of G/N are the equivalence classes of elements of G under the relation $a \sim b$ if and only if $a = xby$ for some x, y in N.]
3. Prove that if $f : G \to H$ is a morphism with kernel N, then $f = pf'$ where $p : G \to G/N$ is the projection and $f' : G/N \to H$ has discrete kernel. If f' is an isomorphism onto $\operatorname{Im} f$, then we shall call f a *projection*.
4. Prove that any morphism f can be factored as $f = f_2 f_1$ where $\operatorname{Ob}(f_1)$ is the identity and f_2 is faithful [for faithful, cf. Exercise 14 of 6.5].
5. Prove that if $f : G \to H$ is faithful, then $f = gr$ where r is a deformation retraction and $\operatorname{Ker} g$ is discrete.
6. Prove that any deformation retraction is a projection.
7. Prove that any $f : G \to H$ can be factored as $f = gr$ where r is a deformation retraction and $\operatorname{Ker} g$ is totally disconnected.
8. Prove that any projection is a deformation retraction followed by projection g with $\operatorname{Ob}(g) = 1$.
9. Prove that the category of groupoids admits difference cokernels, and deduce that any morphisms $G_0 \to G_1$, $G_0 \to G_2$ have a pushout.

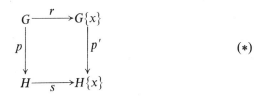

8.4 The van Kampen theorem

Suppose given an adjunction space $W\ {}_f\sqcup Z$ as in the pushout square

$$\begin{array}{ccc} Y & \xrightarrow{f} & W \\ {\scriptstyle i}\downarrow & & \downarrow{\scriptstyle \bar{i}} \\ Z & \xrightarrow[\bar{f}]{} & W\ {}_f\sqcup Z \end{array} \tag{8.4.1}$$

Our object is to determine the groupoid

$$\pi(W\ {}_f\sqcup Z)B$$

for certain (useful) B.

We saw in chapter 7 that the homotopy type of $W\ {}_f\sqcup Z$ depends only on the homotopy types of W and (Z, Y), if Y has reasonable local properties in Z, for example, in the cases studied there, if (Z, Y) has the HEP. To determine $\pi(W\ {}_f\sqcup Z)$ we also need some local conditions—these conditions are essentially in dimensions 0 and 1, and are described in terms of the natural map

$$p : M(f) \cup Z \to W\ {}_f\sqcup Z$$

and its induced morphism of fundamental groupoids.

Suppose that C is a subset of Z representative in Z, that D is a subset of W representative in W, and that $f[C] \subset D$. Let

$$g = f \mid C \cap Y, D, \qquad B = D\ {}_g\sqcup C.$$

Under these conditions we have

8.4.2 (*The van Kampen theorem*) *The following square*

$$\begin{array}{ccc} \pi YC & \xrightarrow{f} & \pi WD \\ {\scriptstyle i}\downarrow & & \downarrow{\scriptstyle \bar{i}} \\ \pi ZC & \xrightarrow[\bar{f}]{} & \pi(W\ {}_f\sqcup Z)B \end{array} \tag{8.4.3}$$

is a pushout if and only if the morphism

$$p : \pi(M(f) \cup Z)A \to \pi(W\ {}_f\sqcup Z)B$$

in which $A = D \cup C$, *is a homotopy equivalence. In particular*, (8.4.3) *is a pushout if* (Z, Y) *has the HEP.*

Proof Let $X = M(f) \cup Z$, $X_2 = X \setminus W$, $X_1 = M(f)$, $X_0 = X_1 \cap X_2$.

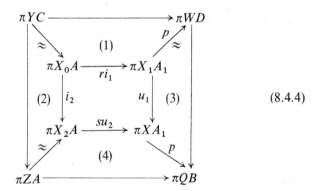

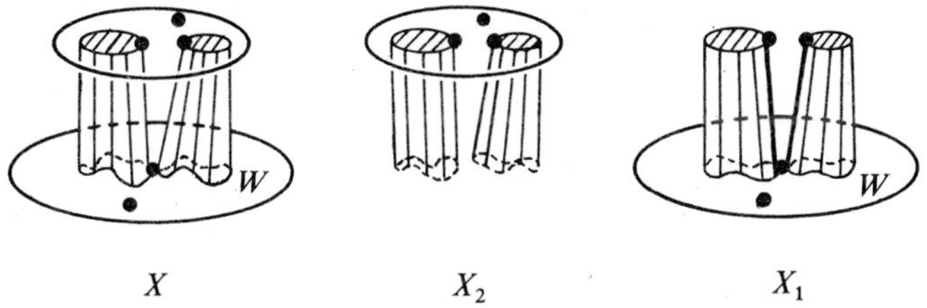

Fig. 8.3

The interiors of X_1, X_2 cover X and so we are in a position to apply 6.7.5 to give us a pushout isomorphic to (8.4.3).

Let $A' = D$, so that in the notation of 6.7.5, $A_1 = D \cup (C \setminus Y)$. Then $p : M(f) \cup Z \to W\,{}_f\!\sqcup Z$ maps A_1 bijectively onto B. Further, A', and hence also A_1, is representative in X_1. Indeed, A_1 meets each path component of W because D is representative in W. Also $C \cap Y$ is representative in Y and each point c of $C \cap Y$ can be joined by the path down the mapping cylinder to fc, which belongs to D; this path is shown as a thick line in X_1 in Fig. 8.3. Notice also that if θc is the class in $\pi X_1 A$ of this path, then $p(\theta c)$ is the identity at fc in πW.

Consider the following diagram in which (i) $Q = W\,{}_f\!\sqcup Z$, (ii) the inner square is the pushout determined by A_1 and the above elements θc as in 6.7.4, (iii) the outer square is (8.4.3).

$$\begin{array}{ccccccc}
\pi YC & & & \longrightarrow & & & \pi WD \\
\downarrow & \searrow \approx & & (1) & & \nearrow p \ \approx & \downarrow \\
 & & \pi X_0 A & \xrightarrow{ri_1} & \pi X_1 A_1 & & \\
 & (2) & \downarrow i_2 & & u_1 \downarrow & (3) & \\
 & & \pi X_2 A & \xrightarrow{su_2} & \pi X A_1 & & \\
\downarrow & \nearrow \approx & & (4) & & \searrow p & \downarrow \\
\pi ZA & & & \longrightarrow & & & \pi QB
\end{array} \qquad (8.4.4)$$

The cell (2) is induced by inclusions and so is commutative. The cell (3) is induced by p and its restrictions, so (3) is commutative. The commutativity of (1) and (4) is a consequence of $p(\theta c) = 1$ $(c \in C \cap Y)$. Thus (8.4.4) is commutative.

Each morphism marked $\approx$ is induced by a homotopy equivalence and is bijective on objects. Therefore these morphisms are isomorphisms. Hence, each of the following statements is equivalent to its successor : (*a*) (8.4.3) is a pushout, (*b*) (8.4.4) is an isomorphism of its inner square to its outer square, (*c*) $p : \pi X A_1 \to \pi Q B$ is an isomorphism.

However, the last morphism is bijective on objects so (*c*) is equivalent to (*d*) $p : \pi X A_1 \to \pi Q B$ is a homotopy equivalence. Since $\pi X A_1$ is a deformation retract of $\pi X A$, (*d*) itself is equivalent to (*e*) $p : \pi X A \to \pi Q B$ is a homotopy equivalence.

This proves the main part of 8.4.2. The last statement of 8.4.2 follows from 7.5.4. □

8.4.2 (*Corollary 1*) *Let W, Z be closed in $W \cup Z$, let $(Z, W \cap Z)$ have the HEP and let B be a set representative in $W \cap Z$, W, Z. Then the square of morphisms induced by inclusions*

$$\begin{array}{ccc} \pi(W \cap Z)B & \longrightarrow & \pi W B \\ \downarrow & & \downarrow \\ \pi Z B & \longrightarrow & \pi(W \cup Z)B \end{array}$$

is a pushout.

Proof This is a consequence of 8.4.2 with $f : Y \to W$ the inclusion. □

This result is, of course, similar to 6.7.2, and is in many cases more convenient to use than the earlier result.

8.4.2 (*Corollary 2*) *Suppose the assumptions of* 8.4.2 (*Corollary 1*) *hold and also $\pi(W \cap Z)B$ is discrete. Then $\pi(W \cup Z)B$ is isomorphic to the free product of groupoids*

$$\pi Z B * \pi W B.$$

Proof From the pushout square of 8.4.2 (Corollary 1) it is easy to deduce that the morphism $\pi Z B \sqcup \pi W B \to \pi(W \cup Z)B$ determined by the two inclusions of Z, W into $W \cup Z$ is universal. □

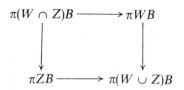

Remark Even this corollary is false without some local assumptions on Y in Z (or in W). For example, let H be the subspace of $\mathbf{R}^2$ which is the union of all circles centre $(1/n, 0)$ and radius $1/n$ for n a positive integer—this space has been called the 'Hawaiian earring'. Let $0 = (0, 0)$ be the base point of H. The space CH is contractible and so the group $\pi(CH, 0)$ is trivial. However, H. B. Griffiths [1, 2] has shown that $\pi(CH \vee CH, 0)$ is non-trivial and in fact can be generated only by an uncountable number of elements. Again, the fundamental group of $H \vee H$ is not the free product $\pi(H, 0) * \pi(H, 0)$. However the space X of Fig. 8.4 formed by joining two Hawaiian ear-rings together does have its fundamental group isomorphic

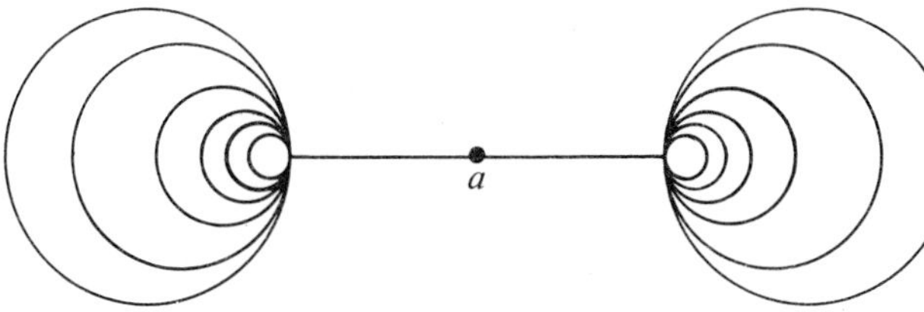

Fig. 8.4

to $\pi(H, 0) * \pi(H, 0)$—this is easy to prove from 8.4.2 (Corollary 2) by taking W, Z to be the left- and right-hand halves of X meeting in $\{a\}$. In this case, the obvious map $X \to H \vee H$ induces a morphism $\pi(X, a) \to \pi(H \vee H, 0)$ which is injective but not surjective (the proof of this statement is not easy—cf. H. B. Griffiths loc. cit.).

Suppose now that we are in the situation of 8.4.2, that (Z, Y) has the HEP, that $C \subset Y$ and $B = D = f[C]$.

8.4.2 (*Corollary 3*) *If further πYC, πWD are discrete then*

$$\bar{f} : \pi ZC \to \pi(W\ {}_f\!\sqcup Z)D$$

is a universal morphism.

Proof This follows from 8.4.2 and the definition of universal morphism. □

We can derive a number of useful results from this. For example, if D consists of a single point d (and the other assumptions of 8.4.2 (Corollary 3) hold) then the fundamental group $\pi(W\ {}_f\!\sqcup Z, d)$ is isomorphic to $U(\pi ZC)$, the universal group of πZC. In particular, if Z is path-connected and $c_0 \in C$, then

$$\pi(W\ {}_f\!\sqcup Z, d) \cong \pi(Z, c_0) * F$$

where F is a free group with one generator for each element of C other than c_0.

We now derive the fundamental group of a cell-complex, first dealing with the 1-dimensional case.

8.4.5 *If K is a connected cell-complex and $v \in K^0$, then the groupoid $\pi K^1 K^0$ is a free groupoid and the fundamental group $\pi(K^1, v)$ is a free group on $r_1 - r_0 + 1$ generators where r_n is the number of n-cells of K, $n = 0, 1$.*

Proof K^1 is obtained by adjoining 1-cells to K^0, that is,

$$K^1 = K^0 {}_f\sqcup (\Lambda \times \mathbf{E}^1)$$

where Λ is a discrete set and $f : \Lambda \times \mathbf{S}^0 \to K^0$ is the attaching map. Let $C = \Lambda \times \mathbf{S}^0$; since K^1 is connected, $f[C] = K^0$. Since $\pi(\Lambda \times \mathbf{S}^0)C$ and $\pi K^0 K^0$ are discrete groupoids (and also $(\Lambda \times \mathbf{E}^1, \Lambda \times \mathbf{S}^0)$ has the HEP) the morphism

$$\bar{f} : \pi(\Lambda \times \mathbf{E}^1)C \to \pi K^1 K^0$$

is universal. But $\pi(\Lambda \times \mathbf{E}^1)C$ is isomorphic to $\Lambda \times I$. So the result follows from 8.2. □

Notice that 8.4.5 also gives the generators of $\pi K^1 K^0$ as follows. For each λ in Λ let ι_λ denote the unique road in $\pi(\Lambda \times \mathbf{E}^1)C$ from $(\lambda, -1)$ to $(\lambda, +1)$. Then the generators of $\pi K^1 K^0$ are the elements $\bar{f}(\iota_\lambda)$, λ in Λ; thus if v_0, v_1 are vertices of K joined by a 1-cell, then the road in $\pi K(v_0, v_1)$ determined by the characteristic map of this 1-cell is one of these generators.

8.4.5 (*Corollary 1*) *The fundamental group of the circle, $\pi(\mathbf{S}^1, 1)$ is isomorphic to $\mathbf{Z}$, with generator the class of the path.*

$$\mathbf{I} \to \mathbf{S}^1$$

$$t \rightsquigarrow e^{2\pi i t}.$$

Proof This is immediate from 8.4.3 and the previous remark, since the given path is a characteristic map for the 1 cell of $\mathbf{S}^1$. □

We now show that $\pi(K^2, v)$ is isomorphic to $\pi(K^1, v)$ with relations for each 2-cell. Let us suppose

$$K^2 = K^1 {}_g\sqcup (M \times \mathbf{E}^2)$$

where $g : M \times \mathbf{S}^1 \to K^1$. Suppose also that K^1 is connected. For each m in M, let $v_m = g(m, e)$ where $e = (1, 0)$ and let

$$\rho_m = g(\iota_m) \in \pi(K^1, v_m)$$

where ι_m is a generator of the fundamental group of $M \times \mathbf{S}^1$ at (m, e). Let v be an element of K^1 and let a_m be an assigned element of $\pi K^1(v, v_m)$ (with $a_m = 1$ if perchance $v_m = v$).

8.4.6 *The fundamental group $\pi(K^2, v)$ is isomorphic to the free group $\pi(K^1, v)$ with the relations*

$$a_m^{-1} \rho_m a_m = 1, \quad m \in M.$$

Proof We first show that if $V = \{v_m : m \in M\} \cup \{v\}$ then $\pi K^2 V$ is the groupoid $\pi K^1 V$ with the relations $\rho_m = 1, m \in M$. Let $C = M \times \{e\}$.

We have a pushout square

$$\begin{array}{ccc} \pi(M \times \mathbf{S}^1)C & \xrightarrow{g} & \pi K^1 V \\ {\scriptstyle i}\downarrow & & \downarrow{\scriptstyle \bar{i}} \\ \pi(M \times \mathbf{E}^2)C & \xrightarrow[\bar{g}]{} & \pi K^2 V \end{array}$$

Suppose $f : \pi K^1 V \to F$ is any morphism such that $f\rho_m = 1$, $m \in M$. Then Im (fg) is discrete. Since $\pi(M \times \mathbf{E}^2)C$ is a discrete groupoid on C, f defines a morphism $\bar{f} : \pi(M \times \mathbf{E}^2)C \to F$ such that $\bar{f}i = fg$. So there is a unique morphism $f' : \pi K^2 V \to F$ such that $f'\bar{i} = f, f'\bar{g} = \bar{f}$. The last condition is redundant, since $\pi(M \times \mathbf{E}^2)C$ is discrete and so $\bar{g}$ is determined by $\mathrm{Ob}(\bar{g}) = \mathrm{Ob}(g) : C \to V$, a surjective function.

This proves that $\pi K^2 V$ is $\pi K^1 V$ with relations $\rho_m = 1$, $m \in M$. The conclusion of 8.4.4 follows from 8.3.3. □

8.4.7 *If K is a cell complex and A a subset of K^2, then the inclusion $\pi K^2 A \to \pi KA$ is an isomorphism.*

Proof We first prove that $\mathbf{S}^n$ is simply-connected for $n > 1$. Let e be a point of $\mathbf{S}^{n-1} = E^n_+ \cap E^n_-$. By 8.4.2 (Corollary 1) we have a pushout square

$$\begin{array}{ccc} \pi(\mathbf{S}^{n-1}, e) & \longrightarrow & \pi(E^n_+, e) \\ \downarrow & & \downarrow \\ \pi(E^n_-, e) & \longrightarrow & \pi(\mathbf{S}^n, e) \end{array}$$

But E^n_+, E^n_- are homeomorphic to $\mathbf{E}^n$ and so are simply-connected. Hence $\pi(E^n_+, e)$, $\pi(E^n_-, e)$ are trivial groups and therefore $\pi(\mathbf{S}^n, e)$ is trivial.

Now consider any adjunction space $W \,{}_f\!\sqcup \mathbf{E}^{n+1}$ where $f : \mathbf{S}^n \to W$ and $n > 1$. Let $e \in \mathbf{S}^n$ and suppose W is path-connected. There is a pushout square.

$$\begin{array}{ccc} \pi(\mathbf{S}^n, e) & \xrightarrow{f} & \pi(W, fe) \\ {\scriptstyle i}\downarrow & & \downarrow{\scriptstyle \bar{i}} \\ \pi(\mathbf{E}^{n+1}, e) & \xrightarrow[\bar{f}]{} & \pi(W \,{}_f\!\sqcup \mathbf{E}^{n+1}, fe) \end{array}$$

Since the two left-hand groups are trivial, $\bar{i}$ is an isomorphism. □

In order to compute the fundamental group of spaces, it is clearly necessary to compute maps $\pi(\mathbf{S}^1, e) \to \pi(K^1, fe)$. The following result is crucial.

8.4.8 *Let $f : \mathbf{S}^1 \to \mathbf{S}^1$ be the map $z \rightsquigarrow z^n$, n an integer. Then the induced morphism $f : \pi(\mathbf{S}^1, e) \to \pi(\mathbf{S}^1, e)$ of additive groups is multiplication by n.*

Proof The result is clearly true if $n = 0$, since f is then constant, or if $n = 1$, since f is then the identity. Suppose $n > 1$; let w be the complex number $e^{2\pi i/n}$ and let $w^r = e^{2\pi ir/n}$, $r = 0, 1, \ldots, n - 1$. Let X_r be the subset of $\mathbf{S}^1$ of points $e^{2\pi i\theta}$, $r/n \leqslant \theta \leqslant (r + 1)/n$, let C_r consist solely of w^r, w^{r+1} and let $C = \{w^r : 0 \leqslant r < n\}$. Since X_r is simply-connected there is a unique element ι_r in $\pi X_r(w^r, w^{r+1})$. The morphism $\pi X_r C_r \to \pi \mathbf{S}^1 C$ induced by inclusion is injective and so we regard $\pi X_r C_r$ as a subgroupoid of $\pi \mathbf{S}^1 C$. A generator a of $\pi(S^1, e)$, where $e = w^0$, is then given by

$$a = \iota_{n-1} + \cdots + \iota_0.$$

The map $f : \mathbf{S}^1 \to \mathbf{S}^1$ determines by restriction $f' : X_r \to \mathbf{S}^1$; clearly $f'\iota_r = a$. Therefore,

$$fa = f(\iota_{n-1} + \cdots + \iota_0) = a + \cdots + a = na.$$

If $n < 0$, let $m = -n$. Then $z \rightsquigarrow z^n$ is the composite of $g : z \rightsquigarrow z^{-1}$ and $z \rightsquigarrow z^m$. But if $b : \mathbf{I} \to \mathbf{S}^1$ is the path $t \rightsquigarrow e^{2\pi it}$, then gb is $t \rightsquigarrow e^{-2\pi it}$, that is, $gb = -b$. Hence, in $\pi(S^1, e)$, $ga = -fa$; therefore $fa = -ma = na$. □

If $f : \mathbf{S}^1 \to \mathbf{S}^1$ is a map such that $f : \pi(\mathbf{S}^1, 1) \to \pi(\mathbf{S}^1, 1)$ is multiplication by n, then we say f is of *degree n*.

Examples 1. Let $K = \mathbf{S}^1 \vee \cdots \vee \mathbf{S}^1$ be a wedge of n circles, with the cell structure $e^0 \cup e_1^1 \cup \cdots \cup e_n^1$. Let v be the vertex of K. Then $\pi(K, v)$ is a free group on n-generators, the generators being the classes of the loops which pass once round one of the circles.

2. The fundamental group of the real projective plane $P^2(\mathbf{R})$ and real projective n-space $P^n(\mathbf{R})$ $(n > 1)$ are the same, by 8.4.7 and the fact that $P^2(\mathbf{R})$ can be identified with the 2-skeleton of $P^n(\mathbf{R})$. Also, $P^2(\mathbf{R}) = \mathbf{S}^1 {}_f\sqcup \mathbf{E}^2$ where $f : \mathbf{S}^1 \to \mathbf{S}^1$ is of degree 2 [5.3]. It follows that the fundamental group of $P^2(\mathbf{R})$ is the group $\mathbf{Z}/2\mathbf{Z} = \mathbf{Z}_2$.

3. We can also state that the fundamental groups of $\mathbf{S}^2$ and $\mathbf{S}^1 \times \mathbf{S}^1$ are 0 and $\mathbf{Z} \times \mathbf{Z}$ respectively. It follows that no two of the spaces $\mathbf{S}^2$, $\mathbf{S}^1 \times \mathbf{S}^1$, $P^n(\mathbf{R})$ are of the same homotopy type; *a fortiori*, no two of these spaces are homeomorphic.

4. The Klein bottle has a cell structure $K = e^0 \cup e_1^1 \cup e_2^1 \cup e^2$. From Fig. 4.4, p. 90 it is clear that, if $\{a, b\}$ is a set of generators of $\pi(K^1, v)$ as

given in 8.4.5, then the relation determined by the 2-cell of K is $abab^{-1}$. Thus $\pi(K, v)$ is a free group on two generators a, b with the relation $abab^{-1} = 1$.

EXERCISES

1. Prove that 8.4.7 is a consequence of the cellular approximation theorem.
2. Prove that the spaces $P^n(\mathbf{H})$ are simply connected.
3. Prove that $\mathbf{S}^1$ is not a retract of $P^n(\mathbf{R})$, $n > 1$.
4. Let $X = Y \cup Z$ where Z, Y are path-connected and (X, Y) has the HEP. Let $a_0, a_1, \ldots, a_n$ be points one in each path-component of $Y \cap Z$. Let i, j be the inclusions of $Y \cap Z$ into Y, Z respectively. Let $\alpha_r \in \pi Y(a_0, a_r)$, $\beta_r \in \pi Z(a_0, a_r)$, $r = 0, \ldots, n$, with $\alpha_0 = 1$, $\beta_0 = 1$. Let F be a free group on elements γ_r, $r = 0, \ldots, n$ with the relation $\gamma_0 = 1$. Prove that $\pi(X, a_0)$ is isomorphic to the free product of the groups $\pi(Y, a_0)$, $\pi(Z, a_0)$ and F with the relations

$$\alpha_r^{-1}(i\rho_r)\alpha_r = \gamma_r(\beta_r^{-1}(j\rho_r)\beta_r)\gamma_r^{-1}$$

for all $\rho_r \in \pi(Y \cap Z, a_r)$ and $r = 0, \ldots, n$. [Here γ_r corresponds to the element $(u\beta_r^{-1})(v\alpha_r)$ of $\pi(X, a_0)$ where u, v are the inclusions of Y, Z respectively into X.]
5. Let K, L be 1-dimensional (finite) cell-complexes. Prove that if $\varphi : \pi KK^0 \to \pi LL^0$ is any morphism, then there is a map $f : K \to L$ such that $\pi f = \varphi$. Prove also that if $f, g : K \to L$ are cellular maps such that $\pi f \simeq \pi g : \pi KK^0 \to \pi LL^0$, then f is homotopic to g.
6. Extend the results 8.4.5, 8.4.6, 8.4.7 to (infinite) CW-complexes. Prove that if G is any group, then there is a CW-complex K and vertex x of K such that $\pi(K, x)$ is isomorphic to G. Deduce that if G is any groupoid then there is a CW-complex K such that πKK^0 is isomorphic to G. [You may assume that if G is any group then there is a free group F and free subgroup R of F such that G is isomorphic to F/R.]
7. Prove that $\mathbf{R}^2$ and $\mathbf{R}^n$ for $n > 2$ are not homeomorphic.
8. Let $p : z \rightsquigarrow z^n + a_{n-1}z^{n-1} + \cdots + a_1 z + a_0$ and $q : z \rightsquigarrow z^n$ be polynomials with $a_i \in \mathbf{C}$. For $r > 0$ let $C_r = \{z \in \mathbf{R}^2 : |z| = r\}$. Prove that for r large enough, p and q restrict to homotopic maps $C_r \to \mathbf{R}^2 \setminus \{0\}$. Prove that for any $r > 0$ this restriction of q is essential and hence show that the polynomial p has a root.

NOTES

Most of the results on groupoids are due to Higgins [1, 2]. The proof of associativity of the multiplication of the words of $U_\sigma(G)$ is borrowed from the treatment of free products of groups in Lang [1]. The construction of arbitrary direct limits of categories and of groupoids is dealt with briefly, and with a different emphasis, in Gabriel–Zisman [1]. For applications of the fundamental group to knot theory, see Crowell–Fox [1], Neuwirth [1], Fox [1].

9. Covering spaces

The notion of a covering space is of interest for several reasons. First, the construction of a covering space $\tilde{X}$ from a given space X (which we give in 9.5) is useful since the structure of $\tilde{X}$ is in a sense simpler than that of X ($\tilde{X}$ has a smaller fundamental group than X). Second, the method of studying a given space X by means of particular kinds of maps into X is of wide importance in topology—the covering maps onto X exemplify this, and indeed form a particular example of the fibre maps whose ramifications extend through topology, differential geometry, analysis, and even algebra. Third, and most important from the point of view of this book, the topology is here very nicely modelled in the algebra of groupoids, so that with this chapter we have completed also the elements of the algebra of groupoids. This algebra is of recent origin and has been developed by P. J. Higgins [1, 2], although the notion of a groupoid itself goes back to a paper of Brandt in 1926. The algebra given here is likely to prove important and suggestive.

We shall use additive notation for groupoids in this chapter.

9.1 Covering maps and covering homotopies

Throughout this chapter all spaces will be assumed *locally path-connected.* This assumption is not essential for all the results (for an alternative approach see Hilton–Wylie [1] or Spanier [3]) but it does seem to lead to a smoother theory; also the study of covering maps for spaces which are not locally path-connected is something of a curiosity.

Definitions Let $p : \tilde{X} \to X$ be a map of topological spaces. A subset U of X is called *canonical* (with respect to p) if U is open, path-connected, and each path-component of $p^{-1}[U]$ is open in $\tilde{X}$ and is mapped by p homeomorphically onto U; each path-component of $p^{-1}[U]$ is also called *canonical.* The map p is called a *covering map* if each x in X has a canonical neighbourhood; and in such case $\tilde{X}$ is called a *covering space* of X. The covering map p is called *connected* if both $\tilde{X}$ and X are path-connected.

The local picture of a covering map is thus that of Fig. 9.1(a).

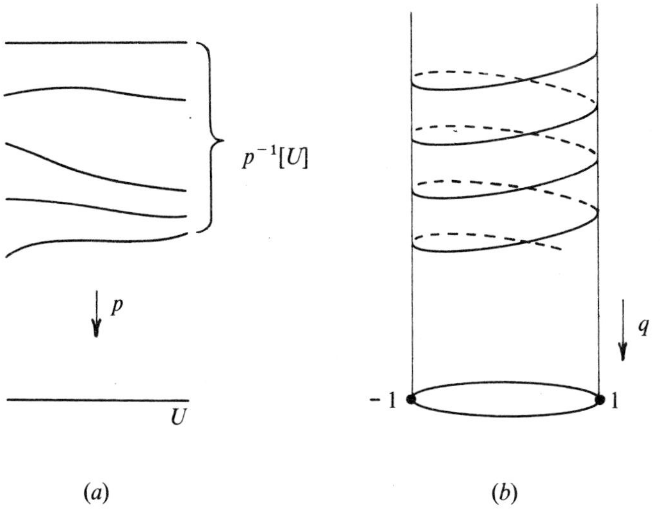

Fig. 9.1

EXAMPLES 1. Consider the map $p : \mathbf{R} \to \mathbf{S}^1$, $t \leadsto e^{2\pi i t}$. The sets

$$U_1 = \mathbf{S}^1 \setminus \{1\}, \qquad U_{-1} = \mathbf{S}^1 \setminus \{-1\}$$

form an open cover of $\mathbf{S}^1$. The path-components of $p^{-1}[U_1]$ (resp. $p^{-1}[U_{-1}]$) are the open intervals $]n, n + 1[$ (resp. $]n - \frac{1}{2}, n + \frac{1}{2}[$) for all $n \in \mathbf{Z}$. So p is a covering map. A good picture of p is obtained by writing $p = qr$ where $r : \mathbf{R} \to \mathbf{R}^2 \times \mathbf{R}$ is $t \leadsto (pt, t)$ and $q : \mathbf{R}^2 \times \mathbf{R} \to \mathbf{R}^2$ is the projection. Thus the image of r is the helix of Fig. 9.1(b).

2. Let $f : \mathbf{S}^1 \to \mathbf{S}^1$ be the map $z \leadsto z^n$, where n is a non-zero integer. It is easily proved by using the sets U_1, U_{-1} of Example 1 that f is a covering map.

3. If $p : \tilde{X} \to X$ is a covering map, then for each x in X the subspace $p^{-1}[x]$ of $\tilde{X}$ is a discrete space—this follows from the fact that if U is a canonical set then the path-components of $p^{-1}[U]$ are open in $\tilde{X}$. It follows that the fundamental map $p : \mathbf{R}^{n+1}_* \to P^n(\mathbf{R})$ is not a covering map.

4. However, let $i : \mathbf{S}^n \to \mathbf{R}^{n+1}_*$ be the inclusion so that $h = pi$ is the Hopf map. Let U_i be the subset of $P^n(\mathbf{R})$ of points px such that the ith-coordinate of x is non-zero. The sets U_i, $i = 1, \ldots, n + 1$ form an open-cover of $P^n(\mathbf{R})$. The path-components of $h^{-1}[U_i]$ are two open hemispheres of $\mathbf{S}^n$

each of which is mapped homeomorphically by h onto U_i (for example, $h^{-1}[U_{n+1}] = \mathbf{S}^n \setminus \mathbf{S}^{n-1}$). Therefore, h is a covering map.

Let $p : \tilde{X} \to X$ be a covering map. If the diagram of maps

$$\begin{array}{ccc} & & \tilde{X} \\ & \overset{f}{\nearrow} & \big\downarrow p \\ Y & \xrightarrow[h]{} & X \end{array}$$

is commutative, then we say f is a *lifting* of h, or that *f covers h*. We shall see that the central properties of covering maps are concerned with liftings; for example, we have the following *covering homotopy property*.

9.1.1 *Let $p : \tilde{X} \to X$ be a covering map and suppose given a commutative diagram of maps (in which i is $z \rightsquigarrow (z\ 0)$)*

$$\begin{array}{ccc} Z & \xrightarrow{f} & \tilde{X} \\ {\scriptstyle i}\big\downarrow & & \big\downarrow{\scriptstyle p} \\ Z \times [0, r] & \xrightarrow[H]{} & X \end{array}$$

so that H is a homotopy of pf. Then H is covered by a unique homotopy $F : Z \times [0, r] \to \tilde{X}$ of f.

Proof Since Z is locally path-connected, each path-component of Z is open in Z. Therefore, it is sufficient to work in each path-component of Z at a time, that is, to assume Z is path-connected.

Step 1. Suppose that Im H is contained in a canonical subset U of X. Since Z is path-connected, Im f is contained in a path-component V of $p^{-1}[U]$. Let p_V be the inverse of the homeomorphism $p \mid V, U$. Then the map

$$\begin{aligned} F : Z \times [0, r] &\to \tilde{X} \\ (z, t) &\rightsquigarrow p_V H(z, t) \end{aligned}$$

is a homotopy of f which covers H. Since $Z \times [0, r]$ is path-connected, it is also the only such homotopy of f.

Step 2. Suppose that the homotopy H is a sum

$$H = H_n + \cdots + H_1$$

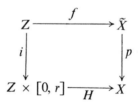

such that the image of each H_i is contained in a canonical subset of X. By Step 1, we can define inductively homotopies $F_i : f_i \simeq f_{i+1}$, $i = 1, \ldots, n$, such that $f_1 = f$ and F_i covers H_i. Clearly,

$$F = F_n + \cdots + F_1$$

is a homotopy of f which covers H. If F' is any other homotopy of f which covers H, then the given subdivision of H determines a subdivision $F' = F'_n + \cdots + F'_1$. It follows by induction that $F'_i = F_i$, $i = 1, \ldots, n$ and so $F' = F$.

Step 3. Let $z \in Z$. We use Step 2 to show that the result is true with Z replaced by some neighbourhood M^z of z.

For each t in $[0, r]$ there are open neighbourhoods M_t, N_t of z, t respectively such that $H[M_t \times N_t]$ is contained in a canonical subset of X. The set $\{N_t : t \in [0, r]\}$ is an open cover of $[0, r]$ which, by compactness, has a finite subcover $\{N_t : t \in A\}$ say. Let

$$M = \bigcap_{t \in A} M_t$$

and let M^z be an open path-connected neighbourhood of z contained in M.

Let l be the Lebesgue number of the cover $\{N_t : t \in A\}$ and let n be an integer such that $0 < r/n < l/2$. For each $i = 1, \ldots, n$ the interval $L_i = [(i-1)r/n, ir/n]$ is contained in some N_t, $t \in A$, and so $H[M^z \times L_i]$ is contained in a canonical subset of X.

Let H_i^z be the homotopy

$$\begin{aligned} M^z \times [0, r/n] &\to X \\ (w, t) &\rightsquigarrow H(w, (i-1)r/n + t) \end{aligned}$$

and let

$$H^z = H \mid M^z \times [0, r] = H_n^z + \cdots + H_1^z.$$

By Step 2, H^z is covered by a unique homotopy F^z of $f \mid M^z$.

Step 4. Let $F : Z \times [0, r] \to \tilde{X}$ be the function $(z, t) \rightsquigarrow F^z(z, t)$. The uniqueness part of the above argument shows that for each $z \in Z$

$$F \mid M^z \times [0, r] = F^z.$$

Since M^z is a neighbourhood of z, it follows that F is continuous. Clearly, F is the unique homotopy of f covering H. □

Our main use of the covering homotopy property is to prove the following *path lifting property*.

9.1.2 *Let $p : \tilde{X} \to X$ be a covering map, let $\tilde{x} \in \tilde{X}$ and let $x = p\tilde{x}$. Then paths a, b in X with initial point x lift uniquely to paths $\tilde{a}$, $\tilde{b}$ with initial point $\tilde{x}$; and a is equivalent to b if and only if $\tilde{a}$ is equivalent to $\tilde{b}$.*

Proof The existence and uniqueness of $\tilde{a}$ (and also $\tilde{b}$) is immediate from 9.1.1 by taking Z to consist of a single point z and defining

$$fz = \tilde{x}, \qquad H(z, t) = at, \qquad t \in [0, r], \quad r = |a|.$$

Suppose next that a, b are paths from x to y of length r which are homotopic rel end points. Then, twisting the usual way of writing a homotopy, there is a map

$$H : \mathbf{I} \times [0, r] \to X$$

such that

$$\begin{array}{lll} H(s, 0) = x, & H(s, r) = y, & s \in \mathbf{I} \\ H(0, t) = at, & H(1, t) = bt, & t \in [0, r]. \end{array}$$

Let $f : \mathbf{I} \to \tilde{X}$ be the constant map with value $\tilde{x}$. Then H is a homotopy of pf and so, by 9.1.1, H is covered by a unique homotopy F of f.

The paths in $\tilde{X}$

$$t \rightsquigarrow F(0, t), \qquad t \rightsquigarrow F(1, t)$$

have initial point $\tilde{x}$ and cover a, b respectively; so these paths are $\tilde{a}$, $\tilde{b}$ respectively. The path $s \rightsquigarrow F(s, r)$ covers the constant path at y and so (by uniqueness) is itself constant; the path $s \rightsquigarrow F(s, 0)$ is constant by definition of F. Therefore, F (suitably twisted) is a homotopy rel end points $\tilde{a} \sim \tilde{b}$.

Finally, if there are real numbers $s, s' \geqslant 0$ such that $s + a$, $s' + b$ are homotopic rel end points, then $s + \tilde{a}$, $s' + \tilde{b}$ cover $s + a$, $s' + b$ respectively, and hence $\tilde{a}$ is equivalent to $\tilde{b}$. □

Remark In fact, we shall use only the consequence 9.1.2 of 9.1.1 and not 9.1.1 itself. A simpler proof of 9.1.1 can be given in the case $Z = [0, q]$ by subdividing the rectangle $[0, q] \times [0, r]$ into rectangles so small that each is mapped by H into a canonical subset of X. But the interest of 9.1.1 is that, provided the word uniqueness is omitted, it applies to many other situations. For example, the fundamental map $\mathbf{K}_*^{n+1} \to P^n(\mathbf{K})$ satisfies this weaker covering homotopy property—this follows from Exercise 8 of 5.3 and results of A. Dold [1] (see also 2.7.12 of Spanier [3]).

EXERCISES

1. Which of the following maps are covering maps?

(i) $\mathbf{R} \to \mathbf{R}$, $x \rightsquigarrow x^2$.
(ii) $\mathbf{R} \to \mathbf{R}$, $x \rightsquigarrow x^3$.
(iii) $\mathbf{R}_* \to \mathbf{R}_*$, $x \rightsquigarrow x^2$.
(iv) $\mathbf{S}^1 \to \mathbf{S}^1$, $e^{i\theta} \rightsquigarrow e^{2i|\theta|}$ $(-\pi < \theta \leqslant \pi)$.
(v) $]0, 3[\to \mathbf{S}^1$, $t \rightsquigarrow e^{2\pi i t}$.

2. Let $f : Y' \to Y$, $g : X \to X'$ be homeomorphisms and let $p : Y \to X$ be a covering map. Prove that $gpf : Y' \to X'$ is a covering map.
3. Let $p : \tilde{X} \to X$ be a covering map, let $f : Y \to \tilde{X}$ be a map, let $A \subset Y$ and let $G : A \times \mathbf{I} \to \tilde{X}$ be a homotopy of $f \mid A$. Prove that, if $H : Y \to X$ is a homotopy of pf which agrees with pG on A, then H is covered by a homotopy F of f such

that F extends G. Deduce that for any map $u : A \to \tilde{X}$, p induces an injection $p_* : Y/\!/u \to Y/\!/pu$ (here $Y/\!/u$ is defined in chapter 7).
4. Prove that 9.1.1 is true without the assumption that Z is locally path-connected.
5. Let $p : \tilde{X} \to X$ be a covering map and let $f : Y \to X$ be a map. Let

$$\tilde{X}_f = \{(y, x) \in Y \times \tilde{X} : fy = p\tilde{x}\}$$

and let $p_f : \tilde{X}_f \to Y$ be the restriction to $\tilde{X}_f$ of the projection $Y \times \tilde{X} \to Y$. Prove that p_f is a covering map. Prove also that f lifts to $\tilde{X}$ if and only if there is a map $s : Y \to \tilde{X}_f$ such that $p_f\, s = 1$.
6. Let $p : \tilde{X} \to X$ be a covering map and let $i : A \to X$ be the inclusion of the subspace A of X. Prove that if $q = p \mid p^{-1}[A], A$ then there is a homeomorphism $g : \tilde{X}_i \to p^{-1}[A]$ such that $qg = p_i$ (where $\tilde{X}_i$, p_i are as in the previous Exercise).
7. Let G be the sheaf of germs of continuous functions from $\mathbf{R}$ to $\mathbf{R}$ [Exercise 3 of 2.8]. Prove that the projection $p : G \to \mathbf{R}, f^x \rightsquigarrow f(x)$, is not a covering map.
8. Give another proof of 9.1.1 on the following lines. First prove it is true when Z consists of a single point. In the general case, we then have that for each z in Z the map $H \mid \{z\} \times [0, r]$ lifts to a homotopy $F \mid \{z\} \times [0, r]$ of $f \mid \{z\}$. So we have a function $F : Z \times [0, r] \to \tilde{X}$. Prove that F is continuous by supposing that F is non-continuous and obtaining a contradiction by considering the infinum of the set of t such that F is not continuous at some (z, t).
9. Let $p : \tilde{X} \to X$ be a connected covering map which is non-trivial (i.e., is not a homeomorphism). Let $Y = X \times X \times X \times \cdots$ be the countable product of X with itself, and let $\tilde{X}^n$ be the n-fold product of $\tilde{X}$ with itself. Let $\tilde{Y}_n = \tilde{X}^n \times Y$. Define $p_n : \tilde{Y}_n \to Y$ by $(\tilde{x}_1, \ldots, \tilde{x}_n, x_1, x_2, \ldots) \rightsquigarrow (p\tilde{x}_1, \ldots, p\tilde{x}_n, x_1, x_2, \ldots)$. Prove that p_n is a covering map. Let $\tilde{Z} = \bigsqcup_{n \geqslant 1} \tilde{Y}_n$ and let Z be the countably infinite topological sum of Y with itself. Let $q = \bigsqcup p_n : \tilde{Z} \to Z$ and let $r : Z \to Y$ be the obvious projection. Prove that q and r are covering maps but that the composite rq is not a covering map.

9.2 Covering groupoids

In this section we shall show how covering spaces are modelled in the category of groupoids.

Let G be a groupoid. For each object x of G the *star* of x in G, denoted by $\mathrm{St}_G\, x$, is the union of the sets $G(x, y)$ for all objects y of G. Thus $\mathrm{St}_G\, x$ consists of all elements of G with initial point x. When no confusion will arise we abbreviate $\mathrm{St}_G\, x$ to $\mathrm{St}\, x$.

Definition Let $p : \tilde{G} \to G$ be a morphism of groupoids. We say p is a *covering morphism* if for each object $\tilde{x}$ of $\tilde{G}$ the restriction of p

$$\mathrm{St}_{\tilde{G}}\, \tilde{x} \to \mathrm{St}_G\, p\tilde{x}$$

is bijective; in such case, we call $\tilde{G}$ a *covering groupoid* of G. The covering morphism p is called *connected* if both $\tilde{G}$ and G are connected.

For any morphism $p : \tilde{G} \to G$ and object $\tilde{x}$ of $\tilde{G}$ we call the subgroup

$p[\tilde{G}\{\tilde{x}\}]$ of $G\{p\tilde{x}\}$ the *characteristic group* of p at $\tilde{x}$—by an abuse of language, we also refer to this group as the characteristic group of $\tilde{G}$, $\tilde{x}$. If p is a covering morphism, then p maps $\tilde{G}\{\tilde{x}\}$ isomorphically onto this characteristic group; also, if $a \in G\{p\tilde{x}\}$, then there is a unique element $\tilde{a}$ of St $\tilde{x}$ such that $p\tilde{a} = a$, but $\tilde{a}$ will be a loop, that is, $\tilde{a}$ will belong to $\tilde{G}\{\tilde{x}\}$, if and only if a itself belongs to the characteristic group of $\tilde{G}$, $\tilde{x}$.

The sense in which a characteristic group of a covering morphism p characterizes p will be discussed later.

EXAMPLES 1. Recall that I is a simply-connected groupoid with two objects 0, 1 and one element ι of $I(0, 1)$. If 0 denotes ambiguously the zero of I at 0 or 1, we have

$$\text{St}_I\, 0 = \{0, \iota\}, \qquad \text{St}_I\, 1 = \{0, -\iota\}.$$

In the group $\mathbf{Z}_2$ (which has one object 0 say)

$$\text{St}_{\mathbf{Z}_2}\, 0 = \{0, 1\}.$$

Hence if $p : I \to \mathbf{Z}_2$ is the unique morphism such that $p\iota = 1$, then p is a covering morphism. Also each characteristic group of p is trivial.

2. In the group $\mathbf{Z}_3$ (with one object 0)

$$\text{St}_{\mathbf{Z}_3}\, 0 = \{0, 1, -1\}.$$

Hence, although the morphism $p : I \to \mathbf{Z}_3$ which sends ι to 1 and $-\iota$ to -1 is surjective on the elements, p is not a covering morphism.

3. For groups, the only covering morphisms are isomorphisms.

We now show the utility for topology of covering morphisms.

9.2.1 *Let $p : \tilde{X} \to X$ be a covering map, let A be a subset of X and let $\tilde{A} = p^{-1}[A]$. Then the induced morphism*

$$\pi p : \pi\tilde{X}\tilde{A} \to \pi XA$$

is a covering morphism.

Proof Let $\tilde{x} \in \tilde{A}$ and let $p\tilde{x} = x$. For each path a in X with initial point x, let $\tilde{a}$ denote the unique covering path in $\tilde{X}$ with initial point $\tilde{x}$. If the final point of a is in A, then the final point of $\tilde{a}$ is in $\tilde{A}$. Also, the equivalence class of $\tilde{a}$ depends only on the equivalence class of a (by 9.1.2). So the function cls $a \rightsquigarrow$ cls $\tilde{a}$ is inverse to the restriction of p which maps St $\tilde{x} \to$ St x. □

Once more we obtain:

9.2.1 (*Corollary 1*) *The circle $\mathbf{S}^1$ has fundamental group isomorphic to the integers $\mathbf{Z}$.*

Proof Let $p : \mathbf{R} \to \mathbf{S}^1$ be the covering map $t \rightsquigarrow e^{2\pi i t}$ so that

$$p' = \pi p : \pi\mathbf{RZ} \to \pi(\mathbf{S}^1, 1)$$

is a covering morphism. Now $\pi\mathbf{RZ}$ is 1-connected: so for each $n \in \mathbf{Z}$ there is a unique element τ_n of $\pi\mathbf{R}(0, n)$. Of course, τ_1 is represented by the path $\mathbf{I} \to \mathbf{R}$, $t \rightsquigarrow t$, while $\tau_{n+1} - \tau_n$—the unique element of $\pi\mathbf{R}(n, n+1)$—is represented by the path $\mathbf{I} \to \mathbf{R}$, $t \rightsquigarrow t + n$. It follows from the definition of p that

$$p'(\tau_{n+1} - \tau_n) = p'\tau_1.$$

Hence, if $\tau = p'\tau_1$, then

$$\begin{aligned} p'\tau_n &= p'(\tau_n - \tau_{n-1}) + \cdots + p'(\tau_2 - \tau_1) + p'\tau_1 \\ &= n\tau. \end{aligned}$$

Since p' is a covering morphism, $p'\tau_n \neq 0$, and so

$$n\tau \neq 0.$$

Also, if $a \in \pi(\mathbf{S}^1, 1)$, then $a = p'\tau_n$ for some n and hence

$$a = n\tau.$$

This shows that $\pi(\mathbf{S}^1, 1)$ is an infinite cyclic group with generator τ. □

9.2.1 (*Corollary 2*) *For $n > 1$, real projective n-space $P^n(\mathbf{R})$ has fundamental group isomorphic to $\mathbf{Z}_2$.*

Proof The Hopf map $h : \mathbf{S}^n \to P^n(\mathbf{R})$ is a covering map, and by the proof of 8.4.7 $\mathbf{S}^n$ is 1-connected for $n > 1$. Let $\tilde{x}, -\tilde{x}$ be antipodal points of $\mathbf{S}^n$ and let $x = h\tilde{x} = h(-\tilde{x})$. By 9.2.1, the morphism

$$\pi h : \pi\,\mathbf{S}^n\{x, -x\} \to \pi(P^n(\mathbf{R}), x)$$

is a covering morphism. However, the groupoid $\pi\,\mathbf{S}^n\{\tilde{x}, -\tilde{x}\}$ is isomorphic to I. It follows that the group $\pi(P^n(\mathbf{R}), x)$ has two elements, and so is isomorphic to $\mathbf{Z}_2$. □

We give in this section some simple results on covering groupoids.

9.2.2 *Let $p : \tilde{G} \to G$ be a covering morphism such that G is connected. If a, b are any elements of G, then $p^{-1}[a], p^{-1}[b]$ have the same cardinality.*

Proof Suppose $c \in G(x, y)$. It is an easy deduction from the definition of covering morphism that the functions

$$p^{-1}[c] \to p^{-1}[x], \qquad p^{-1}[c] \to p^{-1}[y]$$

which send an element of $p^{-1}[c]$ to its initial and its final point respectively, are both bijections.

Suppose now $a \in G(x, x_1)$, $b \in G(y, y_1)$. Since G is connected, there is an element c in $G(x, y)$ and we deduce

$$\divideontimes\, p^{-1}[a] = \divideontimes\, p^{-1}[x] = \divideontimes\, p^{-1}[y] = \divideontimes\, p^{-1}[b] \quad \square$$

In particular, if $p^{-1}[a]$ has n elements for each a in G, then we call p an *n-fold covering morphism*. We shall see later that for connected covering morphisms, this number n is the index in $G\{p\tilde{x}\}$ of the characteristic group of p at $\tilde{x}$ ($\tilde{x} \in \mathrm{Ob}(\tilde{G})$).

9.2.3 *Let $r : K \to H$, $q : H \to G$ be morphisms such that q is a covering morphism. Then r is a covering morphism if and only if qr is a covering morphism.*

Proof Let $x \in \mathrm{Ob}(K)$ and consider the composite $(qr)' = q'r'$

$$\mathrm{St}_K\, x \xrightarrow{r'} \mathrm{St}_H\, rx \xrightarrow{q'} \mathrm{St}_G\, qrx$$

where q', r' are induced by q, r. We are given q' is a bijection. Therefore $(qr)'$ is a bijection if and only if r' is a bijection. $\square$

It is not hard to prove for spaces the result corresponding to one part of 9.2.3. The other part is more tricky (it is false in general) and will be left till an Exercise in 9.5.

9.2.4 *If $r : Z \to Y$, $q : Y \to X$ are maps of spaces such that qr and q are covering maps, then r is a covering map.*

Proof Let $x \in X$. Let U be a canonical neighbourhood of x for the map q, and let V be a canonical neighbourhood of x for the map qr. Let W be an open, path-connected neighbourhood of x contained in $U \cap V$. Clearly, W is canonical for both q and qr (since any open, path-connected subset of a canonical set is again canonical).

For each y in $q^{-1}[x]$, let W_y be the path-component of $q^{-1}[W]$ which contains y—these sets W_y are disjoint and open in Y. Clearly,

$$r^{-1}q^{-1}[W] = \bigcup \{r^{-1}[W_y] : y \in q^{-1}[x]\}$$

and the sets $r^{-1}[W_y]$ are disjoint and open in Z. Therefore, each path-component W' of $r^{-1}[W_y]$ is also a path-component of $r^{-1}q^{-1}[W]$. But $qr \mid W', W$ and $r \mid W_y, W$ are homeomorphisms. Therefore $r \mid W', W_y$ is a homeomorphism. $\square$

EXERCISES

1. Let $f : \mathbf{S}^1 \to \mathbf{S}^1$ be the map $z \rightsquigarrow z^n$, where n is a non-zero integer. Prove that the induced map $\pi f : \pi(\mathbf{S}^1, 1) \to \pi(\mathbf{S}^1, 1)$ is multiplication by n.
2. Let $p : \tilde{G} \to G$ be a covering morphism. Prove that if G is connected and $\tilde{G}$ is non-empty then $\mathrm{Ob}(p)$ is surjective.

3. Prove that a 1-fold covering morphism is an isomorphism.
4. Let $p : \tilde{X} \to X$ be a covering map of spaces such that X is path-connected. Prove that the sets $p^{-1}[x]$ have the same cardinality for all x in X, and that if $\tilde{X}$ is non-empty then p is surjective.
5. Let $r : Z \to Y$, $q : Y \to X$ be covering maps such that $q^{-1}[x]$ is finite for each x in X. Prove that qr is a covering map.
6. Show that the notion of covering morphism for groupoids extends to functors of categories in such a way that if $p : \tilde{X} \to X$ is a covering map of spaces, then $Pp : P\tilde{X} \to PX$ is a covering functor of categories.

9.3 On lifting sums and morphisms

Let $p : \tilde{G} \to G$ be a covering morphism. Because of the analogy with the fundamental groupoid of covering spaces, we say that an element $\tilde{a}$ of $\tilde{G}$ *covers*, or is a *lifting*, of $p\tilde{a}$; similarly, we say a sum $\tilde{a}_n + \cdots + \tilde{a}_1$ in $\tilde{G}$ covers, or is lifting, of $p\tilde{a}_n + \cdots + p\tilde{a}_1$. The basic property of covering morphisms is that not only elements of G but also sums in G can be lifted into $\tilde{G}$.

9.3.1 *Let $\tilde{x}$ be an object of $\tilde{G}$ and let $p\tilde{x} = x$. If*

$$a = a_n + \cdots + a_1$$

belongs to St x, *then there are unique elements $\tilde{a}_n, \ldots, \tilde{a}_1$ of $\tilde{G}$ such that*
(a) $p\tilde{a}_i = a_i$, $i = 1, \ldots, n$
(b) $\tilde{a} = \tilde{a}_n + \cdots + \tilde{a}_1$ *is defined and belongs to* St $\tilde{x}$.

Proof Since $p : \text{St}\,\tilde{y} \to \text{St}\,p\tilde{y}$ is bijective for each object $\tilde{y}$ of $\tilde{G}$, the elements $\tilde{a}_i$ are uniquely defined by the inductive conditions (i) $p\tilde{a}_i = a_i$, (ii) $\tilde{a}_1 \in \text{St}\,\tilde{x}$, (iii) $\tilde{a}_i \in \text{St}\,\tilde{x}_i$ where $\tilde{x}_1 = \tilde{x}$ and $\tilde{x}_i$ is the final point of $\tilde{a}_{i-1}$ for $i > 1$. □

As mentioned in the last section, the characteristic group of $\tilde{G}$, $\tilde{x}$ (that is, the subgroup $p[\tilde{G}\{\tilde{x}\}]$ of $G\{p\tilde{x}\}$) plays an important role in the theory. The relationship of these groups for various $\tilde{x}$ is described in the next result.

First we need a definition: subgroups C of $G\{x\}$, D of $G\{y\}$ are called *conjugate* (in G) if there is an element c of $G(x, y)$ such that

$$c + C - c = D$$

This relation implies easily that $-c + D + c = C$.

9.3.2 *Let C be the characteristic group of $\tilde{G}$, $\tilde{x}$.*
(a) If D is the characteristic group of $\tilde{G}$, $\tilde{y}$, and $\tilde{x}$, $\tilde{y}$ lie in the same component of $\tilde{G}$, then C and D are conjugate.
(b) If D is a subgroup $G\{y\}$ and D is conjugate to C, then D is the characteristic group of $\tilde{G}$, $\tilde{y}$ for some $\tilde{y}$.

Proof (*a*) Let c be an element of $\tilde{G}(\tilde{x}, \tilde{y})$ and let $p\tilde{c} = c$. If $a \in C$, then a is covered by an element $\tilde{a}$ of $\tilde{G}\{\tilde{x}\}$. By 9.3.1, $c + a - c$ is covered by $\tilde{c} + \tilde{a} - \tilde{c}$, which is an element of $G\{x\}$. Therefore $c + a - c \in D$. This proves (*a*).
(*b*) Suppose $c + C - c = D$ where $c \in G(x, y)$. Let $\tilde{c}$ be an element of $\operatorname{St} \tilde{x}$ covering c, let $\tilde{y}$ be the final point of $\tilde{c}$, and let D' be the characteristic group of $\tilde{G}, \tilde{y}$. We prove that $D = D'$.

If $b \in D$, then $b = c + a - c$, where $a \in C$. It follows that b is covered by an element $\tilde{c} + \tilde{a} - \tilde{c}$ of $\tilde{G}\{\tilde{y}\}$, whence $b \in D'$. Conversely, if $b \in D'$, then b is covered by an element $\tilde{b}$ of $\tilde{G}\{\tilde{y}\}$, whence $b = c + a - c$ where $a = p(-\tilde{c} + \tilde{b} + \tilde{c}) \in C$. □

We shall now show the sense in which the characteristic group of $\tilde{G}, \tilde{x}$ *is* characteristic. It is convenient to work in the category of pointed groupoids: a *pointed groupoid* G, x consists of a groupoid G and object x of G. A *pointed morphism* $G, x \to H, y$ of pointed groupoids consists of G, x and H, y and a morphism $f : G \to H$ such that $fx = y$; such a pointed morphism is usually called a morphism $G, x \to H, y$ and is often denoted simply by f. The *characteristic group* of a pointed morphism $f : G, x \to H, y$ is the characteristic group of f at x.

If $p : \tilde{G}, \tilde{x} \to G, x$ is a pointed morphism, then we say p is a *covering morphism* if the morphism $p : \tilde{G} \to G$ is a covering morphism. Similarly, if we are given a commutative diagram of pointed morphisms

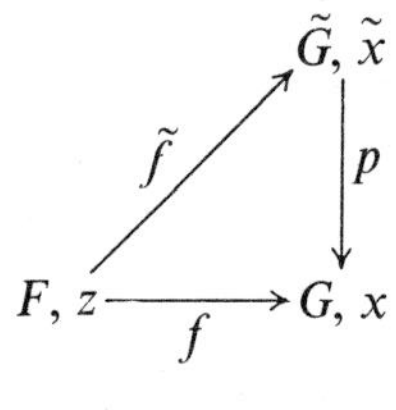

then we say $\tilde{f}$ is a *lifting* of f.

9.3.3 *Let* $p : \tilde{G}, \tilde{x} \to G, x$ *be a covering morphism, and* $f : F, z \to G, x$ *a morphism such that* F *is connected. Then* f *lifts to a morphism* $\tilde{f} : F, z \to \tilde{G}, \tilde{x}$ *if and only if the characteristic group of* f *is contained in that of* p*; and if this lifting exists, then it is unique.*

Proof Suppose first that $\tilde{f}$ exists; the relation $p\tilde{f} = f$ implies

$$f[F\{z\}] \subset p[\tilde{G}\{\tilde{x}\}]$$

and this proves the necessity of the condition.

Suppose, conversely, that the characteristic group of f is contained in the characteristic group C of p. Since p restricts to an isomorphism $\tilde{G}\{\tilde{x}\} \to C$, the morphism

$$f : F\{x\} \to G\{x\}$$

lifts uniquely to a morphism

$$\tilde{f} : F\{z\} \to \tilde{G}\{\tilde{x}\}.$$

For each object v of F let τ_v be an element of $F(z, v)$ (with $\tau_z = 0$). If $a \in F(u, v)$ then a can be written uniquely as

$$a = \tau_v + a' - \tau_u$$

with $a' \in F\{z\}$; and if, further, $b \in F(v, w)$, then $(b + a)' = b' + a'$. Now each element $f\tau_v$ is covered by a unique element $\tilde{f}\tau_v$ of St $\tilde{x}$; so we define

$$\tilde{f}a = \tilde{f}\tau_v + \tilde{f}a' - \tilde{f}\tau_u$$

and it follows that

$$\begin{aligned} \tilde{f}b + \tilde{f}a &= \tilde{f}\tau_w + \tilde{f}b' + \tilde{f}a' - \tilde{f}\tau_u \\ &= \tilde{f}\tau_w + \tilde{f}(b + a)' - \tilde{f}\tau_u \\ &= \tilde{f}(b + a). \end{aligned}$$

Therefore $\tilde{f}$ is a morphism.

Clearly $\tilde{f}$ lifts f. Also any morphism which lifts f must agree with $\tilde{f}$ on $F\{z\}$ and on the elements τ_v; therefore, such a lifting must coincide with $\tilde{f}$. This proves uniqueness of the lifting. □

9.3.3 (*Corollary 1*) *If $p : \tilde{G}, \tilde{x} \to G, x$ and $q : \tilde{H}, \tilde{y} \to G, x$ are connected covering morphisms with characteristic groups C, D respectively, and if $C \subset D$, then there is a unique covering morphism $r : \tilde{G}, \tilde{x} \to \tilde{H}, \tilde{y}$ such that $p = qr$. If $C = D$ then r is an isomorphism.*

Proof By 9.3.3, p lifts uniquely to a morphism $r : \tilde{G}, \tilde{x} \to \tilde{H}, \tilde{y}$ such that $qr = p$. By 9.2.3, r is a covering morphism. Finally, if $C = D$ the usual universal argument shows that r is an isomorphism. □

9.3.3 (*Corollary 2*) *A 1-connected covering groupoid of G covers every covering groupoid of G.*

Proof If $p : \tilde{G}, \tilde{x} \to G, x$ is a covering morphism and $\tilde{G}$ is 1-connected, then the characteristic group of p is trivial and so contained in any subgroup of $G\{x\}$. □

Because of 9.3.3 (Corollary 2) a 1-connected covering groupoid of G is called a *universal covering groupoid* of G. The existence of such for connected G will be proved later; its uniqueness, as a pointed groupoid, is a consequence of the last part of 9.3.3 (Corollary 1).

EXERCISES

Throughout these exercises we suppose there is given a covering morphism $p : \tilde{G}, \tilde{x} \to G, x$.

1. Let $a, b \in G(x, y)$ and let $c = -a + b$. Let $\tilde{a}, \tilde{b}, \tilde{c}$ be elements of St $\tilde{x}$ covering a, b, c respectively. Prove that $\tilde{b} \in \tilde{G}\{\tilde{x}\}$ if and only if $\tilde{a}, \tilde{b}$ have the same final point.

2. Prove that if Γ is a graph which generates G, then $\tilde{\Gamma} = p^{-1}[\Gamma]$ is a graph which generates $\tilde{G}$. Prove also that if Γ generates G freely, then $\tilde{\Gamma}$ generates $\tilde{G}$ freely.

3. The covering morphism p is called *regular* if it is connected and for each a in $G\{x\}$ the elements of $p^{-1}[a]$ are either all or none of them loops. Prove that p is regular if and only if p is connected and has characteristic group normal in $G\{x\}$.

4. Prove that if p is connected and has characteristic group $G\{x\}$, then p is an isomorphism.

5. A *cover transformation* is a morphism $h : \tilde{G} \to \tilde{G}$ such that $ph = p$. Prove that if p is connected then any cover transformation is an isomorphism; show also that in this case the set of cover transformations is a group under the operation of composition.

6. Prove that if p is regular with characteristic group C, then the group of cover transformations is isomorphic to the quotient group $G\{x\}/C$.

7. A morphism $p : H \to G$ of groupoids is called a *fibration* if, for each object x of H, p restricts to a surjection $\mathrm{St}_H\, x \to \mathrm{St}_G\, px$. Prove that if $f : F \to G$ is any morphism of connected groupoids then there is a fibration $p : H \to G$ and an equivalence $f' : F \to H$ such that $f = pf'$. [Choose y in Ob(G). Let Ob(H) consist of pairs (a, g) for $a \in \mathrm{Ob}(F)$, $g \in G(y, fa)$. Let the morphisms in H from (a, g) to (b, h) be determined by the $w \in F(a, b)$ such that $fw = h - g$.]

8. Let $p : H \to G$ be a fibration of groupoids. Prove that if the diagram of morphisms

$$\begin{array}{ccc} F & \xrightarrow{\ f\ } & H \\ {\scriptstyle i}\downarrow & & \downarrow{\scriptstyle p} \\ F \times I & \xrightarrow[\ \varphi\ ]{} & G \end{array}$$

is commutative, where i is given by $a \rightsquigarrow (a, 0)$, then there is a morphism $\psi : F \times I \to H$ such that $\psi i = f$, $p\psi = \varphi$. Prove further that if p is a covering morphism then ψ is unique. [This is the *covering homotopy property* for fibrations and coverings of groupoids: it would be possible, and reasonable, to use this as the definition of fibrations and coverings of groupoids.]

9. Let $\pi_0 X$ denote the set of components of a groupoid X. If $p : H \to G$ is a morphism, and $y \in \mathrm{Ob}(G)$, then the *fibre* F_y of p over y is the subgroupoid $p^{-1}[y]$ of H. Prove that if p is a fibration then there is a functor $\Gamma : G \to \mathscr{S}et$ which on objects is given by $y \rightsquigarrow \pi_0 F_y$.

10. Continuing the previous exercise, let $x \in \mathrm{Ob}(H)$, $y = px$, $F = F_y$. Let x', x'', y' in $\pi_0 F$, $\pi_0 H$, $\pi_0 G$ be the components containing x, x, y respectively,

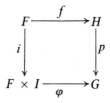

and take x', x'', y' as base points of these sets. Let $\partial : G\{y\} \to \pi_0 F$ be given by $g \rightsquigarrow (\Gamma g)x'$. Prove that the following sequence

$$0 \to F\{x\} \xrightarrow{i'} H\{x\} \xrightarrow{p'} G\{y\} \xrightarrow{\delta} \pi_0 F \xrightarrow{i_*} \pi_0 H \xrightarrow{p_*} \pi_0 G$$

in which (i) p', i' are the restrictions of p and the inclusion $i : F \to H$, (ii) p_*, i_* are induced by p, i (e.g., p_* is cls $a \rightsquigarrow$ cls pa), is exact (cf. Exercise 15 of 7.5). Prove also that $i_* \alpha = i_* \beta$ if and only if there is a g in $G\{y\}$ such that $(\Gamma g)\alpha = \beta$.

11. Let Y be a space, let (X, A) be a pair of spaces and let $i : A \to X$ be the inclusion. Prove that i induces a morphism $i^* : \pi Y^X \to \pi Y^A$ such that cls $F_t \rightsquigarrow$ cls $F_t i$. Prove that i^* is a fibration of groupoids for all Y if and only if (X, A) has the RWHEP [Exercise 9 of 7.5]; hence generalise 7.2.4 to the case (X, A) has the RWHEP.

9.4 Existence of covering groupoids

Our main purpose in this section is to prove that if x is an object of the connected groupoid G, then any subgroup of $G\{x\}$ is the characteristic group of some covering morphism $\tilde{G}, \tilde{x} \to G, x$. We shall deduce this from a more general construction of covering groupoids.

Let G be a groupoid and Γ a functor $G \to \mathscr{S}$et. Let us suppose that the sets Γx, $x \in \mathrm{Ob}(G)$, are disjoint, and have union S. If $s \in S$ then there is a unique object x_s of G such that $s \in \Gamma x_s$; further, we can define G_s, the *group of stability* of s, to be the group of elements g of G such that $(\Gamma g)s$ is defined and equals s. Thus G_s is, roughly, the subgroup of $G\{x_s\}$ of elements which when acting through Γ leave s fixed.

9.4.1 *Under the above assumptions there is a covering morphism $p : \tilde{G} \to G$ such that*
(a) $\mathrm{Ob}(\tilde{G}) = S$, the union of the sets Γx, $x \in \mathrm{Ob}(G)$,
(b) $\mathrm{Ob}(p)$ is the function $s \rightsquigarrow x_s$,
(c) for each $s \in \mathrm{Ob}(\tilde{G})$, the characteristic group of p at s is G_s, the group of stability of s.
Further, $\tilde{G}$ is connected if and only if for each $s, t \in S$ there is a g in G such that $(\Gamma g)s = t$.

Proof For the construction of $\tilde{G}$ it remains only to write down the elements of $\tilde{G}$ and the values of p on these elements.

Let $s \in S$. An element of $\tilde{G}$ with initial point s shall be a pair (g, s) for some element g of G with initial point x_s; the final point of (g, s) shall be $(\Gamma g)s$. The addition in $\tilde{G}$ shall be given by

$$(h, t) + (g, s) = (h + g, s);$$

this is to be defined, of course, if and only if $(\Gamma g)s = t$, in which case, $h + g$ is certainly defined. It is clear that, with this definition, $\tilde{G}$ is a category—in fact, it is a groupoid, the negative (inverse) of (g, s) being $(-g, (\Gamma g)s)$. We define p on elements by $(g, s) \rightsquigarrow g$; clearly, this makes p a morphism of groupoids.

Let $s \in S$; then $\mathrm{St}_{\tilde{G}}\, s$ consists of all pairs (g, s) such that $g \in \mathrm{St}_G\, x_s$. So p maps $\mathrm{St}_{\tilde{G}}\, s$ bijectively onto $\mathrm{St}_G\, ps$. That is, p is a covering morphism. Further, $\tilde{G}\{s\}$ has elements the pairs (g, s) such that $(\Gamma g)s = s$. Hence the characteristic group of p at s is G_s.

Finally, if $s, t \in S$, then $\tilde{G}(s, t)$ is non-empty if and only if there is an element g of G such that $(\Gamma g)s = t$; this proves the last part of the theorem. □

The assumption in the theorem that the sets Γx, $x \in \mathrm{Ob}(G)$ are disjoint is no grave loss of generality. For if this property does not hold, then we can replace Γ by $\Gamma' : G \to \mathscr{S}\mathrm{et}$, where if $x \in \mathrm{Ob}(G)$ then

$$\Gamma' x = \{(x, s) : s \in \Gamma x\}$$

and if $g \in G(x, y)$ then

$$\Gamma' g : \Gamma' x \to \Gamma' y, \qquad (x, s) \rightsquigarrow (y, (\Gamma g)s).$$

Further if $\theta x : \Gamma x \to \Gamma' x$ is the function $s \rightsquigarrow (x, s)$, then θ is a homotopy of functors $\Gamma \simeq \Gamma'$.

9.4.2 *Let x be an object of the connected groupoid G, and let C be a subgroup of $G\{x\}$. Then C is the characteristic group of a connected covering morphism $p : \tilde{G}, \tilde{x} \to G, x$ for which $p^{-1}[x]$ is the set of cosets $G\{x\}/C$.*

Proof If $g \in \mathrm{St}\, x$, then we call the set

$$g + C = \{g + c : c \in C\}$$

a (*left-*) *coset* of C in G. Notice that two cosets $a + C$, $b + C$ are equal if and only if a, b have the same final point and $-a + b \in C$.

We define a functor $\Gamma : G \to \mathscr{S}\mathrm{et}$ as follows. If $y \in \mathrm{Ob}(G)$ then Γy is the set of all cosets $a + C$ for $a \in G(x, y)$; if $g \in G(y, z)$ then $\Gamma g : \Gamma y \to \Gamma z$ is the function $a + C \rightsquigarrow g + a + C$. Clearly, Γ is a functor, and the sets Γy, $y \in \mathrm{Ob}(G)$ are disjoint. So the construction of 9.4.1 gives a covering morphism $p : \tilde{G} \to G$. Further, $\tilde{G}$ is connected since for any cosets $a + C$, $b + C$, the function $\Gamma(b - a)$ sends $a + C$ to $b + C$.

Let s in $\mathrm{Ob}(\tilde{G})$ be the coset $C = 0 + C$ so that $p(s) = x$. The stability group of s is the set of g such that $g + C = C$; this stability group is simply C, and so the characteristic group of p at s is C.

Finally, $p^{-1}[x]$ is the set of cosets $g + C$ for $g \in G\{x\}$. Hence $p^{-1}[x] = G\{x\}/C$. □

The pointed groupoid constructed in 9.4.2 is sometimes written

Tr(G, C). However, since we have uniqueness as well as existence, the particular construction of Tr(G, C) can be forgotten.

9.4.2 (*Corollary 1*) *Any connected groupoid has a universal covering groupoid.*

Proof If G is a connected groupoid and $x \in \mathrm{Ob}(G)$, then a universal covering groupoid of G is Tr(G, 0) where 0 is the trivial subgroup of $G\{x\}$. □

9.4.2 (*Corollary 2*) *A connected covering morphism $\tilde{G}, \tilde{x} \to G, x$ is an n-fold covering if and only if its characteristic group has index n in $G\{x\}$.*

Proof By 9.3.3 (Corollary 3) $\tilde{G}$ is isomorphic to Tr(G, C) where C is the characteristic group of $\tilde{G}, \tilde{x}$. □

EXERCISES

1. Prove that a subgroup C of a free group G is free, and that if G is of rank r and C is of index n in G, then C is of rank $nr - n + 1$. [Construct $\tilde{G} =$ Tr(G, C) and use Exercise 2 of 9.3 and 8.2.2(*c*).]
2. Let R be a ring and I a left ideal of R; then, considering I simply as a subgroup of the abelian group R, we can form the groupoid Tr(R, I). What additional structure can be put on this object by using the multiplication? [This is more of a research project than an exercise!]
3. Let G be a groupoid and let $\mathscr{C}$ov(G) be the category whose objects are covering morphisms $p : \tilde{G} \to G$ and whose morphisms from $p : \tilde{G} \to G$ to $p' : \tilde{G}' \to G$ are determined by p, p' and a morphism $r : \tilde{G} \to \tilde{G}'$ such that $p'r = p$. Prove that $\mathscr{C}$ov(G) is equivalent to the category $\mathscr{F}$un $(G, \mathscr{S}$et$)$ [Exercise 6 of 6.6]. [If $p : \tilde{G} \to G$ is a covering morphism, define a functor $Lp : G \to \mathscr{S}$et which on objects sends $x \rightsquigarrow p^{-1}[x]$, and if $a \in G(x, y)$, $\tilde{x} \in p^{-1}[x]$, then $Lp(a)(\tilde{x})$ is to be the final point of the element $\tilde{a}$ which starts at $\tilde{x}$ and covers a.]

9.5 Lifted topologies

In this section we apply the previous results to prove the existence of liftings of maps and the existence of covering spaces—that is, we prove topological theorems corresponding to 9.3.3 and 9.4.1. However this correspondence is not complete since the existence of a covering map $\tilde{X} \to X$ implies a local condition on X. (We recall that all spaces are, in any case, assumed locally path-connected.)

We introduce some useful language. If $f : H \to G$ is a morphism of groupoids, and $x \in \mathrm{Ob}(G)$, then $\chi_f\{x\}$ denotes $G\{x\}$ if $f^{-1}[x]$ is empty, and otherwise denotes the intersection of the characteristic groups of f at y for all $y \in f^{-1}[x]$. Thus χ_f is a wide, totally disconnected subgroupoid of G.

The notion of characteristic group and of χ_f will be used for maps of spaces. More precisely, if $f : Y \to X$ is a map, then the *characteristic group* of f at y is simply the characteristic group of $\pi f : \pi Y \to \pi X$ at y; and this group is also called the characteristic group of $f : Y, y \to X, fy$ or even, by an abuse of language, the characteristic group of Y, y. Similarly, we write χ_f for $\chi_{\pi f}$, so that χ_f is a subgroupoid of πX.

Let X be a space and χ any wide subgroupoid of πX. A subset U of X is called *weakly χ-connected* if for all x in U the characteristic group at x of the inclusion $U \to X$ is contained in $\chi\{x\}$. Clearly, any subset of a weakly χ-connected set is again weakly χ-connected.

We say X itself is *semi-locally χ-connected* if each x in X has a weakly χ-connected neighbourhood; if χ is simply-connected, this is also expressed as: X is *semi-locally simply-connected.* Our first result shows the utility of these definitions.

9.5.1 *Let $p : \tilde{X} \to X$ be a covering map. Then X is semi-locally χ_p-connected.*

Proof Let U be a canonical subset of X. If $x \in U$ and $\tilde{x} \in p^{-1}[x]$ then the inclusion $i : U, x \to X, x$ lifts to a map $U, x \to \tilde{X}, \tilde{x}$. So the morphism $\pi U, x \to \pi X, x$ lifts, and it follows that the characteristic group of U, x is contained in the characteristic group of p at $\tilde{x}$. This is true for all $\tilde{x}$ in $p^{-1}[x]$ and so the characteristic group of U, x is contained in $\chi_p\{x\}$.

Thus any canonical subset of X is weakly χ_p-connected. By definition of covering map, any x in X has a canonical neighbourhood. □

The following corollary is immediate.

9.5.1 (*Corollary 1*) *If X is path-connected and has a* 1*-connected covering space, then X is semi-locally simply-connected.*

It may be proved using the methods of chapter 8 that the Hawaiian ear-ring [cf. Fig. 8.4 p. 285] is not semi-locally simply-connected (the origin being a 'bad' point) and so this space has no 1-connected covering space.

Suppose now that X is a space and

$$q : \tilde{G} \to \pi X$$

is a covering morphism of groupoids. Let X be semi-locally χ_q-connected, and let

$$\tilde{X} = \mathrm{Ob}(\tilde{G}),\ p = \mathrm{Ob}(q) : \tilde{X} \to X.$$

We shall use q to 'lift' the topology of X to a topology on $\tilde{X}$.

Let $\mathscr{U}$ be the set of all open, path-connected, weakly χ_q-connected subsets of X. If U is any element of $\mathscr{U}$ consider the diagram

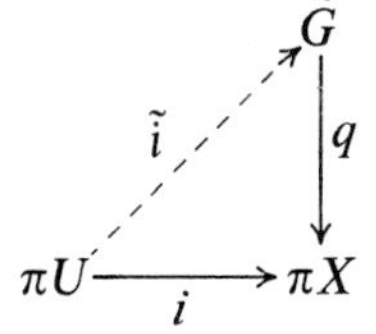

where i is induced by inclusion. By 9.3.3, i lifts to a morphism $\tilde{\imath} : \pi U \to G$; the set $\tilde{\imath}[U]$ is a subset $\tilde{U}$ of $\tilde{X}$ which we call a *lifting* of U. The set of all such liftings of U for all U in $\mathscr{U}$ is written $\tilde{\mathscr{U}}$. Since X is semi-locally χ_q-connected, the set $\tilde{\mathscr{U}}$ covers $\tilde{X}$.

We need a lemma about these liftings.

Let V be path-connected and $g : V \to X$ a map such that $g[V]$ is contained in a set U of $\mathscr{U}$. Let $g' : \pi V \to \tilde{G}$ be a lifting of $\pi g : \pi V \to \pi X$, and let $\tilde{U}$ be a lifting of U.

9.5.2 *If* $\tilde{g} = \mathrm{Ob}(g')$, *and* $\tilde{g}[V]$ *has one point in common with* $\tilde{U}$, *then* $\tilde{g}[V] \subset \tilde{U}$.

Proof Let $\tilde{x} = \tilde{g}v$ be an element of $\tilde{g}[V] \cap \tilde{U}$ and let $x = p\tilde{x} = gv$. Consider the diagram

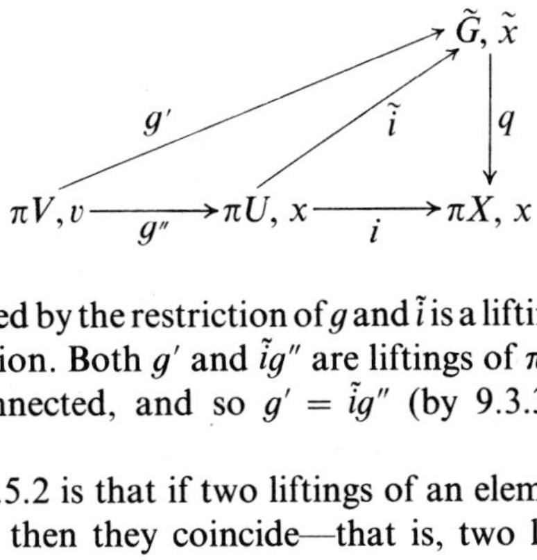

in which g'' is induced by the restriction of g and $\tilde{\imath}$ is a lifting of the morphism i induced by inclusion. Both g' and $\tilde{\imath}g''$ are liftings of $\pi g : \pi V, v \to \pi X, x$. But V is path-connected, and so $g' = \tilde{\imath}g''$ (by 9.3.3). It follows that $\tilde{g}[V] \subset \tilde{U}$. □

A corollary of 9.5.2 is that if two liftings of an element U of $\mathscr{U}$ have a point in common, then they coincide—that is, two liftings of U either coincide or are disjoint.

We now prove that $\tilde{\mathscr{U}}$ is a base for the open sets of a topology on $\tilde{X}$ [cf. 5.6.2]. Let $\tilde{U}$, $\tilde{V}$ be respectively liftings of elements U, V of $\mathscr{U}$, and suppose $\tilde{w} \in \tilde{U} \cap \tilde{V}$. Let $w = p\tilde{w}$ and let W be an open path-connected set such that $w \in W \subset U \cap V$; then W is weakly χ_q-connected and so has a lifting $\tilde{W}$ such that $\tilde{w} \in \tilde{W}$. By 9.5.2, $\tilde{W}$ is contained in both $\tilde{U}$ and $\tilde{V}$ and hence $\tilde{W} \subset \tilde{U} \cap \tilde{V}$.

This completes the proof that $\tilde{\mathscr{U}}$ is a base for the open sets of a topology on $\tilde{X}$. This topology on $\tilde{X}$ will be called the *topology of X lifted by q* or simply the *lifted topology*.

Suppose now that $\tilde{X}$ has this topology.

9.5.3 *Let $f : Z \to X$ be a map. If $\pi f : \pi Z \to \pi X$ lifts to a morphism $f' : \pi Z \to \tilde{G}$, then*

$$\tilde{f} = \mathrm{Ob}(f') : Z \to \tilde{X}$$

is continuous and is a lifting of $f : Z \to X$. All liftings of f arise in this way.

Proof Let $z \in Z$ and let $\tilde{U}$ be a lifting of a set U of $\mathscr{U}$ such that $\tilde{f}z \in \tilde{U}$. Then U is a neighbourhood of fz and so there is an open neighbourhood V of z, which may be assumed path-connected, such that $f[V] \subset U$. Let $j : V \to Z$ be the inclusion. Then $f'\pi(j)$ is a lifting of $\pi(fj)$. By 9.5.2, $\tilde{f}[V] \subset \tilde{U}$. This proves continuity of $\tilde{f}$.

If $\tilde{f} : Z \to \tilde{X}$ is any lifting of f, then $\pi\tilde{f}$ lifts πf, and so all liftings arise in the above way. □

On the other hand, the lifted topology is relevant to the case of a given covering map.

9.5.4 *Let $p : \tilde{X} \to X$ be a covering map. The topology of $\tilde{X}$ is that of X lifted by $\pi p : \pi\tilde{X} \to \pi X$.*

Proof Let U be a canonical subset of X. If $\tilde{U}$ is a path-component of $p^{-1}[U]$, then the inclusion $U \to X$ lifts to a map $U \to \tilde{X}$ with image $\tilde{U}$. It follows that $\tilde{U}$ is a lifting of U. Clearly, these liftings $\tilde{U}$ for all canonical subsets U of X form a base for the open sets of the given topology on $\tilde{X}$.

On the other hand, let $\mathscr{U}$ be the set of open, path-connected, weakly χ_p-connected subsets of X. We know that each canonical subset of X belongs to $\mathscr{U}$. Let $U \in \mathscr{U}$, and let $\tilde{U}$ be a lifting of U. Because each point of U has a canonical neighbourhood contained in U, it is easy to prove that $\tilde{U}$ is open in $\tilde{X}$ and is mapped by p homeomorphically on to U. Therefore U is canonical. Hence the lifted topology coincides with the given topology. □

By combining 9.5.4 with 9.5.3 and the corollaries to 9.3.3 we obtain the following results.

9.5.4 (*Corollary 1*) *If $p : \tilde{X}, \tilde{x} \to X, x$ and $q : \tilde{Y}, \tilde{y} \to X, x$ are connected covering maps with characteristic groups C, D respectively, and if $C \subset D$, then there is a unique map $r : \tilde{X}, \tilde{x} \to \tilde{Y}, \tilde{y}$ such that $p = qr$. Further, r is a covering map, and is a homeomorphism if $C = D$.*

9.5.4 (*Corollary 2*) *A 1-connected covering space of X covers every covering space of X.*

Because of 9.5.4 (Corollary 2) a 1-connected covering space of X is called a *universal covering space* of X. It is a special case of the last part of 9.5.4 (Corollary 1) that any two universal covering spaces of a connected space X are homeomorphic.

We now show that the lifted topology does give rise to a covering space. Let $q : \tilde{G} \to \pi X$ be a covering morphism, let $\tilde{X} = \mathrm{Ob}(\tilde{G})$, $p = \mathrm{Ob}(q)$. Suppose also that X is semi-locally χ_q-connected and that $\mathscr{U}$ is the set of open, path-connected, weakly χ_q-connected subsets of X.

9.5.5 *The lifted topology is the only topology on $\tilde{X}$ such that*
(*a*) *$p : \tilde{X} \to X$ is a covering map*
(*b*) *there is an isomorphism $r : \tilde{G} \to \pi\tilde{X}$ which is the identity on objects and such that the following diagram commutes*

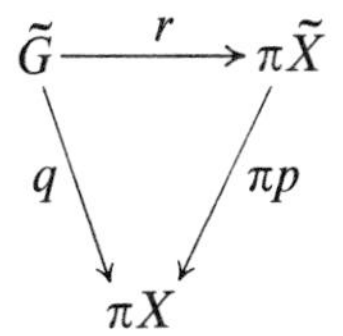

Proof We first prove that if $\tilde{X}$ has the lifted topology then p is a covering map.

Let $\tilde{\mathscr{U}}$ be the set of liftings of elements of $\mathscr{U}$, so that $\tilde{\mathscr{U}}$ is a base for the open sets of the lifted topology on X. If $U \in \mathscr{U}$, then $p^{-1}[U]$ is a union of elements of $\tilde{\mathscr{U}}$; therefore p is continuous. Also, if $\tilde{U} \in \tilde{\mathscr{U}}$, then $p[\tilde{U}] \in \mathscr{U}$; therefore p is open. If $\tilde{U}$ is a lifting of a set U of $\mathscr{U}$ then $p \mid \tilde{U}, U$ is a bijection and hence a homeomorphism; since also $\tilde{U}$ is open in $\tilde{X}$, it follows that $\tilde{U}$ is canonical in $\tilde{X}$ and U is canonical in X. Therefore p is a covering map.

We now define a morphism $r : \tilde{G} \to \pi\tilde{X}$. On objects, r is to be the identity. Let $\alpha \in \tilde{G}(\tilde{x}, \tilde{y})$, and suppose $q\alpha \in \pi X(x, y)$. Let $a : \mathbf{I} \to X$ be a representative of $q\alpha$; then a induces a morphism $\pi a : \pi\mathbf{I} \to \pi X$ such that

$$(\pi a)(\iota) = q\alpha$$

where ι is the unique element of $\pi\mathbf{I}(0, 1)$.

Since $\mathbf{I}$ is 1-connected, πa lifts uniquely to a morphism

$$a' : \pi\mathbf{I}, 0 \to \tilde{G}, \tilde{x}.$$

Notice that $a'(\iota)$ is a lifting of $q\alpha$, so that

$$a'(\iota) = \alpha.$$

By 9.5.3 $\tilde{a} = \mathrm{Ob}(a') : \mathbf{I} \to \tilde{X}$ is continuous, and we define

$$r\alpha = \operatorname{cls} \tilde{a};$$

clearly, $r\alpha \in \pi\tilde{X}(\tilde{x}, \tilde{y})$. Also $r\alpha$ is independent of the choice of representative a in $q\alpha$, since different representatives a_1, a_2 are equivalent and so have equivalent liftings $\tilde{a}_1$, $\tilde{a}_2$.

Suppose that $\beta \in \tilde{G}(\tilde{y}, \tilde{z})$. Then $r(\beta + \alpha)$ and $r\beta + r\alpha$ both lift $q(\beta + \alpha)$. Hence $r(\beta + \alpha) = r\beta + r\alpha$. This proves that r is a morphism and, by definition of r, $(\pi p)r = q$. By 9.2.3, r itself is a covering morphism. This implies that for each $\tilde{x}$, $\tilde{y}$ in $\tilde{X}$

$$r : \tilde{G}(\tilde{x}, \tilde{y}) \to \pi\tilde{X}(\tilde{x}, \tilde{y})$$

is injective. It is also surjective, since any γ in $\pi\tilde{X}(\tilde{x}, \tilde{y})$ is covered by an element $\tilde{\gamma}$ of $\mathrm{St}_{\tilde{G}}\, \tilde{x}$; if $\tilde{\gamma} \in \tilde{G}(\tilde{x}, \tilde{y}')$ then $r\tilde{\gamma} = r\tilde{y}'$ and so $\tilde{y} = \tilde{y}'$. Hence r is an isomorphism. This proves that (*a*) and (*b*) are satisfied by the lifted topology.

The uniqueness of a topology satisfying (*a*) and (*b*) follows from 9.5.4: since r is the identity on objects the topology of X lifts by q and by πp to the same topology on $\tilde{X}$. □

We now discuss covering spaces of cell complexes.

9.5.6 *Let $p : \tilde{X} \to X$ be a finite covering map and let X be a cell complex. Then $\tilde{X}$ can be given a cell structure for which p is a cellular map.*

Proof Let $h_\alpha : \mathbf{E}^{n_\alpha} \to X$ be a characteristic map for an n_α-cell of X. By 9.5.3, 9.5.4, and since $\mathbf{E}^{n_\alpha}$ is 1-connected, h_α lifts to a map $\tilde{h}_\alpha : \mathbf{E}^{n_\alpha} \to \tilde{X}$. Since $\mathbf{E}^{n_\alpha}$ is compact and $\tilde{X}$ is Hausdorff, $\tilde{h}_\alpha$ is a closed map; it is also injective on $\mathbf{B}^{n_\alpha}$ (since h_α is) and so $\tilde{h}_\alpha$ maps $\mathbf{B}^{n_\alpha}$ homeomorphically to a subset $\tilde{e}^{n_\alpha}$ of $\tilde{X}$. That is, the open cells of $\tilde{X}$ are liftings of the open cells of X. By the uniqueness of liftings of maps, these cells of $\tilde{X}$ are disjoint; clearly they cover $\tilde{X}$.

It is clear that $\tilde{X}^n = p^{-1}[X^n]$. It follows that for each lifting $\tilde{h}_\alpha$ of h_α, we have $\tilde{h}_\alpha[\mathbf{S}^{n-1}] \subset \tilde{X}^{n-1}$. The number of liftings of h_α is finite. So the maps $\tilde{h}_\alpha$ are characteristic maps of a cell structure on $\tilde{X}$. The cellularity of p is obvious. □

A similar result is true for any covering space of a cell complex. But if the covering is infinite then the number of cells of $\tilde{X}$ is infinite, and so it is necessary for this to discuss the topology of infinite cell complexes.

We conclude our account of the elements of topology with an entertaining application of these results on covering spaces.

9.5.7 *If $n \geqslant 2$ there is no map $f : \mathbf{S}^n \to \mathbf{S}^1$ such that $f(-x) = -fx$ for all x in $\mathbf{S}^n$.*

Proof Suppose there is such an f and consider the diagram

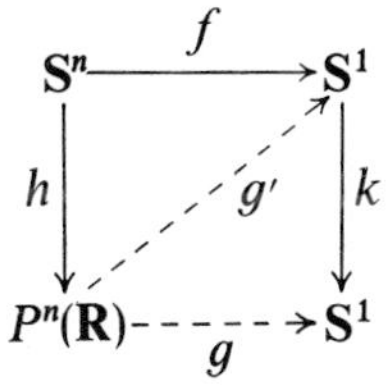

in which h is the Hopf map and k is the map $z \rightsquigarrow z^2$. Both h, k are covering maps which identify antipodal points of the spheres they are defined on.

By the given condition on f, $kf(-x) = kfx$ for all x in $\mathbf{S}^n$. Therefore f defines a map $g : P^n(\mathbf{R}) \to \mathbf{S}^1$ such that $gh = kf$. Since $P^n(\mathbf{R})$, $\mathbf{S}^1$ have fundamental groups $\mathbf{Z}_2$, $\mathbf{Z}$ respectively, any characteristic group of g is 0. So g lifts to a map g' such that $kg' = g$. Notice that

$$k\,g'h = gh = kf$$

so that $g'h$ and f are liftings of the same map. Hence $g'h$ and f agree at all points if they agree at one.

Let $x \in \mathbf{S}^n$. Then $g'hx$, fx are either the same or antipodal points. In the latter case, $g'h(-x) = g'hx = -fx = f(-x)$. So $g'h, f$ agree at either x or $-x$. Hence they agree everywhere, i.e., $g'h = f$. But this implies that $f(-x) = fx$, and so we have a contradiction. □

9.5.7 (*Corollary 1*) *Let $n \geqslant 2$ and let $g : \mathbf{S}^n \to \mathbf{R}^2$ be a map. Then there exists a point x_0 in $\mathbf{S}^n$ such that $g(x_0) = g(-x_0)$.*

Proof Let $h : \mathbf{S}^n \to \mathbf{R}^2$ be the map $x \rightsquigarrow g(x) - g(-x)$. Then $h(-x) = -h(x)$ for all x in $\mathbf{S}^n$.

Suppose there is no x_0 in $\mathbf{S}^n$ such that $g(x_0) = g(-x_0)$. Then $h(x)$ is never zero, and so the map $f : \mathbf{S}^n \to \mathbf{S}^1$, $x \rightsquigarrow h(x)/\|h(x)\|$, is well-defined. But then $f(-x) = -f(x)$ for all x in $\mathbf{S}^n$; this is impossible by 9.5.7. □

The above result with $\mathbf{R}^2$ replaced by $\mathbf{R}^n$ is known as the Borsuk–Ulam Theorem. However, this more general theorem cannot be proved with the methods treated in this book.

EXERCISES

1. Let $p : \tilde{X} \to X$ be a connected covering map. A *cover transformation* $\tilde{X} \to \tilde{X}$ is a map $h : \tilde{X} \to \tilde{X}$ such that $ph = p$. Prove that the set of cover transformations $\tilde{X} \to \tilde{X}$ forms under composition a group isomorphic to the group of cover transformations $\pi\tilde{X} \to \pi\tilde{X}$.
2. Prove that if $q : \tilde{X} \to X$, $r : \tilde{Y} \to Y$ are covering maps, then so also is $q \times r : \tilde{X} \times \tilde{Y} \to X \times Y$.

3. Let $q : \tilde{X} \to \pi X$ be a covering morphism and let X be semi-locally χ_q-connected. Let $\mathscr{U}$ be a set of open, path-connected weakly χ_q-connected subsets of X such that $\mathscr{U}$ is a base for the open sets of X. Let $\tilde{\mathscr{U}}$ be the set of liftings of elements of $\mathscr{U}$. Prove that $\tilde{\mathscr{U}}$ is a base for the open sets of the lifted topology on $\tilde{X} = \mathrm{Ob}(\tilde{G})$.
4. Let $r : Z \to Y$, $q : Y \to X$ be covering maps. Prove that $qr : Z \to X$ is a covering map if and only if X is semi-locally χ_{qr}-connected.
5. Prove that the Borsuk-Ulam Theorem does imply 9.5.7 (corollary 1).
6. Prove that no subspace of $\mathbf{R}^2$ is homeomorphic to $\mathbf{S}^2$.
7. Let $n \geqslant 2$ and let $g : \mathbf{S}^n \to \mathbf{R}^2$ be a map such that $g(-x) = -g(x)$ for all x in $\mathbf{S}^n$. Prove that there is a point x_0 in $\mathbf{S}^n$ such that $g(x_0) = 0$.

NOTES

Other interesting results on covering spaces are given in Spanier [3], which also shows the link-up with fibrations. A different approach to covering spaces is taken in Chevalley [1].

Covering spaces form one of the older parts of algebraic topology—cf., for example, Seifert–Threlfall [1]. By contrast, the development of covering groupoids is recent. Higgins in [1] introduced the groupoid $\mathrm{Tr}(G, C)$ for C a subgroup of the group G, and mentioned that the construction generalized to groupoids. He defined the covering groupoids (but assuming connectedness) in lecture notes and in [2]. An equivalent definition was independently given by Gabriel–Zisman [1] who also show the connection between covering groupoids of a groupoid G and functors $G \to \mathscr{S}\mathrm{et}$. (Exercise 3 of 9.4 is taken from their book.) However, the similar construction of a groupoid $\tilde{G}$ for the case of a group G acting on a set appears certainly in work of Ehresmann in 1958 (and for the case of a groupoid acting on a set in Ehresmann [1] p. 49–50); and, of course, it has long been known that if C is a subgroup of the group G then G operates on the set G/C of cosets of C in G.

Finally, the use of groupoids here, and in algebra, suggests that there is truth in the slogan: 'where there are conjugates there also are groupoids'.

Appendix: Functions and cardinality

A.1 Functions

Let A, B be sets. A subset F of $A \times B$ is called *functional* if for each $a \in A$, there is exactly one $b \in B$ such that $(a, b) \in F$. A *function f from A to B* will consist of the set A, the set B and a functional subset F of $A \times B$; more precisely, f is the triple (A, B, F). The set A is the *domain* of f, the set B is the *codomain* of f, and the functional subset F of $A \times B$ is the *graph* of f. (The codomain is sometimes called the *range* of f, but the word range is also used for what we shall call the *image* of f.) If $a \in A$ then the element b of B such that $(a, b) \in F$ is called the *value* of f on a, and is written $f(a)$, or fa, or even f_a; we also say f *sends*, or *maps*, a to fa. In order to emphasize the fact that f depends on both A and B we often write $f : A \to B$, or $A \xrightarrow{f} B$, for f. When a particular name is not required, a function from A to B is written simply $A \to B$.

Our definition ensures that two functions $f : A \to B$, $f' : A' \to B'$ are equal if and only if they have the same domain (i.e., $A = A'$), the same codomain (i.e., $B = B'$) and the same graph.

Functions are often defined by formulae. Thus the formulae $2x^2 + \sin x$ defines a function $\mathbf{R} \to \mathbf{R}$ whose value on a real number x is $2x^2 + \sin x$; this function we denote by

$$\mathbf{R} \to \mathbf{R}$$

$$x \rightsquigarrow 2x^2 + \sin x$$

and when the domain and codomain can be understood from the context we shall often denote such a function simply by $x \rightsquigarrow 2x^2 + \sin x$. We also allow that the domain of a function specified in this way shall be the maximum domain on which the formula gives a unique answer; thus $x \rightsquigarrow \log x + (x^2 - 1)^{-1}$ denotes a function with domain $\{x \in \mathbf{R} : x > 0$ and $x \neq 1\}$.

A.1.1 Let $f : A \to B$, $g : B \to C$ be functions. The *composite* of these two functions is the function

$$\begin{aligned} gf : A &\to C \\ x &\rightsquigarrow gfx. \end{aligned}$$

Sometimes, when extra clarity is essential, we write $g \circ f$ for gf. If $h : C \to D$ is another function, then we have the associate law

$$h(gf) = (hg)f.$$

The *identity function* on A is the function

$$\begin{aligned} 1, \quad \text{or} \quad 1_A : A &\to A \\ x &\rightsquigarrow x. \end{aligned}$$

If $f : A \to B$, then $f 1_A = f$, $1_B f = f$.

A.1.2 A function $f : A \to B$ is *surjective* (and is a *surjection*) if for each b in B, there is an a in A such that $fa = b$; f is *injective* (and is an *injection*) if for all a, a' in A, $fa = fa'$ implies $a = a'$. Finally, f is *bijective* (and is a *bijection*) if f is injective and surjective. For example, the identity $1 : A \to A$ is a bijection.

A.1.3 Let $f : A \to B$, $g : B \to A$ be functions such that $gf = 1_A$. Then we call g a *left-inverse* of f, and f a *right-inverse* of g. If $a \in A$, then $a = gfa$, and so g is surjective. If a, $a' \in A$ and $fa = fa'$, then $a = gfa = gfa' = a'$; so f is injective.

Suppose further that $fg = 1_B$. Then f is surjective, g is injective. Thus the two relations $gf = 1_A, fg = 1_B$ imply that both f and g are bijective. If these two relations hold, we say g is an *inverse* of f. This inverse is unique because if, further, $g'f = 1_A$ then

$$g' = g'1_B = g'fg = 1_A g = g.$$

If g is *the* inverse of f, then f is the inverse of g, and we write

$$g = f^{-1}, f = g^{-1}.$$

Suppose now $f : A \to B$ is a bijection. Then the subset $\{(fa, a) \in B \times A : a \in A\}$ is functional and so defines a function $g : B \to A$ which sends each b in B to the unique a in A such that $b = fa$. Thus g is the inverse of f.

A.1.4 Let $f : A \to B$ be a function, let $A' \subset A$, $B' \subset B$ and suppose that $a \in A'$ implies $fa \in B'$. The function

$$\begin{aligned} A' &\to B' \\ a &\rightsquigarrow fa \end{aligned}$$

is called the *restriction* (or *cut down*) of f to A, B and is written

$$f \mid A', B'.$$

We also write $f \mid A', B$ (i.e., in case $B' = B$) simply as $f \mid A'$.

In particular, the restriction

$$1_A \mid A' : A' \to A$$

of the identity function is called the *inclusion function* of A' into A. It is clearly injective. We emphasize that if $A' \neq A$, then the inclusion function $A' \to A$ is not the same as the identity function $A' \to A'$, since these functions have different codomains.

A.1.5 Let $f : A \to B$ be a function and let X, Y be any sets. The *image* of X by f is the subset $f[X]$ of B consisting of the elements $f(a)$ for all a in $A \cap X$. That is,

$$f[X] = \{b \in B : \exists a \in A \cap X \quad \text{such that} \quad f(a) = b\}.$$

We note that $f[X] = f[X \cap A]$; but it is convenient to allow $f[X]$ to be defined for any set X, rather than restrict X to be a subset of A. The set $f[A]$ is called the *image of f* and is written $\operatorname{Im} f$.

The *inverse image* of Y by f is the set $f^{-1}[Y]$ of elements a of A such that $f(a) \in Y$. That is,

$$f^{-1}[Y] = \{a \in A : f(a) \in Y\}.$$

In this case, $f^{-1}[Y] = f^{-1}[Y \cap \operatorname{Im} f]$, but again it is convenient to allow $f^{-1}[Y]$ to be defined for any set Y.

The purpose of the square bracket notation $f[X]$ is to avoid ambiguity between the image of a set and the value of the function. For example, $f[\varnothing]$ is always the empty set, but if $A = \{\varnothing\}$, then $f(\varnothing)$ will be an element of B, not necessarily empty.

We shall make use of some abbreviations. For any y, we write $f^{-1}[y]$ for $f^{-1}[\{y\}]$. In some circumstances we omit the square brackets. For example, $f^{-1}f[X]$ means $f^{-1}[f[X]]$, and if $[a, b[$ is an interval of $\mathbf{R}$, then $f[a, b[$ means $f[\,[a, b[\,]$, and $f^{-1}[a, b[$ means $f^{-1}[\,[a, b[\,]$.

Examples In the following examples we consider only functions whose domains are subsets of $\mathbf{R}$.

1. Let f be the function $x \rightsquigarrow x^2$. Then

$$f[-2, 1] = [0, 4], \qquad f^{-1}[-2, 4] = [0, 2].$$

2. Let f be the function $x \rightsquigarrow -\log x$. Then

$$f[0, 3] = [-\log 3, \to[, \qquad f^{-1}[0, 1[=]e^{-1}, 1].$$

A.1.6 We use in this book a number of relations between images and inverse images, and union and intersection. We state these here and leave to the reader their proof and illustration with examples. Let $f : A \to B$, $g : B \to C$ be functions, X, Y sets and $(X_j)_{j \in J}$ a family of sets.

A.1.6(1) $f^{-1}f[X] \supset X \cap A$

A.1.6(2) $ff^{-1}[Y] = Y \cap f[A]$

A.1.6(3) $f[\bigcup_{j \in J} X_j] = \bigcup_{j \in J} f[X_j]$

A.1.6(4) $f[\bigcap_{j \in J} X_j] \subset \bigcap_{j \in J} f[X_j]$

A.1.6(5) $f^{-1}[\bigcup_{j \in J} X_j] = \bigcup_{j \in J} f^{-1}[X_j]$

A.1.6(6) $f^{-1}[\bigcap_{j \in J} X_j] = \bigcap_{j \in J} f^{-1}[X_j]$

A.1.6(7) $(gf)[X] = gf[X]$

A.1.6(8) $(gf)^{-1}[Y] = f^{-1}g^{-1}[Y]$

A.1.6(9) $f^{-1}[B \setminus Y] = A \setminus f^{-1}[Y]$

Remark There are two other definitions of function than the one adopted here. The first, and usual, definition is that a function is a set F of ordered pairs with the property that $(x, y), (x, y') \in F \Rightarrow y = y'$. That is, a function is identified with its graph. From F we can recover the domain of F, namely the set $\{x : \exists y \text{ such that } (x, y) \in F\}$, and the image of F, namely the set $\{y : \exists x \text{ such that } (x, y) \in F\}$. There are definite advantages to this definition, particularly in analysis; for example, the inverse of an injective function is easy to define. However, this definition is not sufficient for our purposes. We really do require that two functions $A \to B$, $A' \to B'$ are the same if and only if they have the same graphs, and $A = A'$, $B = B'$. In fact, when A, B are sets with structures, the functions will be indexed also with these structures.

A compromise between the two definitions, with the advantages of both, is to define a function to be a triple $f = (A, B, F)$ where F is a subset of $A \times B$ such that $(x, y), (x, y') \in F \Rightarrow y = y'$. The domain of f is then a subset of A. A function $f : A \to B$ is then called a *mapping* if its domain is all of A.

This definition is in fact required in one part of chapter 5, and a function of this type will then be written $f : A \rightarrowtail B$ and called a function *out of A to B*. For the rest of the book we require the domain of a function $A \to B$ to be all of A and so we keep to the first definition given.

EXERCISES

1. Let $f : A \to B$, $g : B \to C$ be functions. Prove that

(*a*) If f and g are injective, so is gf.
(*b*) If f and g are surjective, so is gf.
(*c*) If f and g are bijective, so is gf.
(*d*) If gf is surjective, so also is g.
(*e*) If gf is injective, so also is f.

2. Let $f : A \to B$ be a function.

(*a*) Prove that f is surjective if and only if for all C and all functions $g, g' : B \to C$, the relation $gf = g'f$ implies $g = g'$.
(*b*) Prove that f is injective if and only if for all C and all functions $g, g' = C \to A$, the relation $fg = fg'$ implies $g = g'$.
3. Prove the relations given in A.1.6.
4. Let $f: X \to X'$, $g : Y \to Y'$ be functions. We define the *cartesian product* of f and g to be the function

$$\begin{aligned} f \times g : X \times Y &\to X' \times Y' \\ (x, y) &\rightsquigarrow (fx, gy). \end{aligned}$$

Prove that

(*a*) $f \times g$ is injective $\Leftrightarrow f, g$ are injective,
(*b*) $f \times g$ is surjective $\Leftrightarrow f, g$ are surjective.

Prove also that if $f' : X' \to X''$, $g' : Y' \to Y''$ then

$$(g' \times f')(g \times f) = g'g \times f'f,$$

and that $1_X \times 1_Y = 1_{X \times Y}$.
5. Let $f : X \to Y$ be a function, let $A \subset X$ and let $g = f \mid A, f[A]$. Prove that if $U \subset f[A]$, $V \subset Y$, then

$$U = f[A] \cap V \Leftrightarrow g^{-1}[U] = A \cap f^{-1}[V].$$

A.2 Finite, countable and uncountable sets

We shall not attempt an account of all the properties we need of the basic structures of mathematics (that is, such structures as **N**, **Z**, **Q**, **R**). In this section, we begin by stating without proof some results on counting which are particularly relevant. For further details see for example Halmos [1], Flett [1].

For each natural number $n > 0$, S_n denotes the set of natural numbers less than n; S_0 denotes the empty set.

A.2.1 *If there is a bijection* $S_m \to S_n$, *then* $m = n$.

This result implies that the following definition makes sense. Let

$n \in \mathbf{N}$. A set X *has n elements* if there is a bijection $S_n \to X$. Such a bijection $f: S_n \to X$ labels the elements of X as $f(0), \ldots, f(n-1)$ and so, in effect, counts them. By A.2.1, such a counting process leads to a unique answer.

A set X is *finite* if X has n elements for some natural number n.

A.2.2 *A subset X of* $\mathbf{N}$ *is finite if and only if X is bounded above.*

Here by X is *bounded above* is meant that there is a natural number greater than every element of X. In such case, the set X has a greatest element. But $\mathbf{N}$ itself has no greatest element. Therefore, $\mathbf{N}$ is *infinite*, that is, $\mathbf{N}$ is not finite.

Another consequence of A.2.2 and of its proof is that if X is a finite subset of $\mathbf{N}$ with n elements, then every subset Y of X is finite with at most n elements. Further, if Y has n elements, then $Y = X$. An immediate consequence is:

A.2.3 *If X is a finite set with n elements, then every subset Y of X is finite with at most n elements. Further, if Y has n elements, then $Y = X$.*

The process of comparing a set X with S_n generalizes to arbitrary sets. Two sets X and Y have the *same cardinality*, written $\text{※}X = \text{※}Y$, if there is a bijection $X \to Y$. This relation between sets is an equivalence relation [cf. Glossary].

A set X is *countably infinite* if $\text{※}X = \text{※}\mathbf{N}$; X is *countable* if X is finite or countably infinite; and otherwise, X is *uncountable*. We shall prove later that uncountable sets exist.

A.2.4 *Any subset of* $\mathbf{N}$ *is countable.*

Consequently, any subset of a countable set is again countable. But there are non-finite proper subsets of $\mathbf{N}$, for example $\mathbf{N} \setminus \{0\}$, the set of even numbers, the set of prime numbers, and so on; all these must, by A.2.4, be countably infinite. Thus the last part of A.2.3 does not generalize to infinite sets. In fact, though we do not prove this, a set X is infinite if and only if there is a proper subset Y of X such that $\text{※}X = \text{※}Y$.

When X is a finite set, we write $\text{※}X$ for the number of elements of X. It is not hard to prove, and we leave it as an exercise to the reader, that if X and Y are finite then

$$\text{※}(X \times Y) = \text{※}X . \text{※}Y$$

and that if also X and Y are disjoint, then

$$\text{※}(X \cup Y) = \text{※}X + \text{※}Y.$$

It might be thought that if X, Y are countably infinite, then $X \times Y$ is not countable. Surprisingly, this is false.

A.2.5 $\mathbf{N} \times \mathbf{N}$ *and* $\mathbf{N}$ *have the same cardinality.*

Proof Consider the function

$$f : \mathbf{N} \times \mathbf{N} \to \mathbf{N}$$
$$(x, y) \rightsquigarrow 2^x 3^y.$$

Clearly, f is an injection, and so $\mathbf{N} \times \mathbf{N}$ is countable. But $\mathbf{N} \times \mathbf{N}$ is not finite since it contains the infinite set $\mathbf{N} \times \{0\}$. Therefore $\mathbf{N} \times \mathbf{N}$ is countably infinite. □

A more elementary proof of A.2.5 is suggested in Exercise 14 of A.2. We use A.2.5 to prove two important results (A.2.7 and A.2.8). First we need:

A.2.6 *Let X be countable and $f : X \to Y$ a surjection. Then Y is countable, and is finite if X is finite.*

Proof If X is empty, then so also is Y and the result follows. We suppose then that X is non-empty.

We first note that there is a surjection $g : \mathbf{N} \to X$. In fact, if X is infinite, then there is a bijection $g : \mathbf{N} \to X$; if X is finite, there is a bijection $e : S_n \to X$ for some $n > 0$, and the function

$$g : \mathbf{N} \to X$$
$$m \rightsquigarrow \begin{cases} e(m), & m < n \\ e(n-1), & m \geqslant n \end{cases}$$

is a surjection.

The function $h = fg : \mathbf{N} \to Y$ is a surjection. Hence for each y in Y, the set $h^{-1}[y]$ is non-empty, and so has a smallest element $k(y)$. Then $hk(y) = y$, whence the function $k : Y \to \mathbf{N}$ is injective [A.1.6]. So Y is countable by A.2.4.

If X is finite, then $k[Y] \subset S_n$, and so Y is finite. □

We say a family $(X_\lambda)_{\lambda \in L}$ is *countable* if L, the set of indices, is countable; and $(X_\lambda)_{\lambda \in L}$ is *finite* if L is finite.

A.2.7 *The union of a countable family of countable sets is countable.*

Proof Let $(X_\lambda)_{\lambda \in L}$ be a countable family such that each X_λ is countable. Let $X = \bigcup_{\lambda \in L} X_\lambda$, $M = \{\lambda \in L : X_\lambda \neq \varnothing\}$. Then M is countable and $X = \bigcup_{\lambda \in M} X_\lambda$. If $M = \varnothing$, then $X = \varnothing$ and the result follows. So suppose $M \neq \varnothing$.

As shown in the proof of A.2.6, there is a surjection $f : \mathbf{N} \to M$ and for each λ in M there is a surjection $g_\lambda : \mathbf{N} \to X_\lambda$. Consider the function

$$g : \mathbf{N} \times \mathbf{N} \to X$$
$$(m, n) \rightsquigarrow g_{f(m)}(n).$$

We prove that g is a surjection; this implies by A.2.6 and A.2.5 that X is countable.

Let $x \in X$. Then $x \in X_\lambda$ for some $\lambda \in M$; further, $\lambda = f(m)$ for some $m \in \mathbf{N}$ and $x = g_\lambda(n)$ for some $n \in \mathbf{N}$. Thus $x = g_{f(m)}(n)$. □

A.2.8 *The set* **Q** *of rational numbers is countable.*

Proof Any rational number is of the form m/n or $-m/n$ for m, n in **N** and $n \neq 0$. Let $X = \{(m, n) \in \mathbf{N} \times \mathbf{N} : n \neq 0\}$. Then X is a subset of $\mathbf{N} \times \mathbf{N}$ and so is countable.

Let $\mathbf{Q}^{\geqslant 0}$ ($\mathbf{Q}^{\leqslant 0}$) denote the set of non-negative (non-positive) rational numbers. The function $(m, n) \rightsquigarrow m/n$ is a surjection $X \to \mathbf{Q}^{\geqslant 0}$. Therefore $\mathbf{Q}^{\geqslant 0}$ is countable. Similarly, $\mathbf{Q}^{\leqslant 0}$ is countable and so $\mathbf{Q} = \mathbf{Q}^{\geqslant 0} \cup \mathbf{Q}^{\leqslant 0}$ is countable. □

In order to construct uncountable sets from countable ones we need the operation $\mathscr{P}(X)$, the *power* of a set X. By definition, $\mathscr{P}(X)$ is the set of all subsets of X. For example, if X is empty, then $\mathscr{P}(X)$ consists of one element, the empty set. If $X = \{0, 1\}$, then $\mathscr{P}(X)$ consists of $\varnothing$, $\{0\}$, $\{1\}$, X. In the exercises we shall suggest a proof that when X is finite

$$\divideontimes \mathscr{P}(X) = 2^{\divideontimes X}.$$

Here we prove:

A.2.9 *For any set X there is an injection $X \to \mathscr{P}(X)$, but no bijection $X \to \mathscr{P}(X)$.*

Proof The function $x \rightsquigarrow \{x\}$ is an injection $X \to \mathscr{P}(X)$.

Suppose $f : X \to \mathscr{P}(X)$ is a bijection with inverse $g : \mathscr{P}(X) \to X$. We derive a contradiction by a marvellous argument due to G. Cantor.

For each x in $X, f(x)$ is a subset of X. So it makes sense to ask whether or not $x \in f(x)$. Let

$$A = \{x \in X : x \notin f(x)\} \tag{*}$$

and let $a = g(A)$ so that $f(a) = A$. We now ask: does $a \in A$? If $a \in A$, then by (*) $a \notin f(a)$. Since $f(a) = A$, we have a contradiction. On the other hand, if $a \notin A$, then, since $A = f(a)$, we have $a \notin f(a)$. So $a \in A$ by (*), and we still have a contradiction. This shows that a bijection $X \to \mathscr{P}(X)$ cannot exist. □

This result implies that $\mathscr{P}(\mathbf{N})$ is uncountable.

A.2.10 *The set* **R** *of real numbers is uncountable.*

Proof It is sufficient to prove that **R** has an uncountable subset. For this we construct an injection

$$\chi : \mathscr{P}(\mathbf{N}) \to \mathbf{R}$$

For any subset X of $\mathbf{N}$ we define the *characteristic function* of X

$$\chi_X : \mathbf{N} \to \{0, 1\}$$
$$n \rightsquigarrow \begin{cases} 0 & \text{if } n \notin X \\ 1 & \text{if } n \in X. \end{cases}$$

Clearly, $\chi_X = \chi_{X'}$ if and only if $X = X'$. We now define

$$\chi : \mathscr{P}(\mathbf{N}) \to \mathbf{R}$$
$$X \rightsquigarrow \sum_{n=0}^{\infty} \chi_X(n)/2^{2n}.$$

Certainly the series for $\chi(X)$ is convergent, and so $\chi(X)$ is a well-defined real number. In fact, $\chi(X)$ is a binary decimal of the form

$$n_0.0n_10n_20n_30\ldots \qquad (*)$$

It follows that if $\chi(X) = \chi(X')$, then $\chi_X(n) = \chi_{X'}(n)$ for all n in $\mathbf{N}$, whence $X = X'$ (the presence of the 0's in (*) ensures that there is no difficulty with repeated 1's). □

The argument of A.2.10 can be refined to show that $\divideontimes\mathbf{R} = \divideontimes\mathscr{P}(\mathbf{N})$, but we do not need this fact.

The reader should keep well aware of the intuitive meaning of A.2.10. The rational numbers are 'dense' in the real line in the sense that there are rational numbers arbitrarily close to any real number; also between any two rational numbers there is another. Thus the rational numbers appear to fill up the real line. But this is illusory: by A.2.10 there are many more real numbers than rational numbers.

A further point is that very little is known about the irrational numbers, and particularly about the non-algebraic numbers (for whose definition see Exercise 14). It can be argued that by the process of mathematical reasoning we can acquire at most a countable number of facts about $\mathbf{R}$; yet $\mathbf{R}$ is uncountable and so its properties must remain largely unexplored.

A.2.11 *The sets* $\mathbf{R} \times \mathbf{R}$ *and* $\mathbf{R}$ *have the same cardinality.*

Proof Let $A =]0, 1]$. The function

$$f : \mathbf{Z} \times A \to \mathbf{R}$$
$$(n, a) \rightsquigarrow n + a$$

is a bijection. Also $\mathbf{Z}$ is countably infinite (the function $n \rightsquigarrow 2n$ $(n \geqslant 0)$, $n \rightsquigarrow -2n - 1$ $(n < 0)$ is a bijection $\mathbf{Z} \to \mathbf{N}$) and so there is a bijection $g : \mathbf{Z} \times \mathbf{Z} \to \mathbf{Z}$. We shall prove below that there is a bijection $h : A \times$

$A \to A$; the composite

$$\mathbf{R} \times \mathbf{R} \xrightarrow{f^{-1} \times f^{-1}} \mathbf{Z} \times A \times \mathbf{Z} \times A \xrightarrow{1 \times T \times 1} \mathbf{Z} \times \mathbf{Z} \times A \times A$$

$$\xrightarrow{g \times h} \mathbf{Z} \times A \xrightarrow{f} \mathbf{R}$$

in which T is the bijection $(a, n) \rightsquigarrow (n, a)$, is the required bijection $\mathbf{R} \times \mathbf{R} \to \mathbf{R}$.

Let B be the set of all sequences $(m_r)_{r \geq 0}$ of natural numbers. We construct a bijection $i : A \to B$ as follows. Each a in A can be written uniquely as a binary decimal,

$$.a_1 a_2 a_3 \ldots$$

where $a_t = 0$ or 1 and the expression does not end in repeated 0's (that is, it is false that $a_r = 0$ for r sufficiently large). We define $i(a)$ to be the sequence $(m_r)_{r \geq 0}$ such that m_0 is the number of 0's between the decimal point and the first 1, and m_r $(r > 0)$ is the number of 0's between the rth and the $(r + 1)$st 1. For example, if $a = .01100101\ldots$ then $i(a)$ is initially the sequence $1, 0, 2, 1, \ldots$.

There is an obvious bijection $j : B \times B \to B$, where $(l_r) = j((m_r), (n_r))$ is the sequence $m_0, n_0, m_1, n_1, \ldots$, that is, $l_{2r} = m_r$, $l_{2r+1} = n_r$. The composite

$$A \times A \xrightarrow{i \times i} B \times B \xrightarrow{j} B \xrightarrow{i^{-1}} A$$

is the required bijection. □

EXERCISES

1. Prove that the function $x \rightsquigarrow x/(1 + |x|)$ is a bijection $\mathbf{R} \to]-1, 1[$.
2. Let a, b be real numbers such that $a < b$. Construct a bijection $[0, 1] \to [a, b]$.
3. Construct bijections $[0, 1] \to [0, 1[\to]0, 1[$.
4. Let X, Y be disjoint sets such that $\#X = \#Y = \#\mathbf{R}$. Prove that $\#(X \cup Y) = \#\mathbf{R}$.
5. Let X, Y be finite sets with m, n elements respectively. Prove that Y^X, the set of all functions $X \to Y$, has n^m elements.
6. Let A be a set. For each subset X of A define the *characteristic function* of X

$$\chi_X : A \to S_2$$
$$a \rightsquigarrow \begin{cases} 0 & \text{if } a \notin X \\ 1 & \text{if } a \in X. \end{cases}$$

Prove that the function $X \rightsquigarrow \chi_X$ is a bijection $\mathscr{P}(A) \to S_2{}^A$. Deduce that if $\# A = n$, then $\#\mathscr{P}(A) = 2^n$.

7. Continuing the notation of Exercise 6, prove that if X and $X_1, \ldots, X_n$ are subsets of A then

(a) $\chi(A \setminus X) = 1 - \chi(X)$,
(b) $\chi(X_1 \cap \cdots \cap X_n) = \chi(X_1)\,\chi(X_2) \cdots \chi(X_n)$,
(c) $\chi(X_1 \cup \cdots \cup X_n) = 1 - (1 - \chi(X_1))(1 - \chi(X_2)) \cdots (1 - \chi(X_n))$.

Use the characteristic function to verify that

$$X \cap (X_1 \cup \cdots \cup X_n) = (X \cap X_1) \cup \cdots \cup (X \cap X_n).$$

8. We define the *projections*

$$\begin{aligned} p_1 : X \times Y &\to X \\ (x, y) &\rightsquigarrow x \end{aligned} \qquad \begin{aligned} p_2 : X \times Y &\to Y \\ (x, y) &\rightsquigarrow y. \end{aligned}$$

Let $\Delta : Z \to Z \times Z$ be the *diagonal map* $z \rightsquigarrow (z, z)$. Prove that the functions

$$\begin{aligned} \rho : (X \times Y)^z &\to X^z \times Y^z \\ f &\rightsquigarrow (p_1 f, p_2 f) \end{aligned} \qquad \begin{aligned} \sigma : X^z \times Y^z &\to (X \times Y)^z \\ (f, g) &\rightsquigarrow (f \times g)\Delta \end{aligned}$$

satisfy $\rho\sigma = 1$, $\sigma\rho = 1$. Deduce that if l, m, n are natural numbers, then

$$(lm)^n = l^n m^n.$$

9. Let X, Y be disjoint sets. Prove that there is a bijection

$$Z^{X \cup Y} \to Z^X \times Z^Y.$$

Deduce that if l, m, n are natural numbers, then

$$l^{m+n} = l^m l^n.$$

10. Let X, Y, Z be sets. Prove that the exponential map

$$\begin{aligned} e : X^{Z \times Y} &\to (X^Y)^Z \\ f &\rightsquigarrow (z \rightsquigarrow (y \rightsquigarrow f(z, y))) \end{aligned}$$

is a bijection. (The notation means that $e(f)$ is the function such that $e(f)(z)(y) = f(z, y)$.) Deduce that if l, m, n are natural numbers, then

$$l^{nm} = (l^m)^n.$$

11. Let X, Y be finite sets with m, n elements respectively where $m \leqslant n$. Determine the number of bijections $X \to X$ and the number of injections $X \to Y$.
12. Read the proof of A.2.10 given in Dieudonné [1; 2.2.17].
13. Prove that the function

$$\begin{aligned} f : \mathbf{N} \times \mathbf{N} &\to \mathbf{N} \\ (x, y) &\rightsquigarrow \tfrac{1}{2}(x + y)(x + y + 1) + y \end{aligned}$$

is a bijection.
14. A real number α is called *algebraic* if α satisfies an equation

$$\alpha^n + a_1 \alpha^{n-1} + \cdots + a_{n-1}\alpha + a_n = 0$$

where $a_i \in \mathbf{Z}$. Prove that the set of algebraic numbers is countable.
15. Prove that the plane $\mathbf{R} \times \mathbf{R}$ is not the union of countably many lines.

A.3 Products and the axiom of choice

Let $(X_\lambda)_{\lambda \in L}$ be a family of sets and let X be the set of all families $x = (x_\lambda)_{\lambda \in L}$ such that $x_\lambda \in X_\lambda$. Then X is called a *product* of the family $(X_\lambda)_{\lambda \in L}$ and is denoted by

$$\prod_{\lambda \in L} X_\lambda.$$

The element x_λ of X_λ is called the *λth coordinate* of x. Thus we have functions

$$p_\lambda : X \to X_\lambda$$
$$x \rightsquigarrow x_\lambda$$

the function p_λ is called the *λth projection*. (Note that a family $(x_\lambda)_{\lambda \in L}$ such that $x_\lambda \in X_\lambda$ is a function $x : L \to X'$ where $X' = \bigcup_{\lambda \in L} X$, and that x_λ is the same as $x(\lambda)$, the value of x at λ.)

Suppose, in particular, that $L = \{0, 1\}$. The function

$$\prod_{\lambda \in L} X_\lambda \to X_0 \times X_1$$
$$x \rightsquigarrow (x_0, x_1)$$

is a bijection, and this shows that the product $\prod_{\lambda \in L} X_\lambda$ is a reasonable generalization of the cartesian product $X_0 \times X_1$.

If one of the sets X_λ is empty, then so also is $\prod_{\lambda \in L} X_\lambda$. The converse of this statement is the Axiom of Choice.

Axiom of choice. If $\prod_{\lambda \in L} X_\lambda$ *is empty, then one of the sets* X_λ *is empty.*

Or, alternatively, if each set X_λ is non-empty, then $\prod_{\lambda \in L} X_\lambda$ is non-empty.

We shall discuss later why this is an axiom rather than a theorem. For the moment, we illustrate the axiom by a simple consequence.

Let X, Y be sets. We say $※Y \leqslant ※X$ if there is an injection $Y \to X$.

A.3.1 *Let* X, Y *be sets and* $f : X \to Y$ *a surjection. Then* $※Y \leqslant ※X$.

Proof We suppose X is non-empty, so that Y is non-empty. Then for each y in Y the set $f^{-1}[y]$ is non-empty. By the Axiom of Choice there is an element k of $\prod_{y \in Y} f^{-1}[y]$.

Now k is a function $Y \to X$. Since $k(y) \in f^{-1}[y]$ for each $y \in Y$, it follows that $fk = 1_Y$. Therefore, k is injective and so $※Y \leqslant ※X$. □

The reader should compare this proof carefully with that of A.2.6. In A.2.6 an injection $k : Y \to \mathbf{N}$ was constructed explicitly using properties of the natural numbers; no such method is available here and we must rely instead on the Axiom of Choice.

The following theorem was proved by Cantor using the Axiom of Choice. Later, proofs were found not using this axiom, and for this reason the theorem is known as the Schröder–Berstein theorem.

A.3.2 *Let X, Y be sets. If $※X \leqslant ※Y$ and $※Y \leqslant ※X$, then $※X = ※Y$.*

On the other hand, the proof of the following theorem does involve the Axiom of Choice.

A.3.3 *If X and Y are sets, then either $※X \leqslant ※Y$ or $※Y \leqslant ※X$.*

Neither of these theorems is essential for this book, and so we omit the proofs.

For any set X there is a set Y such that $※Y > ※X$ (for example, $Y = \mathscr{P}(X)$). An obvious problem is to determine all infinite cardinalities. With our present notation we can state only the following result.

A.3.4 *There are sets $A_{\mathbf{N}}$, and A_n for each n in $\mathbf{N}$, such that $A_0 = \mathbf{N}$ and*

$$※A_0 < ※A_1 < \cdots < ※A_n < \cdots < ※A_{\mathbf{N}}$$

Further, if X is not finite and has cardinality less than that of $A_{\mathbf{N}}$, then $※X = ※A_n$ for some n in $\mathbf{N}$.

There are also greater cardinalities than that of $A_{\mathbf{N}}$, but to describe these, and also to prove A.3.2, the theory of ordinal numbers is required [cf., for example, the Appendix to Kelley [1]].

We can now ask: where, if anywhere, in the list given in A.3.2 does the cardinality of $\mathbf{R}$ fall? We have already proved that $※\mathbf{R} > \mathbf{N}$; Cantor spent a large portion of his life trying to prove the

Continuum axiom: $※\mathbf{R} = ※A_1$

It is now known that his attempts were bound to fail, although we can describe only roughly why this is so.

In order to decide whether or not this axiom, and the Axiom of Choice, can be proved, it is first necessary to describe precisely what is meant by a proof. This involves setting up carefully the system of logic used in our normal arguments. Second, within this system it is necessary to construct a Set Theory; such a theory is an axiomatic system, with undefined terms such as set, membership, and so on, together with axioms guaranteeing the existence or non-existence of certain sets and governing the use of the undefined terms.

Once Set Theory is set up in this way, we can then ask: do the Axiom of

Choice and the Continuum Axiom follow from the other axioms of Set Theory? It was proved by P. Cohen in 1962 that these two axioms are independent of each other and of the other axioms of Set Theory. This means that for each of these axioms we obtain three theories, in which the axiom is either asserted, or denied, or simply left out (and so there are nine theories from these two axioms!).

Few of these theories have been investigated, and there are other varieties of the Continuum Axiom which are the subject of current research. The Axiom of Choice plays an important role in many branches of mathematics, and we shall use it without further mention. The Continuum Axiom has some (possibly undesirable) consequences in measure theory, but has found no applications in topology. For a further discussion of these topics we refer the reader to Gödel [1], Cohen [1].

A.4 Universal constructions

One purpose of this section is to give an abstract characterization of the product of sets and to introduce the 'dual' notion of sum of sets (the word dual is explained in chapter 6). The kind of argument is of importance in discussing adjunction spaces and computations of the fundamental group. The reader is advised to read through the section quickly and return to master it when he finds the methods are needed.

Let $(X_\lambda)_{\lambda \in L}$ be a family of sets. We have seen that associated with the product $\prod\limits_{\lambda \in L} X_\lambda$ is a family of projections

$$p_\lambda : \prod_{\lambda \in L} X_\lambda \to X_\lambda.$$

The crucial property of these projections is that they determine completely functions into the product.

A.4.1 *Let Y be a set and let there be given a family of functions*

$$f_\lambda : Y \to X_\lambda, \quad \lambda \in L.$$

Then there is a unique function $f : Y \to \prod\limits_{\lambda \in L} X_\lambda$ such that

$$p_\lambda f = f_\lambda, \quad \lambda \in L.$$

Proof An element x of $\prod X_\lambda$ is determined completely by its family of 'coordinates' $x_\lambda = p_\lambda x$. Therefore, we can define f by saying that for each y in Y, $f(y)$ is to have coordinates $f_\lambda(y)$ for each λ in L. This is equivalent to $p_\lambda f(y) = f_\lambda(y)$. □

We now show that this property characterizes the product in a certain sense.

A.4.2 *Let $p'_\lambda : X' \to X_\lambda$, $\lambda \in L$ be a family of functions. Then the following conditions are equivalent.*
(*a*) *There is a bijection $p' : X' \to \prod X_\lambda$ such that*

$$p_\lambda p' = p'_\lambda, \quad \lambda \in L$$

(*b*) *For any Y and any family of functions $f_\lambda : Y \to X_\lambda$, $\lambda \in L$, there is a unique function $f : Y \to X'$ such that*

$$p'_\lambda f = f_\lambda, \quad \lambda \in L.$$

Proof In order to bring out the structure of the argument we make some definitions. All families in what follows will be indexed by L; the family (X_λ) is given.

By an *object* we mean a pair $(Y, (f_\lambda))$ consisting of a set Y and a family $f_\lambda : Y \to X_\lambda$ of functions. Such an object is abbreviated to (Y, f_λ). By a *map* $f : (Y, f_\lambda) \to (Z, g_\lambda)$ of objects we mean a function $f : Y \to Z$ such that

$$g_\lambda f = f_\lambda, \quad \text{all } \lambda \in L.$$

We then have the following properties
(i) The identity $1 : (Y, f_\lambda) \to (Y, f_\lambda)$ is a map.
(ii) If $f : (Y, f_\lambda) \to (Z, g_\lambda)$ and $g : (Z, g_\lambda) \to (W, h_\lambda)$ are maps, then so also is $gf : (Y, f_\lambda) \to (W, h_\lambda)$. (Since $h_\lambda gf = g_\lambda f = f_\lambda$.)
(iii) If $f : (Y, f_\lambda) \to (Z, g_\lambda)$ is a map, and $f : Y \to Z$ is a bijection, then $f^{-1} : (Z, g_\lambda) \to (Y, f_\lambda)$ is a map. (The relation $g_\lambda f = f_\lambda$ implies that $f_\lambda f^{-1} = g_\lambda f f^{-1} = g_\lambda$.)

Let $X = \prod X_\lambda$. The property A.4.1 can now be stated as: *for any object (Y, f_λ) there is exactly one map $f : (Y, f_\lambda) \to (X, p_\lambda)$*. Similarly, the property (*a*) above can be stated: *for any object (Y, f_λ) there is exactly one map $f : (Y, f_\lambda) \to (X', p'_\lambda)$*.

We now show that (*b*) $\Rightarrow$ (*a*). By A.4.1 there is exactly one map $p' : (X', p'_\lambda) \to (X, p_\lambda)$. By (*b*) there is exactly one map $p : (X, p_\lambda) \to (X', p'_\lambda)$. By (ii), $p'p : (X, p_\lambda) \to (X, p_\lambda)$ is a map; but there is only one map $(X, p_\lambda) \to (X, p_\lambda)$ and by (i) this is the identity. Hence $p'p = 1$. Similarly, $pp' : (X', p'_\lambda) \to (X', p'_\lambda)$ is a map and so $pp' = 1$. These relations show that p' is a bijection, while the relations given in (*a*) simply state that p' is a map.

We now show that (*a*) $\Rightarrow$ (*b*). Since p' is a bijection it has an inverse, which we write $p : X \to X'$. By (iii), $p : (X, p_\lambda) \to (X', p'_\lambda)$ is a map. Let (Y, f_λ) be an object. By A.4.1 there is a unique map $f : (Y, f_\lambda) \to (X, p_\lambda)$ and by (ii) $pf : (Y, f_\lambda) \to (X', p'_\lambda)$ is a map. So we have constructed a map as required, but we must show that this is unique.

Suppose then $g : (Y, f_\lambda) \to (X', p'_\lambda)$ is a map. Then $p'g : (Y, f_\lambda) \to (X, p_\lambda)$ is a map and hence $p'g = f$. It follows that $g = pp'g = pf$. □

The property A.4.1 of the product is called a *universal property*, and the method of proof of A.4.2 will be called the *usual universal argument*.

The important thing about a product is that it has projections which satisfy the universal property. There is in many cases no canonical choice for the actual set we take as the product: for example, if L consists of three elements 1, 2, 3, then we can take for the product either $\prod_{\lambda \in L} X_\lambda$ (as defined above) or

$$(X_1 \times X_2) \times X_3, \quad \text{or} \quad X_1 \times (X_2 \times X_3)$$

and there seems no reason to prefer one to another. The fact that in each case we have projections satisfying the universal property ensures that there are bijections between any two of these sets.

We therefore now *define* a product of the family $(X_\lambda)_{\lambda \in L}$ of sets to be a set X with projections $p_\lambda : X \to X_\lambda$, $\lambda \in L$, such that the universal property of A.4.1 holds. The set X is then also written $\prod_{\lambda \in L} X_\lambda$.

In the case L is finite, say $L = \{1, \ldots, n\}$, then a product of $X_1, \ldots, X_n$ is written $X_1 \times \cdots \times X_n$. If $f_\lambda : Y \to X_\lambda$, $\lambda = 1, \ldots, n$ is a family of functions, then the unique function $f : Y \to X_1 \times \cdots \times X_n$ such that $p_\lambda f = f_\lambda$ is written $(f_1, \ldots, f_n)$, and f_λ is called the λth component of f.

Sum of sets

We shall also need the less familiar construction of the *sum* (also called the *disjoint union*, or *coproduct*) of sets. We define this *ab initio* by a universal property.

Let $(X_\lambda)_{\lambda \in L}$ be a family of sets. A *sum* of the family is a set X together with functions $i_\lambda : X_\lambda \to X$, called *injections*, with the following property: for any Y and any family of functions $f_\lambda : X_\lambda \to Y$, $\lambda \in L$, there is a unique function $f : X \to Y$ such that $f\, i_\lambda = f_\lambda$, $\lambda \in L$.

The formal difference between the definitions of product and sum are that we are given maps *from* the product and *into* the *sum*. So we distinguish between the two universal properties by saying that the product is ι-universal (ι for initial) and the sum is φ-universal (φ for final).

A.4.3 *Let $i_\lambda : X_\lambda \to X$ $(\lambda \in L)$ be a sum of the family $(X_\lambda)_{\lambda \in L}$, and let $i'_\lambda : X_\lambda \to X'$ $(\lambda \in L)$ be a family of functions. Then the following conditions are equivalent*

(*a*) *There is a bijection $i' : X \to X'$ such that*

$$i' i_\lambda = i'_\lambda, \quad \lambda \in L.$$

(*b*) *X' and the family $(i'_\lambda)_{\lambda \in L}$ is a sum of the family $(X_\lambda)_{\lambda \in L}$.*

Proof The method of proof is similar to that of A.4.2 and we only outline it.

An *object* (Y, f_λ) is defined to be a pair consisting of a set Y and a family $f_\lambda : X_\lambda \to Y$ of functions. A *map* $f : (Y, f_\lambda) \to (Z, g_\lambda)$ is defined to be a function $f : Y \to Z$ such that

$$ff_\lambda = g_\lambda, \quad \lambda \in L.$$

The properties analogous to (i), (ii), (iii) in the proof of A.4.2 are easily verified.

The object (X, i_λ) is thus a sum if and only if there is exactly one map $(X, i_\lambda) \to (Y, f_\lambda)$ for any object (Y, f_λ). The remainder of the proof is analogous to that of A.4.2 and is left to the reader. □

If $i_\lambda : X_\lambda \to X$ is a sum then we write

$$X = \bigsqcup_{\lambda \in L} X_\lambda.$$

In particular, if $L = \{1, \ldots, n\}$ then we write

$$X = X_1 \sqcup \cdots \sqcup X_n.$$

We must now show that sums exist. Suppose first of all that the family (X_λ) consists of disjoint sets. Let $X = \bigcup_{\lambda \in L} X_\lambda$ and let $i_\lambda : X_\lambda \to X$ be the inclusion. Then, clearly, we have a sum of the family, since first a function $f : X \to Y$ is completely determined by the functions $f_\lambda = f \mid X_\lambda = f i_\lambda$, and second, any such family defines a function f. Thus for families of disjoint sets, we can always take $\bigsqcup_{\lambda \in L} X_\lambda$ to be the union $\bigcup_{\lambda \in L} X_\lambda$.

The situation is different if the sets are not disjoint. For example, two functions $f_1 : X_1 \to Y$, $f_2 : X_2 \to Y$ define a function $f : X_1 \cup X_2 \to Y$ such that $f \mid X_1 = f_1, f \mid X_2 = f_2$ if and only if f_1, f_2 agree on $X_1 \cap X_2$, a condition which is vacuous if $X_1 \cap X_2 = \varnothing$ but not otherwise.

A.4.4 *Any family $(X_\lambda)_{\lambda \in L}$ of sets has a sum.*

Proof The idea of the proof is to replace the given family by a family of disjoint sets and then take the union. In fact we replace X_λ by $X'_\lambda = X_\lambda \times \{\lambda\}$; then X'_λ meets X'_μ if and only if $\lambda = \mu$.

Let $X = \bigcup_{\lambda \in L} X'_\lambda$ and let

$$\begin{aligned} i_\lambda : X_\lambda &\to X \\ x &\rightsquigarrow (x, \lambda). \end{aligned}$$

Let $f_\lambda : X_\lambda \to Y$ be a family of functions. Then the function

$$\begin{aligned} f : X &\to Y \\ (x, \lambda) &\rightsquigarrow f_\lambda(x) \end{aligned}$$

is well-defined and is the only function $X \to Y$ whose composite with each i_λ is f_λ. □

A.4.4 (*Corollary 1*) *If $i_\lambda : X_\lambda \to X$ is a sum, then each i_λ is injective.*

Proof If $i_\lambda : X_\lambda \to X$ is the sum constructed in A.4.4, then it is obvious that each i_λ is injective. If $i'_\lambda : X_\lambda \to X'$ is any sum, then there is a bijection $i' : X \to X'$ such that $i'i_\lambda = i'_\lambda$, and it follows that each i'_λ is injective. □

In the case of a finite family $X_1, \ldots, X_n$, the function $f : X_1 \sqcup \cdots \sqcup X_n \to Y$ such that $f i_\lambda = f_\lambda$, $\lambda = 1, \ldots, n$, is sometimes written

$$(f_1, \ldots, f_n)^t.$$

Equivalence relations

We need in chapter 4 some facts on equivalence relations in addition to those given in the Glossary.

Let R be a relation on a set A. We define a new relation E on A by $a\,E\,b \Leftrightarrow$ there is a sequence $a_1, \ldots, a_n$ of elements of A such that

(a) $a_1 = a, \qquad a_n = b,$

(b) for each $i = 1, \ldots, n - 1,$

$$a_i\,R\,a_{i+1} \quad \text{or} \quad a_{i+1}\,R\,a_i \quad \text{or} \quad a_i = a_{i+1}.$$

A.4.5 *The relation E is the smallest equivalence relation on A containing R.*

Proof We first prove that E is an equivalence relation. Let $a, b, c \in A$. Clearly $a\,E\,a$, since the sequence a, a satisfies (a) and (b). Suppose $a\,E\,b$, and that $a_1, \ldots, a_n$, satisfies (a) and (b). Then the sequence $a_n, \ldots, a_1$ ensures that $b\,E\,a$.

Finally, if $a\,E\,b$ and $b\,E\,c$, then by splicing together two sequences we obtain that $a\,E\,c$.

Suppose E' is an equivalence relation containing R. We have to prove that E is contained in E', i.e., that $a\,E\,b \Rightarrow a\,E'\,b$.

Let $a\,E\,b$, and let $a_1, \ldots, a_n$ be a sequence satisfying (a) and (b). Now E' contains R. So $a_i\,E'\,a_{i+1}$ for each $i = 1, \ldots, n - 1$ (by (b)). Hence $a_1\,E'\,a_n$, i.e., $a\,E'\,b$. □

Because of A.4.5, we call E the *equivalence relation generated by R*. Let $p : A \to A/E$ be the projection $a \rightsquigarrow \text{cls}\, a$. Then p can be characterized by a φ-universal property.

A.4.6 (*a*) *For all* a, b *in* A, $a\,R\,b \Rightarrow pa = pb$.
(*b*) *If* $f : A \to B$ *is any function such that*

$$a\,R\,b \Rightarrow fa = fb \quad \text{all } a, b \text{ in } A$$

then there is a unique function $f^* : A/E \to B$ *such that*

$$f^*p = f.$$

Proof The proof of (*a*) is obvious, since $R \subset E$ and

$$pa = pb \quad \Leftrightarrow \quad \text{cls } a = \text{cls } b \quad \Leftrightarrow \quad a\,E\,b.$$

Let $f : A \to B$ be a function, and let E_f be the relation on A given by

$$a\,E_f\,b \quad \Leftrightarrow \quad fa = fb.$$

Then E_f is an equivalence relation. Suppose also $a\,R\,b \Rightarrow fa = fb$. Then E_f contains R and so E_f contains E. Hence if cls a = cls b, then $fa = fb$. Therefore the function

$$\begin{aligned} f^* : A/E &\to B \\ \text{cls } a &\rightsquigarrow fa \end{aligned}$$

is well-defined, and clearly $f^*p = f$. The uniqueness of f^* follows from the fact that p is surjective. □

The usual universal argument shows that if $p' : A \to A'$ is any function satisfying A.4.6(*a*) and (*b*) (with p replaced by p') then there is a unique bijection $p^* : A/E \to A'$ such that $p^*p = p'$. Also, given a bijection $p^* : A/E \to A'$ for some A', then $p' = p^*p$ satisfies A.4.6(*a*) and (*b*) (with p replaced by p');

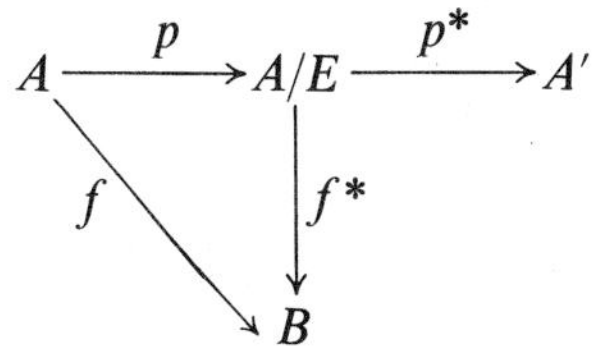

(for if $f : A \to B$ satisfies $a\,R\,b \Rightarrow fa = fb$, then we can construct $f^* : A/E \to B$ such that $f^*p = f$, and then take $f^*(p^*)^{-1} : A' \to B$).

EXERCISES

1. Prove that for any sets X_1, X_2, X_3 there are bijections

$$\begin{aligned} &X_1 \sqcup X_2 \to X_2 \sqcup X_1, \\ &X_1 \sqcup (X_2 \sqcup X_3) \to (X_1 \sqcup X_2) \sqcup X_3 \\ &X_1 \times (X_2 \sqcup X_3) \to (X_1 \times X_2) \sqcup (X_1 \times X_3). \end{aligned}$$

2. Let $f_1 : X_1 \to Y_1, f_2 : X_2 \to Y_2$ be functions. The function

$$f_1 \sqcup f_2 : X_1 \sqcup X_2 \to Y_1 \sqcup Y_2$$

is by definition $(i_1 f_1, i_2 f_2)^t$. Prove that

(*a*) $f_1 \sqcup f_2$ is injective $\Leftrightarrow f_1, f_2$ are injective,
(*b*) $f_1 \sqcup f_2$ is surjective $\Leftrightarrow f_1, f_2$ are surjective.

Prove also that if $g_1 : Y_1 \to Z_1, g_2 : Y_2 \to Z_2$, then

$$(g_1 \sqcup g_2)(f_1 \sqcup f_2) = g_1 f_1 \sqcup g_2 f_2.$$

3. An *n-diagram* is defined to be a pair of functions

$$\{x\} \xrightarrow{j} X \xrightarrow{f} X$$

such that $x \in X$ and j is the inclusion. A *map* of n-diagrams is a commutative diagram,

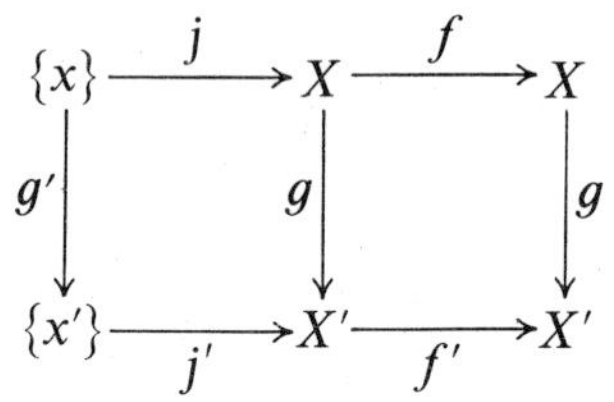

Let $\mathcal{N}$ be the n-diagram $\{0\} \to \mathbf{N} \xrightarrow{s} \mathbf{N}$ where s is the function $n \rightsquigarrow n + 1$. Prove that $\mathcal{N}$ is φ-universal: that is, if $\mathcal{X}$ is any n-diagram, then there is a unique map $\mathcal{N} \to \mathcal{X}$ of n-diagrams. Show that this property expresses the principle of inductive definition of a function [Flett [1; 1.9.1]]. Show that if also $\mathcal{N}'$ is φ-universal, then there are unique maps $\varphi : \mathcal{N} \to \mathcal{N}', \psi : \mathcal{N}' \to \mathcal{N}$ such that $\varphi\psi$, $\psi\varphi$ are the respective identity maps.

Deduce from the φ-universal property the following: if $h : \mathbf{N} \times Y \to Y$ is a function, and $y_0 \in Y$, then there is a unique function $k : \mathbf{N} \to Y$ such that

$$k(0) = y_0, \qquad k(n + 1) = h(n, k(n)), \quad n \in \mathbf{N}.$$

Show that the φ-universal property implies the following form of the principle of induction: if $A \subset \mathbf{N}$ and (i) $0 \in A$, (ii) for all n, $n \in A \Rightarrow n + 1 \in A$, then $A = \mathbf{N}$.

[The most important use of the principle of induction is to define functions. It seems reasonable therefore to axiomatize the integers in a way which immediately allows such a definition. The above universal property (which is due to Verdier) does exactly this.]

Glossary of terms from set theory

Agree Two functions $f, g : X \to Y$ agree at a point x in X if $f(x) = g(x)$; and f, g agree on a subset A of X if $f \mid A = g \mid A$.

Anti-symmetric cf. *Relation.*

Belongs to Synonyms for a belongs to A are: a is a member of A, a is an element of A, a is in A, A contains a, $a \in A$, $A \ni a$.

Bounded Let X be a set and $\leqslant$ a partial order relation on X. Let A be a subset of X. An *upper bound* for A is an element u of X such that $a \in A \Rightarrow a \leqslant u$. A *supremum*, sup A, for A is an upper bound u for A such that if v is any upper bound for A then $u \leqslant v$. The terms *lower bound, infinum* (inf A) are defined as for upper bound and supremum but with $\leqslant$ replaced by $\geqslant$. The set A is *bounded above* if it has an upper bound, *bounded below* if it has a lower bound, and *bounded* if it is bounded above and below. The order relation is *complete* if every non-empty subset of X which is bounded above has a supremum (and this implies that every non-empty subset of X which is bounded below has an infinum).

Cartesian product The Cartesian product of two sets A, B is the set $A \times B$ of all ordered pairs (a, b) for a in A, b in B.

Class There is a necessity to rescue set theory from some rather simple contradictions. One way of doing this is by distinguishing classes and sets. Here class is the general notion and a set is a class which is a member of some other class [cf. Kelley [1]]. A more useful method for category theory is to assume axioms postulating the existence of 'universes' [cf. Sonner [1], Brinkman–Puppe [1]]. Another approach is to axiomatize category theory instead of set theory [cf. Lawvere [1]].

Classifier notation This is the notation $\{x : P(x)\}$; it denotes the set of all x having property $P(x)$ (if such exists; otherwise it has no denotation). We write $\{x \in A : P(x)\}$ for $\{x : P(x)$ and $x \in A\}$. If A can be understood from the context it is common to abbreviate $\{x \in A : P(x)\}$ to $\{x : P(x)\}$. If $a \in A$, then the set $\{x \in A : x \neq a\}$ is written $A^{\neq a}$.

Commutative The following diagrams of functions

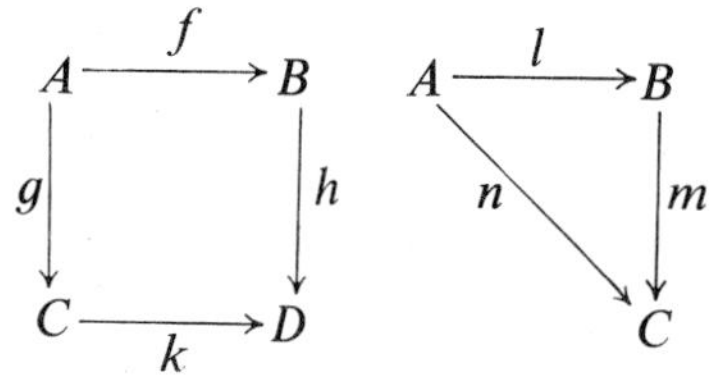

are *commutative* if $hf = kg$ (in the first case) and $n = ml$ (in the second case). This definition is extended to more complicated diagrams in the obvious way.

Complement If A is a subset of B, the complement of A in B is the difference $B \setminus A$. When B can be understood from the context, we refer to $B \setminus A$ simply as the complement of A.

Complete see *Bounded.*

De Morgan laws Let A, B, X be sets, and let $(A_i)_{i \in I}$ be a family of subsets of X. The De Morgan laws are:

$$X \setminus (A \cup B) = (X \setminus A) \cap (X \setminus B), \qquad X \setminus (A \cap B) = (X \setminus A) \cup (X \setminus B)$$

$$X \setminus \bigcup_{i \in I} X_i = \bigcap_{i \in I} (X \setminus X_i), \qquad X \setminus \bigcap_{i \in I} X_i = \bigcup_{i \in I} (X \setminus X_i).$$

Difference If X, A are sets, the difference of X and A is the set $X \setminus A = \{x \in X : x \notin A\}$. The difference satisfies

$$X \setminus (X \setminus A) = X \cap A.$$

Disjoint Two sets A, B are disjoint if $A \cap B = \varnothing$.

Distributivity law If A, B, X are sets, and $(X_i)_{i \in I}$ is a family of sets, then

$$X \cap (A \cup B) = (X \cap A) \cup (X \cap B)$$

$$X \cap \bigcup_{i \in I} X_i = \bigcup_{i \in I} (X \cap X_i)$$

Element An object or member of a set.

Empty set The set $\varnothing$ with no elements.

Equality Two sets are equal if and only if they have the same elements.

Equivalence relation An equivalence relation on a set A is a relation which is symmetric, transitive, and reflexive on A. If $\sim$ is an equivalence relation on A, then for each $a \in A$ the set $\text{cls}\, a = \{b \in A : b \sim a\}$ is called the equivalence class of a. Two sets $\text{cls}\, a$, $\text{cls}\, b$ either coincide, or are disjoint. The set of all such equivalence classes is written $A/\sim$. The function $A \to A/\sim$, $a \rightsquigarrow \text{cls}\, a$ is called the *projection*. We also abbreviate $a \in \text{cls}\, a \in A/\sim$ to $a \in\in A/\sim$.

Factors Let $n : A \to C$ be a function and B a set. It is possible that there are functions $l : A \to B$, $m : B \to C$ such that $n = ml$. The existence of such l, m is expressed by saying n factors through B. Given m, the existence of such an l is expressed by saying n factors through l. Given l, the existence of such an m is expressed by saying n is factors through m. This definition is extended to other categories [cf. chapter 6] than $\mathscr{S}$et in the obvious way.

Family A family $(X_\lambda)_{\lambda \in L}$ with indexing set L is a function $\lambda \rightsquigarrow X_\lambda$ with domain L. We allow the empty family, in which $L = \varnothing$. The restriction of $\lambda \rightsquigarrow X_\lambda$ to a subset of L is called a subfamily of $(X_\lambda)_{\lambda \in L}$.

Greatest element See *Maximal element*.

Inclusion See *Subset*.

Integers The set $\mathbf{Z}$ of integers has elements $0, \pm 1, \pm 2, \ldots$; $\mathbf{Z}$ may be constructed from the set $\mathbf{N}$ of natural numbers.

Intersection The intersection of two sets A, B is the set $A \cap B$ of elements belonging to both A and B; thus

$$A \cap B = \{x : x \in A \quad \text{and} \quad x \in B\}.$$

The intersection of a non-empty family of sets $(X_\lambda)_{\lambda \in L}$ is the set

$$\bigcap_{\lambda \in L} X_\lambda = \{x : x \in X_\lambda \quad \text{for all } \lambda \in L\}.$$

The intersection of an empty family is defined only when we are dealing with families of subsets of a fixed set X, and then the intersection is X itself.

Interval Let X be a set and $\leqslant$ an order relation on X. A subset A of X is an interval if the conditions $x \leqslant z \leqslant y$ and $x, y \in A$ imply $z \in A$. Particular intervals of X are X itself, the empty set and the subsets of X given by

$$[x, y] = \{z : x \leqslant z \leqslant y\}, \qquad [x, \to[= \{z : x \leqslant z\}$$

$$]\leftarrow, x] = \{z : z \leqslant x\}, \qquad [x, y[= \{z : x \leqslant z < y\}$$

$$]x, y] = \{z : x < z \leqslant y\}, \qquad]x, y[= \{z : x < z < y\}$$

$$]x, \to[= \{z : x < z\}, \qquad]\leftarrow, x[= \{z : z < x\}$$

for x, y in X such that $x \leqslant y$. The first three intervals are *closed*, the next two *half-open* and the last three, as well as X and $\varnothing$, are *open*. The *completeness* of the order is equivalent to the fact that any interval of X is of one of the forms given above. An interval $[x, \to[$ is also written $X^{\geqslant x}$; similar notations are used for other intervals involving $\to$ or $\leftarrow$. The *end points* of a non-empty interval $]x, y]$ are x and y; the *end point* of $]\leftarrow, x]$ is x; similar terms are applied to the other intervals.

Least element See *Maximal element.*

Maximal element Let X be a set, $\leqslant$ a partial order relation on X, and A a subset of X. An element a of A is maximal if no element of A is larger than a, i.e., if $a' \in A$ and $a \leqslant a'$ implies $a = a'$. An element a of A is a greatest element if a is larger than every element of A, i.e., if $a' \in A$ implies $a' \leqslant a$. A greatest element is maximal, and is unique. A maximal element need not be a greatest element, nor need it be unique; however it will have both of these properties if $\leqslant$ is an order relation, and in this case the maximal element of A (when it exists) is written max A. The terms minimal, least, min A are defined as for maximal, greatest, max A, but with $\leqslant$ replaced by $\geqslant$, and larger replaced by smaller.

Meet Two sets A, B meet if $A \cap B \neq \varnothing$; in such case we also say A meets B, B meets A.

Natural number The set **N** of natural numbers has elements 0, 1, 2, . . . ; **N** may be described axiomatically [Flett [1]] or constructed explicitly.

Order relation A partial order relation on X is a relation which is transitive, anti-symmetric, and reflexive on X. A partial order is an *order relation* if it is a total relation. A partial order is often written $\leqslant$, in which case $x < y$ means $x \leqslant y$ and $x \neq y$, while $y \geqslant x$ means $x \leqslant y$.

Ordered pair Intuitively, the ordered pair (a, b) consists of the elements a, b taken in order. The crucial property is that $(a, b) = (a', b')$ if and only if $a = a'$, $b = b'$. This property may be derived from the definition

$$(a, b) = \{\{a\}, \{a, b\}\}.$$

Point The word point is a geometric synonym for element, object.

Positive Greater than 0.

Proper see *Subset.*

Rational number A rational number is a ratio m/n for m, n integers and $n \neq 0$. The set **Q** of all rational numbers may be constructed from **Z**.

Real number The set **R** of real numbers satisfies axioms given, for example, in Flett [1]; **R** may be constructed from **Q** by Dedekind sections or Cantor sequences.

Reflexive see *Relation.*

Relation A relation on A is a subset R of $A \times A$. If $(x, y) \in R$ we write also $x\,R\,y$. The relation R is *reflexive* if $x\,R\,x$ for all $x \in A$; *symmetric* if $x\,R\,y \Rightarrow y\,R\,x$ for all $x, y \in A$; *anti-symmetric* if $x\,R\,y$ and $y\,R\,x \Rightarrow x = y$ for all $x, y \in A$; *transitive* if $x\,R\,y$ and $y\,R\,z \Rightarrow x\,R\,z$ for all $x, y, z \in A$; *total on A* if $x\,R\,y$ or $y\,R\,x$ for all $x, y \in A$.

Sequence A family with indexing set a subset of **N**.

Set A set is a 'collection of objects viewed as a whole'. A rigorous treatment of sets requires an axiomatic theory [cf. Kelley [1]]. The set whose elements are exactly $x_1, x_2, \ldots, x_n$ is written $\{x_1, x_2, \ldots, x_n\}$.

Subset The set A is a subset of B if every element of A is an element of B, that is, if $\forall x\; x \in A \Rightarrow x \in B$. Synonyms for A is a subset of B are: $A \subset B$, $B \supset A$, A is contained in B, A is included in B, B contains A. The set A is a proper subset of B if $A \subset B$ and $A \neq B$.

Symmetric see *Relation.*

Total see *Relation.*

Transitive see *Relation.*

Union The union of two sets A, B is the set $A \cup B$ whose elements are those of A and those of B. That is,

$$A \cup B = \{x : x \in A \quad \text{or} \quad x \in B\}.$$

The union of a family $(X_\lambda)_{\lambda \in L}$ of sets is the set

$$\bigcup_{\lambda \in L} X_\lambda = \{x : x \in X_\lambda \quad \text{for some } \lambda \in L\};$$

in particular, the union of the empty family is the empty set.

Zorn's lemma This lemma, which is equivalent to the Axiom of Choice, states that if $\mathcal{T}$ is a partially ordered set such that any ordered subset has an upper bound in $\mathcal{T}$, then $\mathcal{T}$ contains a maximal element.

References

Albert, A.

[1] *Studies in modern algebra*, edited by A. Albert, Mathematical Association of America, Englewood Cliffs (1963).

Arkowitz, M.

[1] 'The generalized Whitehead product' *Pacific J. Math.* 12 (1962) pp. 7–24.

Atiyah, M. F.–Bott, R.–Schapiro, A.

[1] 'Clifford modules' *Topology* 3 Suppl. (1964) pp. 3–38.

Bourbaki, N.

[1] *General topology* Addison-Wesley, Reading (Mass.) (1966).

Brandt, H.

[1] 'Uber eine Verallgemeinerung des Gruppenbegriffes' *Math. Ann.* 96 (1926) pp. 360–6.

Bredon, G. E.

[1] *Sheaf theory* McGraw-Hill, New York (1967).

Brinkman, H. B.–Puppe, D.

[1] *Kategorien und Funktoren* (Lecture notes in mathematics 18) Springer-Verlag, Heidelberg (1966).

Brown, R.

[1] 'Ten topologies for $X \times Y$' *Quart. J. Math. Oxford* (2) 14 (1963) pp. 303–19.

[2] 'Function spaces and product topologies' ibid. 15 (1964) pp. 238–50.

[3] 'On a method of P. Olum' *J. Lond. Math. Soc.* 40 (1965) pp. 303–4.

[4] 'Groupoids and van Kampen's theorem' *Proc. Lond. Math. Soc.* (3) 17 (1967) 385–40.

Cairns, S. S.

[1] *Introductory topology* Ronald Press, New York (1961).

Chevalley, C.

[1] *Theory of Lie groups* Princeton University Press, Princeton (1946).

Cockcroft, W. H.–Jarvis, T. M.

[1] 'An introduction to homotopy theory and duality I' *Bull. Soc. Math. Belgique* 16 (1964) pp. 407–28; 17 (1965) pp. 3–26.

Cohen, P.

[1] *Set theory and the continuum hypothesis* Benjamin, New York (1966).

Crowell, R. H.

[1] 'On the van Kampen theorem' *Pacific J. Math.* 9 (1959) pp. 43–50.

Crowell, R. H.–Fox, R. H.
[1] *Knot theory* Ginn & Co., Boston (1963).
Csazar, A.
[1] *Foundations of general topology* Pergamon Press, Oxford (1963).
Dieudonné, J.
[1] *Foundations of modern analysis* Academic Press, New York (1960).
Dold, A.
[1] 'Partitions of unity in the theory of fibrations' *Ann. Math.* 78 (1963) pp. 223–55.
[2] *Halbexacte homotopiefunktoren* (Lecture notes in mathematics 12) Springer-Verlag, Heidelberg (1966).
[3] *Lectures on algebraic topology* Springer-Verlag, Heidelberg (1968).
Eckmann, B.
[1] 'Homotopy and cohomology theory' *Proc. Inter. Congress of Mathematicians Stockholm 1962* pp. 59–73.
Ehresmann, C.
[1] *Catégories et structures* Dunod, Paris (1965).
Eilenberg, S.–Maclane, S.
[1] 'The general theory of natural equivalences' *Trans. Amer. Math. Soc.* 58 (1945) pp. 231–94.
Eilenberg, S.–Steenrod, N. E.
[1] *Foundations of algebraic topology* Princeton University Press, Princeton (1952).
Flett, T. M.
[1] *Modern analysis* McGraw-Hill, London (1966).
Fox, R. H.
[1] 'A quick trip through knot theory' from *The topology of 3-manifolds and related topics* Prentice-Hall, Englewood Cliffs (1961).
Franklin, S. P.
[1] 'Spaces in which sequences suffice' *Fundamenta Mathematicae* 57 (1965) pp. 107–15.
Freyd, P.
[1] *Abelian categories* Harper Row, New York (1964).
Gabriel, P.–Zisman, M.
[1] *Categories of fractions and homotopy theory* Springer-Verlag, Heidelberg (1967).
Gödel, K.
[1] 'What is Cantor's continuum problem' *Amer. Math. Month.* 54 (1947) pp. 515–25.
Godement, R.
[1] *Topologie algébrique et théorie des faisceaux* Hermann, Paris (1958).
Griffiths, H. B.
[1] 'The fundamental group of two spaces with a common point' *Quart. J. Math. Oxford* (2) 5 (1954) pp. 175–90; (Correction) ibid. 6 (1955) pp. 154–5.
[2] 'Infinite products of semi-groups and local connectivity' *Proc. Lond. Math. Soc.* (3) 6 (1956) pp. 455–80.

Halmos, P. R.
[1] *Naïve set theory* van Nostrand, Princeton (1960).
[2] *Finite dimensional vector spaces* van Nostrand, Princeton (1958).
Higgins, P. J.
[1] 'Presentations of groupoids, with applications to groups' *Proc. Camb. Phil. Soc.* 60 (1964) pp. 7–20.
[2] 'Grushko's theorem' *J. Algebra* 4 (1966) pp. 365–72.
Hilton, P. J.
[1] *An introduction to homotopy theory* Cambridge University Press, London (1953).
[2] *Homotopy theory and duality* Gordon and Beach, New York (1965).
Hilton, P. J.–Wylie, S.
[1] *Homology theory, an introduction to algebraic topology* Cambridge University Press, London (1960).
Hirzebruch, F.
[1] *Topological methods in algebraic geometry* Springer-Verlag, Heidelberg (1966).
Hocking, J. G.–Young, G. S.
[1] *Topology* Addison-Wesley, Reading (Mass.) (1961).
Hoffman, K.–Kunze, R.
[1] *Linear algebra* Prentice-Hall, Englewood Cliffs (1961).
Hu, S. T.
[1] *Homotopy theory* Academic Press, New York (1959).
[2] *Elements of general topology* Holden-Day, San Fransisco (1964).
[3] *Theory of retracts* Wayne State University Press, Detroit (1965).
Husemoller, D.
[1] *Fibre bundles* McGraw-Hill, New York (1966).
James, I. M.
[1] 'The intrinsic join' *Proc. Lond. Math. Soc.* (3) 8 (1958) pp. 507–35.
James, I. M.–Thomas, Emery–Toda, H.–Whitehead, G. W.
[1] 'On the symmetric square of a sphere' *J. Math. Mech.* 12 (1963) pp. 771–6.
Kampen, E. H. van
[1] 'On the connection between the fundamental groups of some related spaces' *Amer. J. Math.* 55 (1933) pp. 261–7.
Kelley, J. L.
[1] *General topology* van Nostrand, Princeton (1955).
Kurosch, A. G.
[1] *General algebra* (trans. from the Russian by K. Hirsch) Chelsea, New York (1963).
Lang, S.
[1] *Algebra* Addison-Wesley, Reading (Mass.) (1965).
Lawvere, F. W.
[1] 'The category of categories as a foundation for mathematics' *Proceedings of the conference on categorical algebra, La Jolla, 1965* Springer-Verlag, Heidelberg (1966).

Lefschetz, S.
[1] *Introduction to topology* Princeton University Press, Princeton (1949).

Maclane, S.
[1] 'Categorical algebra' *Bull. Amer. Math. Soc.* 71 (1965) pp. 40–106.

Massey, W. M.
[1] *Algebraic topology: an introduction* Harcourt, Brace & World, Inc., New York (1967).

Michael, E. A.
[1] '$\aleph_0$-spaces' *J. Math. Mech.* 15 (1966) pp. 983–1002.

Milnor, J. W.
[1] 'Some consequences of a theorem of Bott' *Ann. Math.* 68 (1958) pp. 444–9.
[2] *Morse theory* (Annals of Math. Studies, No. 51) Princeton University Press, Princeton (1963).
[3] *Topology from the differentiable viewpoint* University Press of Virginia, Charlottesville (1965).

Mitchell, B.
[1] *Theory of categories* Academic Press, New York (1965).

Mokobodzki, G.
[1] 'Nouvelle méthode pour demontrer la peracompacité des espaces metrisables' Ann. Inst. Fourier Grenoble 14 (2) (1964) pp. 539–42.

Montgomery, D.–Zippin, L.
[1] *Topological transformation groups* Interscience, New York (1955).

Nagata, J.-I.
[1] *Modern dimension theory* North-Holland, Amsterdam (1965).

Neuwirth, L. P.
[1] *Knot groups* (Annals of Maths. Studies 56) Princeton University Press, Princeton (1965).

Olum, P.
[1] 'Non-abelian cohomology and van Kampen's theorem' *Ann. Math.* 68 (1958) pp. 658–67.

Puppe, D.
[1] 'Homotopic-mengen und ihre induzierten Abildungen *I*' *Math. Zeit.* 69 (1958) pp. 299–344.
[2] 'Bemerkung über die Erweiterung von Homotopien' *Archiv d. Math.* 18 (1967) pp. 81–88.

Seifert, H.
[1] 'Konstruction drie dimensionaler geschlossener Raume' *Berlin Verb. Sachs. Akad. Leipzig, Math.-Phys.* Kl. 83 (1931) pp. 26–66.

Seifert, H.–Threllfall, W.
[1] *Lehrbuch der Topologie* Teubner Verlagsgesellschaft, Stuttgart (1934).

Simmons, G. F.
[1] *Introduction to topology and modern analysis* McGraw-Hill, New York (1963).

Smale, S.
[1] 'A survey of some recent developments in differential topology' *Bull. Amer. Math. Soc.* 69 (1963) pp. 131–45.
Spanier, E. H.
[1] 'Secondary operations on mappings and cohomology' *Ann. Math.* 75 (1962) pp. 260–82.
[2] 'Quasi-topologies' *Duke Math. J.* 30 (1963) pp. 1–14.
[3] *Algebraic topology* McGraw-Hill, New York (1966).
Spanier, E. H.–Whitehead, J. H. C.
[1] 'Carriers and S-theory' in *Algebraic geometry and topology* (*a symposium in honour of S. Lefschetz*) Princeton University Press, Princeton (1957) pp. 330–60.
Steenrod, N. E.
[1] *The topology of fibre bundles* Princeton University Press, Princeton (1951).
[2] 'A convenient category of topological spaces' *Michigan Math. J.* 14 (1967) pp. 133–152.
Steenrod, N. E.–Epstein, D. B.
[1] *Cohomology operations* (Annals of mathematics studies 50) Princeton University Press, Princeton (1962).
Swan, R. G.
[1] *The theory of sheaves* University of Chicago Press, Chicago, (1964).
Todá, H.
[1] *Composition methods in homotopy groups of spheres* (Annals of Mathematics Studies (49) Princeton University Press, Princeton (1962).
Vaidyanathaswamy, R.
[1] *Set topology* Chelsea Publishing Co., New York (1960).
Wall, C. T. C.
[1] 'Topology of smooth manifolds' *J. Lond. Math. Soc.* 40 (1965) pp. 1–20.
Weinzweig
[1] 'The fundamental group of a union of spaces' *Pacific J. Math.* 11 (1961) pp. 763–76.
Whitehead, J. H. C.
[1] 'Combinatorial homotopy *I*' *Bull. Amer. Math. Soc.* 55 (1949) pp. 213–45.

Index

MADE AND PRINTED BY OFFSET IN GREAT BRITAIN BY WILLIAM CLOWES AND SONS LTD., LONDON AND BECCLES